Springer Series in Synergetics

Editor: Hermann Haken

Synergetics, an interdisciplinary field of research, is concerned with the cooperation of individual parts of a system that produces macroscopic spatial, temporal or functional structures. It deals with deterministic as well as stochastic processes.

Volumes 1–35 are listed at the end of the book

Rouslan L. Stratonovich

Nonlinear Nonequilibrium Thermodynamics II

Advanced Theory

With 14 Figures

Springer-Verlag Berlin Heidelberg GmbH

Professor Dr. Rouslan L. Stratonovich

Physics Department
Moscow State University
Lenin Hills
119899 Moscow
Russia

Series Editor:

Professor Dr. Dr. h. c. Hermann Haken

Institut für Theoretische Physik und Synergetik der Universität Stuttgart,
D-70550 Stuttgart, Germany and
Center for Complex Systems, Florida Atlantic University,
Boca Raton, FL 33431, USA

Translated by

A. L. Repiev

Samarkandsky Boulevard 13
Block 3, Appt. 63
109507 Moscow
Russia

ISBN 978-3-662-03072-1   ISBN 978-3-662-03070-7 (eBook)
DOI 10.1007/978-3-662-03070-7

Library of Congress Cataloging-in-Publication Data. Stratonovich, R. L. Nonlinear nonequilibrium Thermodynamics II: advanced theory/ Rouslan L. Stratonovich. p. cm. – (Springer series in synergetics; v. 59) Includes bibliographical references and index.
Nonequilibrium thermodynamics. 2. Nonlinear theories. I. Title. II. Series. QC318.I7S7665   1993   536'.71–dc20   93-39995

© Springer-Verlag Berlin Heidelberg 1994
Originally published by Springer-Verlag Berlin Heidelberg New York in 1994

Typesetting: Camera ready copy from Y. Engelbrecht and P. Veeber using a Springer T$_E$X macro package
SPIN: 10066642   55/3140 - 5 4 3 2 1 0 - Printed on acid-free paper

Preface

This two-volume work gives the first detailed coherent treatment of a relatively young branch of statistical physics — nonlinear nonequilibrium and fluctuational-dissipative thermodynamics. This area of research has taken shape rather recently: its development began in 1959. The earlier theory — linear nonequilibrium thermodynamics — is in principle a simple special case of the new theory. Despite the fact that the title of the book includes the word 'nonlinear', it also covers the results of linear nonequilibrium thermodynamics. The presentation of the linear and nonlinear theories is done within a common theoretical framework that is not subject to the linearity condition.

The author hopes that the reader will perceive the intrinsic unity of this discipline, the uniformity and generality of its constituent parts. This theory has a wide variety of applications in various domains of physics and physical chemistry, enabling one to calculate thermal fluctuations in various nonlinear systems.

The book is divided into two volumes. Fluctuation-dissipation theorems (or relations) of various types (linear, quadratic and cubic, classical and quantum) are considered in the first volume. There one encounters the Markov and non-Markov fluctuation-dissipation theorems (FDTs), theorems of the first, second and third kinds. Nonlinear FDTs are less known than their linear counterparts.

The present second volume of the book deals with the advanced theory. It consists of four chapters. The connection and interdependence of the material in the various chapters of both volumes are illustrated in the accompanying diagram.

Chapter 1 of this second volume is closely related to the Chaps. 5,6 of Vol. I in which the nonequilibrium thermodynamics of arbitrary systems with after-effect is considered. In it the reader will find the non-Markov generating equation from which FDTs of all degrees of nonlinearity can easily be obtained. The quantum generalization of the concept of Markov process is also considered here. Two different definitions of the quantum Markov process are given.

The three other chapters stand somewhat apart. Chapter 2 is devoted to the nonequilibrium thermodynamics of open systems. Just like Chaps. 3 and 4 of the Vol. I it heavily relies on the Markov techniques, but the theory no longer draws to the same degree on the results of equilibrium thermodynamics. Chapter 3 is devoted to wave processes. Nonlinear interaction of waves with reflectors and scatterers is also governed by general nonequilibrium thermodynamic laws, which are essentially a generalization of the well-known Kirchhoff law to coherent waves and nonlinear

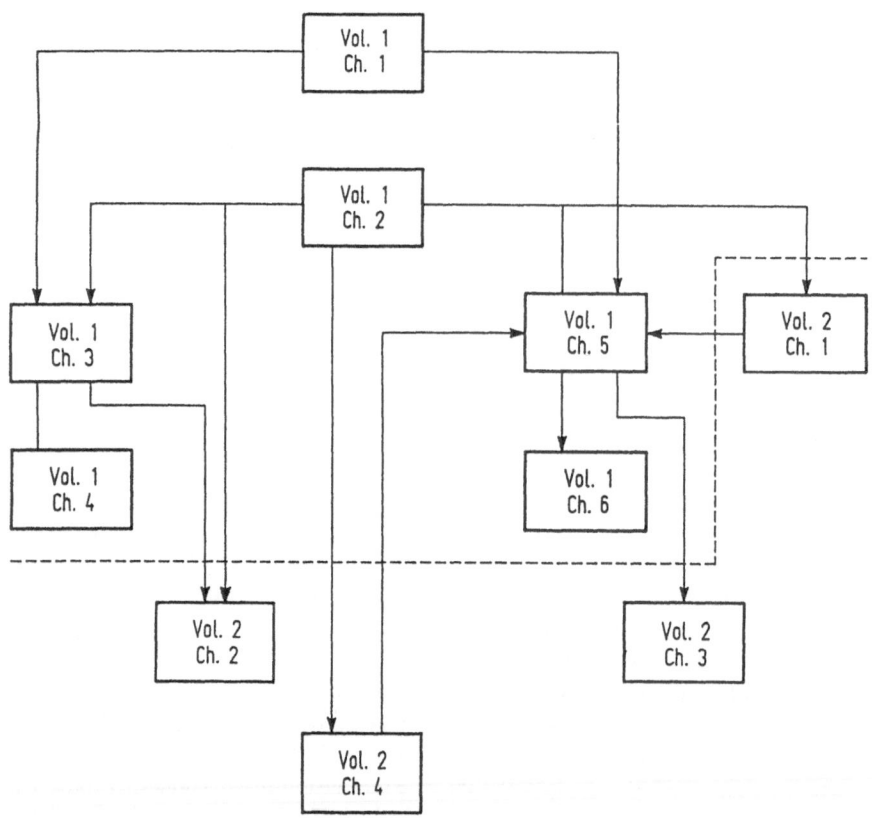

interactions. Linear and nonlinear fluctuation-dissipation relations (FDRs) for co-herent waves interacting with a physical body can be called Kirchhoff-type FDRs.

Chapter 4 treats the motion in phase space used to describe microprocesses in a physical system, and applies the technique of phase space projection operators in nonequilibrium theory. Formally, this topic lies beyond the scope of nonequilibrium thermodynamics, but it is closely related to this theory. Here we solve the problem of how to express the operator of the physical Markov process in terms of micro-dynamics. By methods of projection operators in phase space we can also derive universal thermodynamic relations. This will be demonstrated with reference to linear relations.

The level of rigor and mathematical techniques in the book are those gen-erally accepted in theoretical physics (see, e.g., Course of Theoretical Physics by L.D.Landau and E.M.Lifshits).

The book will appeal to theoretical physicists and applied scientists. The latter may wish to ignore the derivations of some of the universal relations and simply utilize them to handle problems of interest. The theoretical treatment is supported by numerous illustrative examples and applications of the general results to a va-riety of electrical, thermal, mechanical and chemical systems. It is shown that the most complete analysis of specific systems is achieved by the fusion of the Markov and non-Markov techniques of nonequilibrium thermodynamics.

The bibliography, especially in relation to linear nonequilibrium thermodynamics, and the nonequilibrium thermodynamics of open systems, does not claim to be exhaustive.

This second volume of the book was translated from the Russian by Mr. A.P.Repiev.

I would like to thank Professor H.Haken for his suggestion to publish this book in the Springer Series in Synergetics. I am grateful to the first reader of the book Professor Yu.Klimontovich for his useful comments. I also thank Angela Lahee for her careful editing of the manuscript.

Moscow, October 1993   R.L. Stratonovich

Contents

1. Generating Equations of Non-Markov Theory

In the general non-Markov case, just as for the Markov case, we can derive general generating equations, from which, by differentiation, we can obtain various systems of non-Markov FDRs. To begin with, we will consider two generating equations, which correspond to processes occurring under step-wise external forces. They give rise to two-time FDRs. Also derived in this chapter is the generating equation for the case of arbitrary external forces, which give rise to FDRs of the second kind.

The generating equations will be given here in two forms: quantum and non-quantum ones. Quantum relations are of course more difficult to derive.

The chapter considers quasi-classical moments of quantum theory and the related quasi-classical characteristic functional. The generating equation, which the latter obeys, exactly coincides with the nonquantum one. We discuss the possibility of a quantum generalization of the concept of the Markov process. Two different definitions of quantum Markov processes will be given and explored.

1.1 Generating Equations for Step-Driven Processes

1.1.1 Characteristic Function of Simple Relaxation Process

Nonequilibrium processes are called step-driven if they occur under the action of external forces having a step-wise nature:

$$h_\alpha(t) = x_\alpha \eta(t_0 - t) = \begin{cases} x_\alpha & \text{for } t \leq t_0, \\ 0 & \text{for } t > t_0 \end{cases} , \tag{1.1.1}$$

where the values of x are independent of time and characterize the "height of a step".

These processes are discussed in the works of *Bernard* and *Callen* [1.1] and *Stratonovich* [1.2]. They derived the earliest results of nonlinear nonequilibrium thermodynamics. Bernard and Callen considered step-driven processes and came quite close to the derivation of the quadratic FDRs [(5.3.71,90) v.1]. Stratonovich obtained generating equations for step-driven processes.

With the external forces (1.1.1) the situation becomes simpler because in the case of conventional mixing systems, when the constant forces $h_\alpha(t) = x_\alpha$ act for infinitely long time, an equilibrium sets in that corresponds to those forces, i.e. for $t \leq t_0$ the system is described by the equilibrium (for forces x_α) Gibbs distribution

$$w_x(z) = C_x \exp[-\beta\mathcal{H}(z) + \beta x B(z)] . \qquad (1.1.2)$$

After the forces x_α are switched off at time t_0, this distribution becomes nonequilibrium and in the system a relaxation process (which we will call simple) sets in – namely, the process of relaxation of internal parameters from the values corresponding to the probability density (1.1.2) to values corresponding to a new equilibrium probability density

$$w_0(z) = C_0 \exp(-\beta\mathcal{H}(z)) . \qquad (1.1.3)$$

Let us introduce a one-time characteristic function describing the above-mentioned relaxation process

$$\Theta_x(iu, t - t_0) = \int \exp(iuB(z))w(z,t)dz , \quad t \geq t_0 . \qquad (1.1.4)$$

Naturally, it turns out to be dependent on time t_0 of switching-off of the forces and on their value in the past.

Using [(2.1.10) v.1] we can write (1.1.4) as follows:

$$\Theta_x(v, t - t_0) = \exp\left[\sum_{n=1}^{\infty} \frac{1}{n!} \sum_{\alpha_1 \ldots \alpha_n} k_{\alpha_1 \ldots \alpha_n}(t - t_0, x)v_{\alpha_1} \ldots v_{\alpha_n}\right] , \qquad (1.1.5)$$

where

$$k_{\alpha_1 \ldots \alpha_n}(t - t_0, x) = \langle B_{\alpha_1}(t), \ldots, B_{\alpha_n}(t)\rangle_h \qquad (1.1.6)$$

are one-time nonequilibrium correlators. If we put $n = 1$ in (1.1.6), we will obtain a nonequilibrium mean, which is known to obey

$$A_1 = G_{1,2}h_2 + \frac{1}{2}G_{1,23}h_2h_3 + \frac{1}{6}G_{1,234}h_2h_3h_4 + \ldots \qquad (1.1.7)$$

(see Sect. 5.2.1 v.1). Substituting (1.1.1) into (1.1.7) gives

$$\begin{aligned}
k_\alpha(t - t_0, x) &= x_\beta \int_{-\infty}^{t_0} G_{\alpha,\beta}(t, t_2)dt_2 \\
&+ \frac{1}{2}x_\beta x_\gamma \int_{-\infty}^{t_0}\int_{-\infty}^{t_0} G_{\alpha,\beta\gamma}(t, t_2, t_3)dt_2dt_3 + \ldots .
\end{aligned} \qquad (1.1.8)$$

Similarly to (1.1.7), we can write expressions for higher nonequilibrium correlators

$$\begin{aligned}
\langle B_1, \ldots, B_n\rangle_{h(\tau)} &= \langle B_1, \ldots, B_n\rangle_0 + G_{1\ldots n, n+1}h_{n+1} \\
&+ \frac{1}{2}G_{1\ldots n; n+1, n+2}h_{n+1}h_{n+2} + \ldots ,
\end{aligned} \qquad (1.1.9)$$

where $\langle B_1, \ldots, B_n\rangle_0$ is the equilibrium correlator corresponding to the probability density (1.1.3). The functions $G_{1\ldots n, n+1\ldots}$ that appear in this expression can be called superadmittances. Equating the times $t_1 = t_2 = \ldots = t_n = t$ in (1.1.9) and substituting (1.1.1), we obtain

$$k_{\alpha_1 \ldots \alpha_n}(t - t_0, x') = k^0_{\alpha_1 \ldots \alpha_n} + x_\beta \int_{-\infty}^{t_0} G_{\alpha_1 \ldots \alpha_n, \beta}(t, \ldots, t; t')dt' + \frac{1}{2}x_\beta x_\gamma$$

$$\times \int_{-\infty}^{t_0}\int_{-\infty}^{t_0} G_{\alpha_1 \ldots \alpha_n, \beta\gamma}(t, \ldots, t; t', t'')dt'dt'' + \ldots \qquad (1.1.10)$$

$(k^0_{\alpha_1 \dots \alpha_n} = \langle B_1, \dots, B_n \rangle_0$ for $t_1 = \dots = t_n = t)$. Thus, from (1.1.5,6,8,10) we will have

$$\ln \Theta_x(v, t - t_0)$$
$$= \ln \Theta_0(v) + \sum_{m,n=1}^{\infty} \frac{1}{n!m!} v_{\alpha_1} \dots v_{\alpha_n} x_{\beta_1} \dots x_{\beta_m}$$
$$\times \int_{-\infty}^{t_0} \dots \int_{-\infty}^{t_0} G_{\alpha_1 \dots \alpha_n, \beta_1 \dots \beta_m}(t, \dots, t; t'_1, \dots, t'_m) dt'_1 \dots dt'_m, \quad (1.1.11)$$

where $\Theta_0(v)$ is the x-independent characteristic function for the probability density (1.1.3). Consequently, the logarithm of the characteristic function (1.1.4) up to the x-independent term is the generating function for the integrals

$$\int_{-\infty}^{t_0} \dots \int_{-\infty}^{t_0} G_{\alpha_1 \dots \alpha_n, \beta_1 \dots \beta_m}(t, \dots, t; t'_1, \dots, t'_m) dt'_1 \dots dt'_m. \quad (1.1.12)$$

1.1.2 Derivation of the Generating Equation

For the moment we will confine ourselves to the nonquantum case. In addition to the one-time nonequilibrium characteristic function (1.1.4), we can also introduce the two-time equilibrium characteristic function

$$\Theta_{eq}(iu, iu', t_1 - t_2) = \langle \exp[iu_\alpha B_\alpha(z(t_1)) + iu'_\alpha B_\alpha(z(t_2))] \rangle_0, \quad t_1 \geq t_2, (1.1.13)$$

which corresponds to the complete absence of external forces: $h(t) \equiv 0$. We will take into account that at time $t_2 \leq t_1$ the Gibbs probability density (1.1.3) is effective and that the dynamic variables $z(t_1)$ can be expressed, using the Hamilton equations, in terms of the dynamic variables at a previous moment

$$z(t_1) = \varphi_{t_1 - t_2}(z(t_2)), \quad t_1 \geq t_2. \quad (1.1.14)$$

Therefore, we can write (1.1.13) as

$$\Theta_{eq}(v, v', t_1 - t_2) = \int \exp[vB(\varphi_{t_1 - t_2}(z^{(2)}))$$
$$+ v'B(z^{(2)})]C_0 \exp(-\beta\mathcal{H}(z^{(2)}))dz^{(2)}. \quad (1.1.15)$$

At the same time, using (1.1.14) (at $t_1 = t, t_2 = t_0$) and the probability density (1.1.2) taken at t_0, we reduce (1.1.4) to the form

$$\Theta_x(v, t - t_0) = \int \exp[vB(\varphi_{t - t_0}(z))]C_x \exp[-\beta\mathcal{H}(z) + \beta x B(z)]dz. \quad (1.1.16)$$

Comparing (1.1.15) with (1.1.16) gives the expression

$$\Theta_x(v, t - t_0) = (C_x/C_0)\Theta_{eq}(v, \beta x, t - t_0) \quad (1.1.17)$$

which relates the nonequilibrium characteristic function to the two-time equilibrium function. If we put $v = 0$ in (1.1.17) and take into account that $\Theta_x(0, t - t_0) = 1$ and $\Theta_{eq}(0, v', t - t_0)$ is nothing but the one-time equilibrium characteristic function $\Theta_0(v')$ given, say, by [(2.2.28) v.1], we will have

$$C_0/C_x = \Theta_0(\beta x)\,. \tag{1.1.18}$$

Therefore, (1.1.17) can be written as

$$\ln \Theta_{kTv'}(v, t - t_0) = \ln \Theta_{eq}(v, v', t - t_0) - \ln \Theta_0(v')\,. \tag{1.1.19}$$

This is the desired generating equation. Using (1.1.11) and the expression

$$
\begin{aligned}
&\ln \Theta_{eq}(v, v', t - t_0) \\
&= \sum_{n,m} \frac{1}{n!m!} \sum_{\alpha_1\ldots\alpha_n} \sum_{\beta_1\ldots\beta_m} k_{\alpha_1\ldots\alpha_n,\beta_1\ldots\beta_m}(t - t_0) v_{\alpha_1} \ldots v_{\alpha_n} v'_{\beta_1} \ldots v'_{\beta_m}\,,
\end{aligned}
$$

where

$$k_{\alpha_1\ldots\alpha_n,\beta_1\ldots\beta_m}(t - t_0) = \langle B_{\alpha_1}(t), \ldots, B_{\alpha_n}(t), B_{\beta_1}(t_0), \ldots, B_{\beta_m}(t_0)\rangle_0\,, \tag{1.1.20}$$

we can obtain, from (1.1.19), various FDRs relating the integrals (1.1.12) and the two-time equilibrium correlators (1.1.20). For this purpose, we have to differentiate (1.1.19) several times with respect to v and v' and then set v and v' to zero. Specifically, we can easily get

$$(kT)^2 \int_{-\infty}^{t_0} \int_{-\infty}^{t_0} G_{\alpha,\beta\gamma}(t; t', t'') dt' dt'' = \langle B_\alpha(t), B_\beta(t_0), B_\gamma(t_0)\rangle_0\,, \tag{1.1.21a}$$

$$
\begin{aligned}
(kT)^2 &\int_{-\infty}^{t_0} \int_{-\infty}^{t_0} G_{\alpha\beta,\gamma\delta}(t, t; t', t'') dt' dt'' \\
&= \langle B_\alpha(t), B_\beta(t), B_\gamma(t_0), B_\delta(t_0)\rangle_0\,,
\end{aligned} \tag{1.1.21b}
$$

and also many other relations. Note that we can also verify the validity of (1.1.21a) by using [(5.3.78) v.1]. Significantly, from (1.1.19) we can also derive the relations with five and more subscripts.

1.1.3 Generating Equation Obtained from Time Reversibility

Suppose that the condition of time reversibility $\mathcal{H}(\varepsilon z) = \mathcal{H}(z)$ or $w(\varepsilon B) = w(B)$ (see Sect. 3.2.1 v.1) is satisfied. Then the equilibrium characteristic function satisfies the following condition of time-reversal invariance

$$\Theta_{eq}(\varepsilon v', \varepsilon v, \tau) = \Theta_{eq}(v, v', \tau)\,, \quad \tau \geq 0\,. \tag{1.1.22}$$

To derive this, we will have to take into account that, due to the time invariance, the means

$$
\begin{aligned}
&\langle \exp[vB(t_1) + v'B(t_2)]\rangle_0\,, \\
&\langle \exp[v\tilde{B}(\tilde{t}_1) + v'\tilde{B}(\tilde{t}_2)]\rangle_0\,,
\end{aligned} \tag{1.1.23}
$$

corresponding to the forward and reverse times are equal to the same function:

$$\langle \exp[vB(t_1) + v'B(t_2)]\rangle_0 = \Theta_{eq}(v, v', t_1 - t_2) \quad \text{for} \quad t_1 \geq t_2\,, \tag{1.1.24}$$

$$\langle \exp[v\tilde{B}(\tilde{t}_1) + v'\tilde{B}(\tilde{t}_2)]\rangle_0 = \Theta_{eq}(v, v', \tilde{t}_1 - \tilde{t}_2) \quad \text{for} \quad \tilde{t}_1 \geq \tilde{t}_2\,, \tag{1.1.25}$$

An earlier instant in reverse time is a later one in forward time. Using also the time signatures ε_α of internal parameters, we will have

$$\begin{aligned} \tilde{B}_\alpha(\tilde{t}_1) &= \varepsilon_\alpha B_\alpha(t_2)\,, \\ \tilde{B}_\alpha(\tilde{t}_2) &= \varepsilon_\alpha B_\alpha(t_1)\,, \end{aligned} \quad (\tilde{t}_1 - \tilde{t}_2 = t_1 - t_2)\,. \tag{1.1.26}$$

Substituting (1.1.26) into (1.1.25) gives

$$\langle \exp[v_\alpha \varepsilon_\alpha B_\alpha(t_2) + v'_\alpha \varepsilon_\alpha B_\alpha(t_1)] \rangle_0 = \Theta_{eq}(v, v', t_1 - t_2)\,, \quad t_1 \geq t_2\,. \tag{1.1.27}$$

But the left-hand side of this, by (1.1.24), is nothing but $\Theta_{eq}(\varepsilon v', \varepsilon v, t_1 - t_2)$. Therefore, (1.1.27) yields (1.1.22). Using (1.1.19), we can derive from (1.1.22) the generating equation involving the nonequilibrium characteristic functions

$$\ln \Theta_{kTv'}(v, t - t_0) + \ln \Theta_0(v') = \ln \Theta_{kT\varepsilon v}(\varepsilon v', t - t_0) + \ln \Theta_0(v)\,. \tag{1.1.28}$$

Differentiating this we can obtain various relations, for example the following ones:

$$\int_{-\infty}^{t_0} G_{\alpha,\beta}(t;t')dt' = \varepsilon_\alpha \varepsilon_\beta \int_{-\infty}^{t_0} G_{\beta,\alpha}(t;t')dt'\,, \tag{1.1.29a}$$

$$kT \int_{-\infty}^{t_0}\int_{-\infty}^{t_0} G_{\alpha,\beta\gamma}(t;t',t'')dt'dt'' = \varepsilon_\alpha \varepsilon_\beta \varepsilon_\gamma \int_{-\infty}^{t_0} G_{\beta\gamma,\alpha}(t,t;t')dt'\,, \tag{1.1.29b}$$

$$\int_{-\infty}^{t_0}\int_{-\infty}^{t_0} G_{\alpha\beta,\gamma\delta}(t,t;t',t'')dt'dt''$$
$$= \varepsilon_\alpha \varepsilon_\beta \varepsilon_\gamma \varepsilon_\delta \int_{-\infty}^{t_0}\int_{-\infty}^{t_0} G_{\gamma\delta,\alpha\beta}(t,t;t',t'')dt'dt''\,. \tag{1.1.29c}$$

The first of these is equivalent to the reciprocal relation, which is concerned only with the nonfluctuational behavior of a system, i.e. with the admittances $G_{1,2\ldots n}$. Other reciprocal relations concern also the superadmittances $G_{12,3\ldots}, G_{123,4\ldots}, \cdots$, i.e. not only dissipational but also fluctuational characteristics. From (1.1.28), we can then obtain relations for five-subscript and other superadmittances. The general {generalizing [(5.3.45) v.1] } relation has the form

$$L_{\alpha_1\ldots\alpha_n,\beta_1\ldots\beta_m}(t_{12}) = \varepsilon_{\alpha_1}\ldots\varepsilon_{\beta_m} L_{\beta_1\ldots\beta_m,\alpha_1\ldots\alpha_n}(t_{12})\,, \quad t_{12} > 0\,, \tag{1.1.30}$$

where

$$L_{\alpha_1\ldots\alpha_n,\beta_1\ldots\beta_m}(t_{12})$$
$$= \int_{-\infty}^{t_2}\ldots\int_{-\infty}^{t_2} G_{\alpha_1\ldots\alpha_n,\beta_1\ldots\beta_m}(t_1,\ldots,t_1;t'_1,\ldots,t'_m)dt'_1\ldots dt'_m\,. \tag{1.1.31}$$

1.1.4 The First Generating Equation in the Quantum Case

The derivation of the generating equation given in Sect. 1.1.2 is not suitable in the quantum case since with the operator character of the parameters $\hat{B}_\alpha$ and the Hamiltonian $\hat{\mathcal{H}}$ the expression

$$\exp(-\beta\hat{\mathcal{H}} + \beta x \hat{B}) = \exp(-\beta\hat{\mathcal{H}})\exp(\beta x \hat{B}) \tag{1.1.32}$$

in the general case becomes invalid, and so more complicated reasoning is required. We will introduce the notation

$$\hat{D}(\lambda) = \exp(\lambda\hat{\mathcal{H}})\exp(-\lambda(\hat{\mathcal{H}} - x\hat{B}))\,. \qquad (1.1.33)$$

Now, differentiating (1.1.33) with respect to λ gives

$$d\hat{D}(\lambda)/d\lambda = \exp(\lambda\hat{\mathcal{H}})[\hat{\mathcal{H}} - (\hat{\mathcal{H}} - x\hat{B})]\exp[-\lambda(\hat{\mathcal{H}} - x\hat{B})] \qquad (1.1.34)$$

or

$$\begin{aligned} d\hat{D}(\lambda)/d\lambda &= \exp(\lambda\hat{\mathcal{H}})x\hat{B}\exp[-\lambda(\hat{\mathcal{H}} - x\hat{B})] \\ &= \exp(\lambda\hat{\mathcal{H}})x\hat{B}\exp(-\lambda\hat{\mathcal{H}})\hat{D}(\lambda)\,. \end{aligned} \qquad (1.1.35)$$

By virtue of [(5.2.41) v.1], the factor $\exp(\lambda\hat{\mathcal{H}})x\hat{B}(t)\exp(-\lambda\hat{\mathcal{H}})$ in (1.1.35) can be interpreted as

$$\exp(\lambda\hat{\mathcal{H}})x\hat{B}(t)\exp(-\lambda\hat{\mathcal{H}}) = x\hat{B}(t - i\hbar\lambda)\,. \qquad (1.1.36)$$

Integrating (1.1.35) subject to the initial condition

$$\hat{D}(0) = \hat{1} \qquad (1.1.37)$$

gives

$$\exp(\beta\hat{\mathcal{H}})\exp(-\beta\hat{\mathcal{H}} + \beta x\hat{B}) \equiv \hat{D}(\beta) = \overleftarrow{\exp}\left[\int_0^\beta d\lambda\, x\hat{B}(t - i\hbar\lambda)\right]\,. \qquad (1.1.38)$$

Here the arrow over 'exp' has the same meaning as in [(5.2.44) v.1], where it orders the operators $\hat{B}(t - i\hbar\lambda)$ { see [(5.2.45) v.1]}. The only difference is that in [(5.2.44) v.1] the ordering is in time, and in (1.1.38) in λ : the larger λ the farther to the left stands the appropriate operator.

In the quantum case (1.1.16) takes the form

$$\Theta_x(v, t - t_0) = \text{Tr}\{C_x \exp[-\beta\hat{\mathcal{H}} + \beta x\hat{B}(t_0)]\exp v\hat{B}(t)\}\,, \quad t > t_0\,. \qquad (1.1.39)$$

Substituting (1.1.38), or rather

$$\exp(-\beta\hat{\mathcal{H}} + \beta x\hat{B}(t_0)) = \exp(-\beta\hat{\mathcal{H}})\overleftarrow{\exp}\left[\int_0^\beta d\lambda\, x\hat{B}(t_0 - i\hbar\lambda)\right]\,, \qquad (1.1.40)$$

we get

$$\Theta_x(v, t - t_0) = C_x C_0^{-1}\text{Tr}\left\{\hat{\rho}_0\,\overleftarrow{\exp}\left[\int_0^\beta d\lambda\, x\hat{B}(t_0 - i\hbar\lambda)\right]\exp(v\hat{B}(t))\right\}\,, \qquad (1.1.41)$$

where

$$\hat{\rho}_0 = C_0\exp(-\beta\hat{\mathcal{H}}) \qquad (1.1.42)$$

is the equilibrium density matrix without any external forces. As in Sect. 1.1.2, letting $v = 0$ and using (1.1.41) we can easily prove the equality (1.1.18), which enables C_x/C_0 to be eliminated from (1.1.41) (it can also be expressed through the free energy $F(a)$). We thus arrive at

$$\Theta_x(v, t - t_1)$$
$$= \left\langle \overleftarrow{\exp} \left[\int_0^\beta d\lambda x \hat{B}(t_1 - i\hbar\lambda) \right] \exp(v\hat{B}(t)) \right\rangle_0 \exp(\beta F_x - \beta F_0) . \quad (1.1.43)$$

Here we have substituted t_1 for t_0. Expression (1.1.43) serves as a quantum generalization of the generating equation (1.1.19). Considering that

$$B(t_1 - i\hbar\lambda) = \exp(-i\hbar\lambda p_1)B(t_1) \quad (p_1 = \partial/\partial t_1), \quad (1.1.44)$$

we can write it as

$$\Theta_x(v, t - t_1) \exp(\beta F_0 - \beta F_x)$$
$$= \left\langle \overleftarrow{\exp} \left[\int_0^\beta d\lambda \exp(-i\hbar\lambda p_1)x B(t_1)_\lambda \right] \exp(v B(t)) \right\rangle_0 . \quad (1.1.45)$$

Here the subscript λ of $B(t_1)$ (a second argument) is the index ordering the operators. This means that at all λ the operator $B(t_1)_\lambda$ coincides with $B(t_1)$, but the value of λ in $B(t_1)_\lambda$ points to the place of the operator: the smaller λ the farther to the right stands the respective operator $B(t_1)_\lambda$.

The exponential function in (1.1.45) can be expanded into a Taylor series. In the process, the operators with ordering indices may be handled as numerical quantities. Referring to the expression

$$J = \left[\int_0^\beta d\lambda \exp(-i\hbar\lambda p)D(t)_\lambda \right]^2$$
$$= \int_0^\beta \int_0^\beta d\lambda_1 d\lambda_2 \exp[-i\hbar(\lambda_1 p_1 + \lambda_2 p_2)]D(t_1)_{\lambda_1} D(t_2)_{\lambda_2} \quad (1.1.46)$$

we will see how we can develop the power terms with operators ordered in λ_i. We thus have

$$J = \left\{ \int_0^\beta d\lambda_1 \int_0^{\lambda_1} d\lambda_2 + \int_0^\beta d\lambda_2 \int_0^{\lambda_2} d\lambda_1 \right\}$$
$$\times \exp[-i\hbar(\lambda_1 p_1 + \lambda_2 p_2)]D(t_1)_{\lambda_1} D(t_2)_{\lambda_2} \quad \text{for} \quad t_1 = t_2 = t . \quad (1.1.47)$$

As stated above, the smaller λ_i the more to the right must stand the respective operator that has on it the subscript λ_i. Therefore,

$$J = \beta^2 \Phi(p_1, p_2)D(t_1)D(t_2) \quad + \quad \beta^2 \Phi(p_2, p_1)D(t_2)D(t_1)$$
$$\text{for} \cdot t_1 = t_2 = t , \quad (1.1.48)$$

where

$$\Phi(p_1, p_2) = (kT)^2 \int_0^\beta d\lambda_1 \int_0^{\lambda_1} d\lambda_2 \exp(-i\hbar\lambda_1 p_1 - i\hbar\lambda_2 p_2) . \quad (1.1.49)$$

We need not write the subscripts λ_1 and λ_2 in (1.1.48) since the explicit ordering of the operators has already been carried out. Substituting in the second term of (1.1.48) t_2, p_2 for t_1, p_1 and vice versa, we can write (1.1.48) as

$$J = 2\beta^2 \Phi(p_1, p_2) D(t_1) D(t_2) \qquad (1.1.50)$$

for $t_1 = t_2 = t$. Similarly, introducing

$$\Phi(p_1, \ldots, p_m)$$
$$= (kT)^m \int_0^\beta d\lambda_1 \int_0^{\lambda_1} d\lambda_2 \ldots \int_0^{\lambda_{m-1}} d\lambda_m \exp(-i\hbar\lambda_1 p_1 - \ldots - i\hbar\lambda_m p_m),$$
$$m = 1, 2, 3, \ldots, \qquad (1.1.51)$$

we can write arbitrary power terms ordered in λ as follows:

$$\left[\int_0^\beta d\lambda \exp(-i\hbar\lambda p_1) x B(t_1)_\lambda \right]^m = \beta^m P_{1\ldots m} x_{\alpha_1} \ldots x_{\alpha_m} (\Phi_{1\ldots m} B_1 \ldots B_m). \qquad (1.1.52)$$

Changing the notation we obtain

$$\left[\int_0^\beta d\lambda \exp(-i\hbar\lambda p_1) x B(t_1)_\lambda \right]^m$$
$$= m! \beta^m x_{\alpha_1} \ldots x_{\alpha_m} (\Phi_{1\ldots m} B_1 \ldots B_m)_{t_2 = \ldots = t_m = t_1}, \qquad (1.1.53)$$

where, as usual, $B_1 = B_{\alpha_1}(t_1), \ldots$ and, in addition, $\Phi_{1\ldots m} = \Phi(p_1, \ldots, p_m)$. To be more specific, we will write some of the lower functions (1.1.51):

$$\Phi(p_1) = y_1^{-1}(1 - \exp(-y_1)) = 1/\Theta^+(p_1), \qquad (1.1.54a)$$

$$\Phi(p_1, p_2) =$$
$$y_2^{-1}[y_1^{-1}(1 - \exp(-y_1)) - (y_1 + y_2)^{-1}(1 - \exp(-y_1 - y_2))] \qquad (1.1.54b)$$

$(y_1 = i\beta\hbar p_1)$. In the nonquantum limit as $\hbar \longrightarrow 0, \Phi(p_1, \ldots, p_m)$ goes over into $1/m!$.

Expanding the exponential function in (1.1.45) into a Taylor series and using (1.1.53), we obtain the generating equation of the form

$$\Theta_x(v, t - t_1) \exp(\beta F_0 - \beta F_x)$$
$$= \left\langle \beta^m \sum_{m=0}^\infty x_{\alpha_1} \ldots x_{\alpha_m} (\Phi_{1\ldots m} B_1 \ldots B_m) \exp(v B(t)) \right\rangle_0. \qquad (1.1.55)$$

1.1.5 Some Uses of Quantum Generating Equations

In the quantum case, instead of (1.1.21) we can derive from (1.1.55)

$$\int_{-\infty}^{t_1} \int_{-\infty}^{t_1} G_{\alpha,\alpha_1\alpha_2}(t; t', t'') dt' dt'' = \beta^2 P_{12} \Phi(p_1, p_2) \langle B_{\alpha_1}(t_1), B_{\alpha_2}(t_2), B_\alpha(t) \rangle_0,$$

$$\int_{-\infty}^{t_1} \int_{-\infty}^{t_1} G_{\alpha\beta,\alpha_1\alpha_2}(t; t, t', t'') dt' dt''$$
$$= \beta^2 P_{12} \Phi(p_1, p_2) \langle B_{\alpha_1}(t_1), B_{\alpha_2}(t_2), B_\alpha(t) B_\beta(t) \rangle_0, \qquad (1.1.56)$$

for $t_2 = t_1 < t$, where $\Phi(p_1, p_2)$ is given by (1.1.54b).

Now let us consider another special case. Setting $v = 0$ in (1.1.45) yields

$$\sum_{m=1}^{\infty} \frac{\beta^m}{m!} \sum_{\alpha_1 \dots \alpha_m} x_{\alpha_1} \dots x_{\alpha_m} P_{1\dots m}[\Phi_{1\dots m}\langle B_1 \dots B_m \rangle_0]_{t_2=\dots=t_m=t_1}$$
$$= \exp[-\beta F(a+x) + \beta F(a)]. \tag{1.1.57}$$

Consequently in the quantum case the numbers $m_{\alpha_1 \dots \alpha_n}$, which are determined by the free energy, i.e. by

$$\exp[-\beta F(a + kTv) + \beta F(a)] = \sum_{n=0}^{\infty} \frac{1}{n!} \sum_{\alpha} m_{\alpha_1 \dots \alpha_n} v_{\alpha_1} \dots v_{\alpha_n} \tag{1.1.57a}$$

(see [(2.2.28) v.1]) and which, in the nonquantum case, coincide with the one-time equilibrium moments $\langle B_{\alpha_1} \dots B_{\alpha_n} \rangle_{\text{eq}}$, are related to the equilibrium moments by

$$m_{\alpha_1 \dots \alpha_n} = P_{1\dots n}[\Phi(p_1, \dots, p_n)\langle B_{\alpha_1}(t_1) \dots B_{\alpha_n}(t_n) \rangle_0]_{t_2=\dots=t_n=t_1}. \tag{1.1.58}$$

Using the equation

$$P_{1\dots n}(\Phi_{1\dots n}\langle B_1 \dots B_n \rangle_0) = P_{1\dots(n-1)}(\Phi_{1\dots(n-1)}\langle B_1 \dots B_n \rangle_0), \tag{1.1.59}$$

proved in Sect. A1.2, the last equation becomes

$$m_{\alpha_1 \dots \alpha_n} = M^{\text{q.c}}_{\alpha_1 \dots \alpha_n}(t_1, \dots, t_1), \tag{1.1.60}$$

where

$$M^{\text{q.c.}}_{1\dots n} = P_{1\dots(n-1)}(\Phi_{1\dots(n-1)}\langle B_1 \dots B_n \rangle_0). \tag{1.1.61}$$

Expressions (1.1.61) can be referred to as quasi-classical moments. We see that in the quantum case they are related to free energy by the same equations by which the moments are related to free energy in the nonquantum case.

1.1.6 Second Quantum Generating Equation

Using the same procedure as in the derivation of (1.1.45) we can obtain the equation

$$\langle \hat{L} \rangle_{h(\tau)} = \left\langle \overleftarrow{\exp} \left[\int_0^\beta d\lambda \exp(-i\hbar\lambda p_2) x B(t_2)_\lambda \right] \hat{L} \right\rangle_0 \exp(\beta F_x - \beta F_0), \tag{1.1.62}$$

where $\hat{L}$ is the arbitrary operator expressed, say, in terms of $B_\alpha(t)$, and the external force $h(\tau)$ on the left-hand side is given by the expression

$$h_\alpha(\tau) = x_\alpha \eta(t_2 - t), \tag{1.1.63}$$

similar to (1.1.1). This means that the mean here is for a step-driven process.

Using (1.1.36), we can represent (1.1.62) in the form

$$
\langle \hat{L} \rangle_{h(\tau)}
= \left\langle \overleftarrow{\exp} \left[\int_0^\beta d\lambda (\exp(\lambda \mathcal{H}) x B(t_2) \exp(-\lambda \mathcal{H}))_\lambda \right] \hat{L} \right\rangle_0 \exp(\beta F_x - \beta F_0).
$$

$$(1.1.64)$$

We apply the operation of complex conjugation to both sides of this expression. Using

$$
[\mathrm{Tr}(AB \ldots Z)]^* = \mathrm{Tr}[(AB \ldots Z)^+] = \mathrm{Tr}(Z^+ \ldots B^+ A^+) \tag{1.1.65}
$$

yields

$$
\langle L^+ \rangle_{h(\tau)}
= \left\langle L^+ \overrightarrow{\exp} \left[\int_0^\beta d\lambda (\exp(-\lambda \mathcal{H}) x B(t_2) \exp(\lambda \mathcal{H}))_\lambda \right] \right\rangle_0 \exp(\beta F_x - \beta F_0)
$$

$$(1.1.66)$$

where $\overrightarrow{\exp}$ means that the smaller λ, the farther to the left stands the operator $(\ldots)_\lambda$.

Now let us introduce

$$
\Psi_{t_1 - t_2}(y, x) = \left\langle \overrightarrow{\exp} \left[\int_0^\beta d\mu \exp(i\hbar \mu p_1) y B(t_1)_\mu \right] \right\rangle_{h(\tau)} \tag{1.1.67}
$$

or, by (1.1.36),

$$
\Psi_{t_1 - t_2}(y, x) = \left\langle \overrightarrow{\exp} \left[\int_0^\beta d\mu (\exp(-\mu \mathcal{H}) y B(t_1) \exp(\mu \mathcal{H}))_\mu \right] \right\rangle_{x(t_2)}. \tag{1.1.68}
$$

In the subscript $x(t_2)$ the time t_2 is the instant when the forces are switched off, and the operator ordering is in μ.

Next we apply (1.1.62) to (1.1.67), the operator L being taken to be equal to the averaged exponent. As a result, we obtain

$$
\begin{aligned}
\Psi_{t_1 - t_2}(y, x) &= \left\langle \overleftarrow{\exp} \left[\int_0^\beta d\lambda \exp(-i\hbar \lambda p_2) x B(t_2)_\lambda \right] \right. \\
&\quad \times \left. \overrightarrow{\exp} \left[\int_0^\beta d\mu \exp(i\hbar \mu p_1) y B(t_1)_\mu \right] \right\rangle_0 \exp(\beta F_x - \beta F_0).
\end{aligned}
$$

$$(1.1.69)$$

We will now go over to the reverse time $\tilde{t} = -t + \mathrm{const}$, for which, instead of $B(t)$, we will have $\tilde{B}(\tilde{t})$. The time-reversed form of (1.1.68) has a similar form

$$
\Psi_{\tilde{t}_1 - \tilde{t}_2}(\tilde{y}, \tilde{x}) = \left\langle \overrightarrow{\exp} \left[\int_0^\beta d\mu (\exp(-\mu \mathcal{H}) \tilde{y} \tilde{B}(\tilde{t}_1) \exp(\mu \mathcal{H}))_\mu \right] \right\rangle_{\tilde{x}(\tilde{t}_2)}. \tag{1.1.70}
$$

If the time reversibility condition that follows from [(5.3.41) v.1] is met, the left-hand sides of (1.1.68) and (1.1.70) include the same function $\Psi_\tau(y,x)$ of different arguments. And the time-reversed form of (1.1.66), where $\langle L^+ \rangle_{h(\tau)} = \langle L^+ \rangle_{x(t_2)}$, is

$$\langle L^+ \rangle_{\tilde{h}(\tilde{\tau})} \equiv \langle L^+ \rangle_{\tilde{x}(\tilde{t}_2)}$$
$$= \left\langle L^+ \overrightarrow{\exp} \left[\int_0^\beta d\lambda (\exp(-\lambda\mathcal{H}) \tilde{x} \tilde{B}(\tilde{t}_2) \exp(\lambda\mathcal{H}))_\lambda \right] \right\rangle_0 f(\tilde{x}) \qquad (1.1.71)$$

$[f(\tilde{x}) = \exp(\beta F_{\tilde{x}} - \beta F_0)]$. We apply to both sides of (1.1.70) the operation of complex conjugation. From (1.1.65,70) we will have

$$\Psi^*_{\tilde{t}_1 - \tilde{t}_2}(\tilde{y}, \tilde{x})$$
$$= \left\langle \overleftarrow{\exp} \left[\int_0^\beta d\mu (\exp(\mu\mathcal{H}) \tilde{y} \tilde{B}(\tilde{t}_1) \exp(-\mu\mathcal{H}))_\mu \right] \right\rangle_{\tilde{x}(\tilde{t}_2)} . \qquad (1.1.72)$$

Taking the operator L^+ in (1.1.71) to be equal to the exponential expression averaged on the right-hand side of (1.1.72) yields

$$\Psi^*_{\tilde{t}_1 - \tilde{t}_2}(\tilde{y}, \tilde{x}) f^{-1}(\tilde{x}) = \left\langle \overleftarrow{\exp} \left[\int_0^\beta d\mu (\exp(\mu\mathcal{H}) \tilde{y} \tilde{B}(\tilde{t}_1) \exp(-\mu\mathcal{H}))_\mu \right] \right.$$
$$\left. \times \overrightarrow{\exp} \left[\int_0^\beta d\lambda (\exp(-\lambda\mathcal{H}) \tilde{x} \tilde{B}(\tilde{t}_2) \exp(\lambda\mathcal{H}))_\lambda \right] \right\rangle_0 . \qquad (1.1.73)$$

Now recall that, as stated in Sect. 5.3.2 v.1,

$$\tilde{D}(\tilde{t}) = \varepsilon_D D^*(t(\tilde{t}))$$
$$= \varepsilon_D D^*(-\tilde{t} + \text{const}), \qquad (1.1.74)$$

where D is an arbitrary operator, including B_α. By (1.1.74), we have the equalities

$$\tilde{B}_\alpha(\tilde{t}_1) = \varepsilon_\alpha B^*_\alpha(t_2),$$
$$\tilde{B}_\alpha(\tilde{t}_2) = \varepsilon_\alpha B^*_\alpha(t_1), \qquad (1.1.75)$$

which generalize (1.1.26) to the quantum case. Therefore, (1.1.73) becomes

$$\Psi^*_{\tilde{t}_1 - \tilde{t}_2}(\tilde{y}, \tilde{x}) f^{-1}(\tilde{x}) = \left\langle \overleftarrow{\exp} \left[\int_0^\beta d\lambda (\exp(\lambda\mathcal{H}) \varepsilon \tilde{y} B(t_2) \exp(-\lambda\mathcal{H}))_\lambda \right] \right.$$
$$\left. \times \overrightarrow{\exp} \left[\int_0^\beta d\mu (\exp(-\mu\mathcal{H}) \varepsilon \tilde{x} B(t_1) \exp(\mu\mathcal{H}))_\mu \right] \right\rangle_0^* . \qquad (1.1.76)$$

(here we have replaced $\lambda \rightleftarrows \mu$).

The complex conjugation operation on the right-hand side of (1.1.76) comes from (1.1.75). Letting $\tilde{y} = \varepsilon x, \tilde{x} = \varepsilon y, \tilde{t}_1 - \tilde{t}_2 = t_1 - t_2 = \tau$ and again using (1.1.36) yields

$$\Psi_\tau(\varepsilon x, \varepsilon y) f^{-1}(y) = \left\langle \overleftarrow{\exp} \left[\int_0^\beta d\lambda \exp(-i\hbar\lambda p_2) x B(t_2)_\lambda \right] \right.$$

$$\left. \times \overrightarrow{\exp} \left[\int_0^\beta d\mu \exp(i\hbar\mu p_1) y B(t_1)_\mu \right] \right\rangle_0 . \qquad (1.1.77)$$

The equilibrium means in (1.1.69) and (1.1.77) coincide. It follows from this that

$$\Psi_\tau(y, x) \exp(\beta F_0 - \beta F_x) = \Psi_\tau(\varepsilon x, \varepsilon y) \exp(\beta F_0 - \beta F_y) \qquad (1.1.78)$$

or

$$\ln \Psi_\tau(y, x) - \beta F_x = \ln \Psi_\tau(\varepsilon x, \varepsilon y) - \beta F_y . \qquad (1.1.79)$$

Either of the formulas (1.1.78, 79) constitutes a quantum generating equation, as sought for; this follows from the time reversibility.

1.1.7 Simplest Uses of the Last Generating Equation

To derive corollaries from (1.1.79), we will expand (1.1.67) into a series. To this end, instead of (1.1.67) we can take

$$\Psi^*_{t_1-t_2}(y, x) = \left\langle \overleftarrow{\exp} \left[\int_0^\beta d\lambda \exp(-i\hbar\lambda p_1) y B(t_1)_\lambda \right] \right\rangle_{x(t_2)} \qquad (1.1.80)$$

similar to (1.1.72). Putting here $t_2 = t_0$, expanding the exponential function in (1.1.80) into a series and using (1.1.52) gives

$$\Psi^*_{t_1-t_0}(y, x) = \sum_{m=0}^\infty \beta^m y_{\alpha_1} \cdots y_{\alpha_m} \Phi_{1\ldots m} \langle B_{\alpha_1}(t_1) \ldots B_{\alpha_m}(t_m) \rangle_{x(t_0)} \qquad (1.1.81)$$

for $t_m = \ldots = t_2 = t_1$. The nonequilibrium moments here can be expressed by the conventional relationships of Sect. 2.1.2 v.1 in terms of the nonequilibrium correlators (1.1.10). The notation in (1.1.10) retains its meaning in the quantum case as well. We will now write several simple equations

$$\langle B_1 \rangle_{x(t_0)} = k_{\alpha_1}(t_1 - t_0, x)$$

$$= x_{\alpha_1} \int_{-\infty}^{t_0} G_{1,2} dt_2 + \tfrac{1}{2} x_{\alpha_2} x_{\alpha_3} \int_{-\infty}^{t_0} \int_{-\infty}^{t_0} G_{1,23} dt_2 dt_3 + \ldots,$$

$$\langle B_1 B_2 \rangle_{x(t_0)} = k_{\alpha_1\alpha_2}(t_1 - t_0, x) + k_{\alpha_1}(t_1 - t_0, x) k_{\alpha_2}(t_1 - t_0, x)$$

$$= G_{12} + x_{\alpha_3} \int_{-\infty}^{t_0} G_{12,3} dt_3 + \tfrac{1}{2} x_{\alpha_3} x_{\alpha_4} \int_{-\infty}^{t_0} \int_{-\infty}^{t_0} (G_{12,34}$$

$$+ G_{1,3} G_{2,4} + G_{1,4} G_{2,3}) dt_3 dt_4 + \ldots \qquad (1.1.82)$$

Expressions (1.1.82) must be substituted into (1.1.81), and (1.1.81) into (1.1.79) or, equivalently, into

$$\ln \Psi^*_\tau(y, x) - \beta F_x = \ln \Psi^*_\tau(\varepsilon x, \varepsilon y) - \beta F_y . \qquad (1.1.83)$$

Differentiating both sides of (1.1.83) several times with respect to x and several times with respect to y and then letting $x = y = 0$, we will obtain relations of various orders. Thus, differentiating (1.1.83) once with respect to y_α and once with respect to x_β gives

$$\Phi(p) \int_{-\infty}^{t_0} G_{\alpha,\beta}(t, t_2)dt_2 = \Phi(p)\varepsilon_\alpha\varepsilon_\beta \int_{-\infty}^{t_0} G_{\beta,\alpha}(t, t_2)dt_2 \,, \tag{1.1.84}$$

This is the expression (1.1.29a) if we omit the operator $\Phi(p)$. Further, if we differentiate (1.1.83) with respect to $y_\alpha, y_\beta, x_\gamma$ and equate x, y to zero, we will have

$$\int_{-\infty}^{t_0} [\Phi(p_1, p_2)G_{\alpha\beta,\gamma}(t_1, t_2; t_3) + \Phi(p_2, p_1)G_{\beta\alpha,\gamma}(t_2, t_1; t_3)]dt_3$$

$$= kT\varepsilon_\alpha\varepsilon_\beta\varepsilon_\gamma\Phi(p) \int_{-\infty}^{t_0}\int_{-\infty}^{t_0} G_{\gamma,\alpha,\beta}(t; t_2', t_3')dt_2'dt_3' \tag{1.1.85}$$

for $t_1 = t_2 = t$.

Lastly, differentiating (1.1.83) with respect to $y_\alpha, y_\beta, x_\gamma, x_\delta$ at zero gives

$$\Phi(p_1, p_2) \int_{-\infty}^{t_0}\int_{-\infty}^{t_0} [G_{\alpha\beta,\gamma\delta}(t_1, t_2; t_3, t_4) + G_{\beta\alpha,\gamma\delta}(t_1, t_2; t_3 t_4)]dt_3 dt_4$$

$$= \varepsilon_\alpha\varepsilon_\beta\varepsilon_\gamma\varepsilon_\delta\Phi(p_1, p_2) \int_{-\infty}^{t_0}\int_{-\infty}^{t_0} [G_{\gamma\delta,\alpha\beta}(t_1, t_2; t_3 t_4)$$

$$+ G_{\delta\gamma,\alpha\beta}(t_1, t_2; t_3, t_4)]dt_3 dt_4 \tag{1.1.86}$$

for $t_1 = t_2 = t$. Relationships (1.1.85,86) are a quantum generalization of the corresponding relations (1.1.29b,c).

1.2 Non-Markov Generating Equations. General Case

1.2.1 Derivation of Auxiliary Formula. Nonquantum Case

We now want to derive a generating equation for the case when a system is subject to the action of arbitrary external forces $h_\alpha(t)$. The Hamiltonian of the system is

$$\mathcal{H}(z, h(t)) = \mathcal{H}_0(z) - \sum_\alpha B_\alpha(z)h_\alpha(t) \tag{1.2.1}$$

and hence it varies with time explicitly

$$\mathcal{H}(z, t) = \mathcal{H}(z, h(t)) \,. \tag{1.2.2}$$

Such systems are called nonconservative. It is easily verified that their energy changes in time.

Taking a total derivative with respect to time yields

$$\frac{d\mathcal{H}(z(t), t)}{dt} = \sum_i \left(\frac{\partial \mathcal{H}}{\partial q_i}\dot{q}_i + \frac{\partial \mathcal{H}}{\partial p_i}\dot{p}_i\right) + \frac{\partial \mathcal{H}}{\partial t} \,. \tag{1.2.3}$$

Using the Hamilton equations

$$\dot{q}_i = \partial \mathcal{H} / \partial p_i \,,$$
$$\dot{p}_i = -\partial \mathcal{H} / \partial q_i \,, \tag{1.2.4}$$

gives

$$d\mathcal{H}(z(t),t)/dt = \partial \mathcal{H}(z,t)/\partial t \,. \tag{1.2.5}$$

For the case (1.2.1,2) this equation becomes

$$d\mathcal{H}(z(t),t)/dt = -\sum_\alpha B_\alpha(z)\dot{h}_\alpha(t) \,. \tag{1.2.6}$$

Suppose that the forces $h_\alpha(t)$ are only nonzero within the interval $a < t < b$ and are identically zero elsewhere:

$$h_\alpha(t) \equiv 0 \quad \text{for} \quad t \leq a \quad \text{and} \quad t \geq b = -a > 0 \,. \tag{1.2.7}$$

As follows from (1.2.6) the Hamiltonian will then remain constant in the region $t \leq a$, and also in the region $t \geq -a$.

Let us now find the change of the Hamiltonian over the interval $a < t < b$. Using (1.2.6), we will have

$$\mathcal{H}(z(b),b) - \mathcal{H}(z(a),a) = -\int_a^b B_\alpha(z(t))\dot{h}_\alpha(t)dt \,. \tag{1.2.8}$$

If then we carry out integration by parts and take into account that $h_\alpha(a) = h_\alpha(b) = 0$, this relation can be written as

$$\mathcal{H}(z(b),b) - \mathcal{H}(z(a),a) = \int_a^b \dot{B}_\alpha(z(t))h_\alpha(t)dt \,. \tag{1.2.9}$$

Due to (1.2.1,7), $\mathcal{H}[z(b),b]$ is nothing but $\mathcal{H}_0(z(b))$, and $\mathcal{H}(z(a),a) = \mathcal{H}_0(z(a))$. Further, $B_\alpha(z(t))$ is nothing but $B_\alpha(t)$, so that

$$\mathcal{H}_0(z(b)) = \mathcal{H}_0(z(a)) + \int_a^b J_\alpha(t)h_\alpha(t)dt \,. \tag{1.2.10}$$

Here $J_\alpha = \dot{B}_\alpha$. The Hamiltonians in this equation correspond to zero external forces.

1.2.2 Characteristic Functional for Fluxes Under External Forces

Let us introduce the characteristic functional

$$\Theta[iu(t), h(t)] = \left\langle \exp\left\{ i \int_a^b J_\alpha(t)u_\alpha(t) \right\} \right\rangle_{h(\tau)} \,, \tag{1.2.11}$$

describing the statistical behavior of the fluxes $J_\alpha(t); a < t < b$, under external forces that meet the condition (1.2.7). Using the designation $y_a^b = \{y(t), a < t < b\}$ or

$$y_a^b = \lim_{N \to \infty} \{y(a), y(t_1), \ldots, y(t_N)\}$$

with $t_j = a + (b-a)j/N$, the functional $\Theta[iu(t), h(t)]$ can be written as $\Theta\left[iu_a^b, h_a^b\right]$.

By analogy with (1.1.14), the fluxes $J_\alpha(t) = J_\alpha(z(t))$ can be expressed through the initial values $z(a)$ of dynamic variables and through the forces $h_\alpha(t')$ acting over the interval $t' \in (a, t)$:

$$J_\alpha(t) = F_{\alpha, t-a}\left[z(a), h_a^t\right] . \tag{1.2.12}$$

After (1.2.12) has been substituted into (1.2.11), we should average over $z(a)$. Since up to the time $t = a$ no external forces were acting, $w(z(a))$ corresponds to the equilibrium and we can take for $w(z(a))$ the Gibbs probability density

$$w(z(a)) = C_0 \exp[-\beta \mathcal{H}_0(z(a))] . \tag{1.2.13}$$

Accordingly, the characteristic functional (1.2.11) becomes

$$\Theta\left[v_a^b, h_a^b\right] = C_0 \int \exp\left\{ \int_a^b v(t) F_{t-a}[z(a), h_a^t] dt - \beta \mathcal{H}_0(z(a)) \right\} dz(a) . \tag{1.2.14}$$

Replacing here v_a^b by $(v - \beta h)_a^b$ gives

$$\Theta\left[(v - \beta h)_a^b, h_a^b\right]$$
$$= C_0 \int \exp\left\{ \int_a^b v(t) J(t) dt - \beta \int_a^b h(t) J(t) dt - \beta \mathcal{H}_0(z(a)) \right\} dz(a) , \tag{1.2.15}$$

where $J(t)$ is defined by (1.2.12). Using (1.2.10) yields

$$\Theta\left[(v - \beta h)_a^b, h_a^b\right]$$
$$= C_0 \int \exp\left\{ \int_a^b v(t) F_{t-b}[z(b), h_b^t] dt - \beta \mathcal{H}_0(z(b)) \right\} dz(b) . \tag{1.2.16}$$

We have here used the fact that the Jacobian of the canonical transformation $z(a) \to z(b)$, just like that of any other canonical transformation, is unity and that the formula

$$J(t) = F_{t-t_0}\left[z(t_0), h_{t_0}^t\right] \tag{1.2.17}$$

is valid for any t_0 including $t_0 > t$.

1.2.3 Nonquantum Generating Equation

In passing to the reverse time $\tilde{t} = c - t$, it would be convenient to choose c such that a is transformed into b and vice versa. To obtain this, we must set $c = a + b$. Then the initial moment is a and the final moment is b in both forward and reverse time. We suppose that the condition $\mathcal{H}_0(\varepsilon z) = \mathcal{H}(z)$, i.e. $\mathcal{H}_0(q, -p) = \mathcal{H}(q, p)$ of time reversibility is satisfied. Hence the derivatives $\partial \mathcal{H}_0 / \partial q_i$ are even functions of p, and $\partial \mathcal{H}_0 / \partial p_i$ are odd functions of p:

$$\frac{\partial \mathcal{H}_0(q, -p)}{\partial q_i} = \frac{\partial \mathcal{H}_0(q, p)}{\partial q_i},$$

$$\left. \frac{\partial \mathcal{H}_0(q, p)}{\partial p_i} \right|_{p=-p'} = -\frac{\partial \mathcal{H}_0(q, p')}{\partial p'_i}. \tag{1.2.18}$$

For the Hamiltonian (1.2.1), the transformation (1.1.14) has the form

$$z(t) = \varphi_{t-t_0}\left(z(t_0), h_{t_0}^t \right). \tag{1.2.19}$$

It is determined by the dynamical Hamilton equations (1.2.4) or by

$$\dot{q}_i = \frac{\partial \mathcal{H}_0}{\partial p_i} - \sum_\alpha \frac{\partial B_\alpha}{\partial p_i} h_\alpha(t),$$

$$\dot{p}_i = -\frac{\partial \mathcal{H}_0}{\partial q_i} + \sum_\alpha \frac{\partial B_\alpha}{\partial q_i} h_\alpha(t). \tag{1.2.20}$$

Let the division of the interval $[t_0, t]$ be made by the points $t_j, j = 1, \ldots, N$ ($t_N = t$, $t_j < t_k$ if $j < k$). For small $\tau = t_{j+1} - t_j$ we obtain from (1.2.20) the transformation

$$q_i(t_{j+1}) = q_i(t_j) + \tau \left[\frac{\partial \mathcal{H}_0(q(t_j), p(t_j))}{\partial p_i(t_j)} - \sum_\alpha \frac{\partial B_\alpha}{\partial p_i(t_j)} h_\alpha(t) \right],$$

$$p_i(t_{j+1}) = p_i(t_j) - \tau \left[\frac{\partial \mathcal{H}_0(q(t_j), p(t_j))}{\partial q_i(t_j)} - \sum_\alpha \frac{\partial B_\alpha}{\partial q_i(t_j)} h_\alpha(t) \right]. \tag{1.2.21}$$

This transformation is nothing but the transformation

$$z(t_{j+1}) = \varphi_\tau(z(t_j), h(t_j)). \tag{1.2.22}$$

Let us introduce $q'_i(t) = q_i(t)$, $p'_i(t) = -p_i(t)$, i.e. $z'(t) = \varepsilon z(t)$. Substituting $q_i(t) = q'_i(t)$, $p_i(t) = -p'_i(t)$ into (1.2.21) and using (1.2.18) gives

$$q'_i(t_{j+1}) = q'_i(t_j) - \tau \frac{\partial \mathcal{H}_0(q', p')}{\partial p'_i} - \tau \sum_\alpha \left. \frac{\partial B_\alpha(q', p)}{\partial p_i} \right|_{p=-p'} h_\alpha(t_j),$$

$$p'_i(t_{j+1}) = -p'(t_j) - \tau \frac{\partial \mathcal{H}_0(q', p')}{\partial q'_i} + \tau \sum_\alpha \left. \frac{\partial B_\alpha(q, p)}{\partial q_i} \right|_{q=q'} h_\alpha(t_j) \tag{1.2.23}$$

with $q' = q'(t_j)$, $p' = p'(t_j)$. Now we consider the formula

$$B_\alpha(\varepsilon z') = \varepsilon_\alpha B_\alpha(z') \tag{1.2.24}$$

or

$$B_\alpha(q', -p') = \varepsilon_\alpha B_\alpha(q', p')$$

and consequently

$$\frac{\partial B_\alpha(q, -p')}{\partial q_i}\bigg|_{q=q'} = \varepsilon_\alpha \frac{\partial B_\alpha(q', p')}{\partial q'_i},$$

$$\frac{\partial B_\alpha(q', p)}{\partial p_i}\bigg|_{p=-p'} = -\varepsilon_\alpha \frac{\partial B_\alpha(q', p')}{\partial p'_i}. \tag{1.2.25}$$

Substituting (1.2.25) into (1.2.23) and comparing (1.2.23) with (1.2.21) gives

$$\varepsilon z(t_{j+1}) = \varphi_{-\tau}(\varepsilon z(t_j), \varepsilon h(t_j)). \tag{1.2.26}$$

Applying (1.2.26) sequentially for all elementary intervals, beginning with $[t_0, t_1]$, and passing to the limit $\max_j (t_{j+1} - t_j) \longrightarrow 0$, we get

$$\varepsilon z(t) = \varphi_{t_0-t}(\varepsilon z(t_0), \varepsilon h^t_{t_0}) \tag{1.2.27}$$

in the same way as (1.2.19) can be obtained from (1.2.22). For the variables $J_\alpha(t) = J_\alpha(z(t))$ we have

$$J_\alpha(t) = J_\alpha[\varphi_{t-t_0}(z(t_0), h^t_{t_0})]. \tag{1.2.28}$$

Comparing this with (1.2.17) yields

$$F_{\alpha, t-t_0}[z(t_0), h^t_{t_0}] = J_\alpha[\varphi_{t-t_0}(z(t_0), h^t_{t_0})]. \tag{1.2.29}$$

Owing to (1.2.27), the equation $J_\alpha(t) = J_\alpha(z(t))$ implies

$$J_\alpha(t) = J_\alpha[\varepsilon \varphi_{t_0-t}(\varepsilon z(t_0), \varepsilon h^t_{t_0})]. \tag{1.2.30}$$

We have $J_\alpha = \dot{B}_\alpha$. Since the differentiation with respect to time changes the time parity, the formula (1.2.24) has the consequence

$$J_\alpha(\varepsilon z) = -\varepsilon_\alpha J_\alpha(z). \tag{1.2.31}$$

Therefore

$$J_\alpha(t) = -\varepsilon_\alpha J_\alpha[\varphi_{t_0-t}(\varepsilon z(t_0), \varepsilon h^t_{t_0})] \tag{1.2.32}$$

or

$$J_\alpha(t) = -\varepsilon_\alpha F_{\alpha, t_0-t}(\varepsilon z(t_0), \varepsilon h^t_{t_0}) \tag{1.2.33}$$

according to (1.2.29). Applying (1.2.33) instead of (1.2.17) at $t_0 = b$, we have

$$\Theta\left[(v - \beta h)^b_a, h^b_a\right] = C_0 \int \exp\left\{-\int_a^b v(t)\varepsilon F_{b-t}[\varepsilon z, \varepsilon h^t_b]dt - \beta \mathcal{H}_0(z)\right\} dz \tag{1.2.34}$$

instead of (1.2.16).

Further, using the freedom to choose the argument, in (1.2.14) we can take $-\varepsilon_\alpha v_\alpha(a + b - t)$, $\varepsilon_\alpha h_\alpha(a + b - t)$ instead of $v(t), h(t)$ respectively. Then

$$\Theta\left[-\varepsilon v^a_b, \varepsilon h^a_b\right]$$

$$= C_0 \int \exp\left\{-\int_a^b v(a+b-t)\varepsilon F_{t-a}\left[z, \varepsilon h^{a+b-t}_b\right] dt - \beta \mathcal{H}_0(z)\right\} dz. \tag{1.2.35}$$

Introducing a new integration variable $\tilde{t} = a + b - t$ gives

$$\Theta\left[-\varepsilon v_b^a, \varepsilon h_b^a\right] = C_0 \int \exp\left\{-\int_a^b v(\tilde{t})\varepsilon F_{b-\tilde{t}}\left[z, \varepsilon h_b^{\tilde{t}}\right] d\tilde{t} - \beta \mathcal{H}_0(z)\right\} dz \quad (1.2.36)$$

with $z = z(a)$. We see that the right-hand side of (1.2.36) coincides with the right-hand side of (1.2.34) after passing to a new integration variable $z' = \varepsilon z$. Therefore the left-hand sides are equal to each other

$$\Theta\left[(v - \beta h)_a^b, h_a^b\right] = \Theta\left[-\varepsilon v_b^a, \varepsilon h_b^a\right].$$

Writing the functional in the usual form and putting $b = -a > 0$, we have

$$\Theta\left[v(t) - \beta h(t), h(t)\right] = \Theta\left[-\varepsilon v(-t), \varepsilon h(-t)\right]. \quad (1.2.37)$$

We can now pass to the limit as $b = -a \longrightarrow \infty$. Equation (1.2.37) is the nonquantum generating equation that we sought. It was firstly derived by *Bochkov* and *Kuzovlev* [1.3] in 1977.

If we introduce the generating functional

$$\Pi[y(t), h(t)] = \beta^{-1} \ln \Theta[\beta y(t), -h(t)] \quad (1.2.38)$$

we will have, instead of (1.2.37), equivalent equation

$$\Pi[y(t) + h(t), h(t)] = \Pi[-\varepsilon y(-t), \varepsilon h(-t)]. \quad (1.2.39)$$

It is obviously analogous to the generating equation [(3.2.50) v.1] of the Markov theory.

Before we leave this subsection, we will provide a more symmetric form of (1.2.39). Substituting $y - h/2$ for y gives

$$\Pi[y(t) + h(t)/2, h(t)] = \Pi[-\varepsilon y(-t) + \varepsilon h(-t)/2, \varepsilon h(-t)]. \quad (1.2.40)$$

Since

$$\varepsilon_\alpha h_\alpha(-t) = h^{\text{t.c.}}(t), \quad (1.2.41a)$$
$$-\varepsilon_\alpha J_\alpha(-t) = J^{\text{t.c.}}(t) \quad (1.2.41b)$$

(where t.c. means time conjugate) and hence

$$-\varepsilon_\alpha y_\alpha(-t) = y^{\text{t.c.}}(t), \quad (1.2.42)$$

we can write (1.2.40) as

$$\Pi[y(t) + h(t)/2, h(t)] = \Pi[y^{\text{t.c.}}(t) + h^{\text{t.c.}}(t)/2, h^{\text{t.c.}}(t)] \quad (1.2.43)$$

The functional $\Pi[y + h/2, h]$ is thus invariant under time reversal.

1.2.4 Consequence of the Generating Equation: the H-Theorem

Setting in (1.2.37) $v'(t) = 0$, we obtain

$$\Theta[-\beta h(t), h(t)] = 1\,. \qquad (1.2.44)$$

Hence, according to (1.2.11)

$$\left\langle \exp\left\{-\beta \int_a^b J_\alpha(t)h_\alpha(t)dt\right\} - 1 \right\rangle_{h(\tau)} = 0\,. \qquad (1.2.45)$$

It follows that

$$\left\langle \exp\left(-\beta \int_a^b Jhdt\right) + \beta \int_a^b Jhdt - 1 \right\rangle_{h(\tau)} = \beta \left\langle \int_a^b Jhdt \right\rangle_{h(\tau)}\,. \qquad (1.2.46)$$

Since $\exp(x) - 1 - x \geq 0$ at any x, from (1.2.46) we get

$$\left\langle \int_a^b J(t)h(t)dt \right\rangle \geq 0\,, \qquad (1.2.47)$$

or, by (1.2.10),

$$\langle \mathcal{H}_0(b) \rangle - \langle \mathcal{H}_0(a) \rangle \geq 0\,,$$

i.e.

$$\Delta U \equiv U(b) - U(a) \geq 0\,, \qquad (1.2.48)$$

where $U = \langle \mathcal{H}_0 \rangle$ is the internal energy of the system.

We have thus proved that, if the initial distribution is an equilibrium one, the internal energy increases (not decreases) under external forces. From the external bodies producing the external forces some energy is transferred into the system under consideration with the result that the external work ΔW is nonpositive. This assumes that the energy being dissipated evolves as heat and stays in the system and that all interactions with external bodies reduce to the action of external forces $h(t)$. If we further assume that there is some heat exchange with the surrounding medium, then according to the first law of thermodynamics $\Delta Q = \Delta U + \Delta W$, the theorem we have just proved will be given, not by (1.2.48), but by

$$\Delta W \leq 0 \quad \text{or} \quad \Delta U - \Delta Q \geq 0\,. \qquad (1.2.49)$$

This theorem belongs to the H-theorems, several of which were discussed in the first volume (see Sects 4.6.3 and 5.1.8).

1.2.5 Derivation of the Simplest FDRs from the Generating Equation

We would now like to illustrate the uses of the generating equation (1.2.37) for the derivation of FDRs referring to some simple relations.

From (1.2.11) and [(2.1.10) v.1]

$$
\ln \Theta[v(t), h(t)]
$$
$$
= \sum_{m=1}^{\infty} \frac{1}{m!} \int \ldots \int \langle J_{\alpha_1}(t_1), \ldots, J_{\alpha_m}(t_m) \rangle_{h(\tau)} v_{\alpha_1}(t_1) \ldots v_{\alpha_m}(t_m) dt_1 \ldots dt_m.
$$

$$(1.2.50)$$

The integrands here are nonequilibrium correlators for variable external forces $h(t)$.

(a) we will start with the equation (1.2.44). If we then take its logarithm and substitute (1.2.50), we get

$$
-\beta \int \langle J_{\alpha_1}(t_1) \rangle_{h(\tau)} h_\alpha(t_1) dt_1
$$
$$
+\frac{1}{2}\beta^2 \int \langle J\alpha_1(t_1), J_{\alpha_2}(t_2) \rangle_{h(\tau)} h_{\alpha_1}(t_1) h_{\alpha_2}(t_2) dt_1 dt_2
$$
$$
-\frac{1}{6}\beta^3 \langle J_1, J_2, J_3 \rangle_{h(\tau)} h_1 h_2 h_3 + \ldots = 0.
$$

$$(1.2.51)$$

If we take the variational derivative of (1.2.51) with respect to $h_\beta(t)$ and then let $h(\tau) = 0$, we will obtain

$$
\langle J_\beta(t) \rangle_0 = 0.
$$

$$(1.2.52)$$

Then differentiating (1.2.51) with respect to h_1 and h_2 and letting $h(t) = 0$, we have

$$
-\delta \langle J_1 \rangle_h / \delta h_2 - \delta \langle J_2 \rangle_h / \delta h_1 + \beta \langle J_1, J_2 \rangle = 0 \quad \text{for} \quad h(t) \equiv 0.
$$

$$(1.2.53)$$

Owing to [(5.2.8) v.1] this is equivalent to the FDR

$$
Y_{12} = kT(Y_{1,2} + Y_{2,1}) \quad (Y_{12} = \langle J_1, J_2 \rangle_0),
$$

$$(1.2.54)$$

which is nothing but the nonquantum form of [(5.3.92) v.1].

Triple differentiation of (1.2.51) with respect to h_1, h_2, h_3 at $h = 0$, using [(5.2.8) v.1] and the notation of Sect. 5.3.7 v.1, gives

$$
-Y_{1,23} - Y_{2,13} - Y_{3,12} + \beta Y_{12,3} + \beta Y_{13,2} + \beta Y_{23,1} - \beta^2 Y_{123} = 0.
$$

$$(1.2.55)$$

This equation can be derived from the FDRs [(5.3.75,99) v.1].

It should be noted that the equations (1.2.44,47,54,55) also hold when the condition of time-reversal invariance is not met, i.e. when $\mathcal{H}_0(\varepsilon z)$ differs from $\mathcal{H}_0(z)$. Actually, in the absence of time reversibility when (1.2.37) is not valid, we can set $v(t) \equiv 0$ in (1.2.16) and get (1.2.44) since, in so doing, the right-hand side of (1.2.16) becomes unity. The above-mentioned equations thus follow solely from the Gibbs formula (1.2.13).

(b) Consider now equation (1.2.37). If we take its logarithm, substitute (1.2.50), differentiate with respect to v_1 and set $v = 0$, we will have

$$\langle J_{\alpha_1}(t_1)\rangle_{h(\tau)} - \beta \int \langle J_{\alpha_1}(t_1), J_{\alpha_2}(t_2)\rangle h_{\alpha_2}(t_2) dt_2 + O(h^2)$$
$$= -\varepsilon_{\alpha_1}\langle J_{\alpha_1}(-t_1)\rangle_{\varepsilon h(-\tau)}. \tag{1.2.56}$$

Taking the variational derivative with respect to h_2 of the expressions on both sides of (1.2.56) gives

$$\delta\langle J_1\rangle_{h(\tau)}/\delta h_2 - \beta\langle J_1, J_2\rangle_0 = -\varepsilon_{\alpha_1}\delta\langle J_{\alpha_1}(-t_1)\rangle_{\varepsilon h(-\tau)}/\delta h_2 \tag{1.2.57}$$

for $h \equiv 0$. From [(5.2.8) v.1] we can write this as

$$Y_{\alpha\beta}(t_1, t_2) = kT[Y_{\alpha,\beta}(t_1, t_2) + \varepsilon_\alpha\varepsilon_\beta Y_{\alpha,\beta}(-t_1, t_2)]. \tag{1.2.58}$$

Comparison with (1.2.54) gives

$$Y_{\beta,\alpha}(t_2, t_1) = \varepsilon_\alpha\varepsilon_\beta Y_{\alpha,\beta}(-t_1, -t_2), \tag{1.2.59}$$

i.e. the reciprocal relation [(5.3.94) v.1].

From the generating equation (1.2.37) we can also derive other nonquantum FDRs including relations that have not been obtained earlier, namely those with more than four subscripts.

1.2.6 Quantum Case. Auxiliary Formulas

The dynamic variables $\hat{z} = (\hat{q}_i, \hat{p}_i)$ or their functions $\hat{D} = D(\hat{z})$ in the Heisenberg formalism with the time-dependent Hamiltonian $\hat{\mathcal{H}}(t)$ obey

$$d\hat{D}/dt = (i/\hbar)(\hat{\mathcal{H}}(t)\hat{D} - \hat{D}\hat{\mathcal{H}}(t)). \tag{1.2.60}$$

If $\hat{D}$ also varies with time explicitly: $\hat{D} = D(\hat{z}, t)$, then instead of (1.2.60) we will have

$$d\hat{D}/dt = (i/\hbar)(\hat{\mathcal{H}}(t)\hat{D} - \hat{D}\hat{\mathcal{H}}(t)) + \partial\hat{D}/\partial t. \tag{1.2.61}$$

Taking here the Hamiltonian $\hat{H}(t)$ as $\hat{D}$, we obtain

$$d\hat{\mathcal{H}}(t)/dt = \partial\hat{\mathcal{H}}(t)/\partial t. \tag{1.2.62}$$

In the quantum case (1.2.1) becomes

$$\hat{\mathcal{H}}(t) = \hat{\mathcal{H}}_0 - \sum_\alpha \hat{B}_\alpha h_\alpha(t). \tag{1.2.63}$$

Hence, by (1.2.62),

$$d\hat{\mathcal{H}}(t)/dt = -\sum_\alpha \hat{B}_\alpha \dot{h}_\alpha(t) \tag{1.2.64}$$

which is analogous to (1.2.6). Integrating (1.2.64) and using $h(a) = h(b) = 0$ gives

$$\hat{\mathcal{H}}_0(b) = \hat{\mathcal{H}}_0(a) - \int_a^b \hat{B}_\alpha \dot{h}_\alpha(t) dt$$
$$= \hat{\mathcal{H}}_0(a) + \int_a^b \hat{J}_\alpha h_\alpha(t) dt, \tag{1.2.65}$$

where $\hat{\mathcal{H}}_0(t) = \hat{\mathcal{H}}_0(z(t))$, $\hat{J}_\alpha = \dot{\hat{B}}_\alpha$.

The theory of Sect. 1.2.1 thus covers the quantum case as well. The Gibbs formula (1.2.13) now becomes

$$\hat{\rho}_0(a) = C_0 \exp(-\beta\hat{\mathcal{H}}_0(a)) . \tag{1.2.66}$$

We will also consider the Gibbs density matrix for $t \geq b$:

$$\hat{\rho}_0^*(b) = C_0 \exp(-\beta\hat{\mathcal{H}}_0^*(b)) . \tag{1.2.67}$$

Matrix (1.2.66) can be thought of as a time-independent forward-time initial density matrix in the Heisenberg representation, and the matrix (1.2.67) as its time-reversed counterpart.

Substituting (1.2.65) into (1.2.67) gives

$$\hat{\rho}(b) = C_0 \exp[-\beta\hat{\mathcal{H}}_0(a) - \beta\hat{E}] , \tag{1.2.68}$$

where

$$\hat{E} = \int_a^b \hat{J}_\alpha(t) h_\alpha(t) dt . \tag{1.2.69}$$

Using the equation

$$\exp(-\beta\hat{\mathcal{H}}_0 - \beta\hat{E})$$
$$= \exp(-\beta\hat{\mathcal{H}}_0) \overleftarrow{\exp} \left[-\int_0^\beta d\lambda (\exp(\lambda\hat{\mathcal{H}}_0)\hat{E}\exp(-\lambda\hat{\mathcal{H}}_0))_\lambda \right] , \tag{1.2.70}$$

which is similar to (1.1.38) with (1.1.36) taken into account where $\hat{\mathcal{H}}$ can be identified with $\hat{\mathcal{H}}_0$, and applying (1.2.70) to (1.2.68), we arrive at

$$\hat{\rho}_0(b) = C_0 \exp[-\beta\hat{\mathcal{H}}_0(a)] \overleftarrow{\exp} \left[-\int_0^\beta d\lambda (\exp(\lambda\hat{\mathcal{H}}_0)\hat{E}\exp(-\lambda\hat{\mathcal{H}}_0))_\lambda \right] , \tag{1.2.71}$$

where $\hat{\mathcal{H}}_0 = \hat{\mathcal{H}}_0(a)$ and by (1.2.66), the expression appearing before the exponential function is simply ρ_0.

1.2.7 General Quantum Equation
with an Arbitrary Operator Expression

Since from (1.2.7) for $t < a$, a forward-time system is not subject to any external forces, this establishes an equilibrium state described by the Heisenberg density matrix (1.2.66). The latter determines the means

$$\langle \ldots \rangle_{h(\tau)} = \text{Tr}\{\hat{\rho}_0(a)\ldots\}_{h(\tau)} . \tag{1.2.72}$$

The subscripts here are the external forces acting on the system for $t > a$, since they affect the evolution of the operators being averaged. Using (1.2.72), we average such an operator

$$\overleftarrow{\exp} \left\{ -\int_0^\beta d\lambda [\exp(\lambda\hat{\mathcal{H}}_0(a))\hat{E}\exp(-\lambda\hat{\mathcal{H}}_0(a))]_\lambda \right\} \hat{L}[\hat{B}(t)] , \tag{1.2.73}$$

where $\hat{L}[\hat{B}(t)] = \hat{L}[\hat{B}(t), -\infty < t < \infty]$ is a certain functional operator expression, in which the ordering of the operators $\hat{B}_\alpha(t)$ is also defined. This means that a rule is specified by which an operator $\hat{L}$ is placed in correspondence with a set of noncommutative operators $\hat{B}_\alpha(t)$ with α and t assuming various values. Some examples of the operator expression will be considered later. Substituting (1.2.73) into (1.2.72), we will obtain the mean:

$$m = \text{Tr}\left\{\hat{\rho}_0(a) \overleftarrow{\exp}\left\{-\int_0^\beta d\lambda[\exp(\lambda\hat{\mathcal{H}}_0(a))\hat{E}\exp(-\lambda\hat{\mathcal{H}}_0(a))]_\lambda\right\}\hat{L}[\hat{B}(t)]\right\}_{h(\tau)}.$$
(1.2.74)

According to (1.2.71), we get

$$m = \text{Tr}\{\hat{\rho}_0(b)\hat{L}[\hat{B}(t)]\}_{h(\tau)}.$$
(1.2.75)

Specifically, if $\hat{L}[\hat{B}] = \hat{1}$, from the normalization condition of the matrix $\rho_0(b)$, we immediately obtain $m = 1$, i.e.,

$$\left\langle\overleftarrow{\exp}\left[-\int_0^\beta d\lambda[\exp(\lambda\hat{\mathcal{H}}_0(a))\hat{E}\exp(-\lambda\hat{\mathcal{H}}_0(a))]_\lambda\right]\right\rangle_{h(\tau)} = 1.$$
(1.2.76)

Let us now return to the case of an arbitrary $\hat{L}[\hat{B}(t)]$. The matrix $\rho_0(b)$ which enters into (1.2.76) can be treated as an unchanging matrix in the Heisenberg formalism in reverse time. In that case, the operators $\hat{B}_\alpha(t)$, or rather the related operators

$$\hat{\tilde{B}}_\alpha(\tilde{t}) = \varepsilon_\alpha\hat{B}_\alpha^*(-\tilde{t})$$
(1.2.77)

(see (1.1.74)), should be regarded as varying in reverse time. The last expression is equivalent to

$$\hat{B}_\alpha(t) = \varepsilon_\alpha\hat{\tilde{B}}_\alpha^*(-t).$$
(1.2.78)

Substituting (1.2.78) into (1.2.75) gives

$$\begin{aligned} m &= \text{Tr}\{\hat{\rho}_0(b)\hat{L}[\varepsilon\tilde{B}^*(-t)]\}_{h(\tau)} \\ &= \text{Tr}\{\hat{\rho}_0(b)\hat{L}[\varepsilon\tilde{B}(-t)]\}_{h(\tau)}^*. \end{aligned}$$
(1.2.79)

Note that $\hat{\rho}_0^*(b)$ coincides with the time-reversed density matrix $\hat{\tilde{\rho}}_0(b)$ due to the relations [(5.3.25) v.1] and (1.2.67). Instead of t in (1.2.79) we can equally well write $\tilde{t}$. The subscript $h(\tau)$ in (1.2.79) indicates the fixed function, namely, the behavior of acting forces considered in the forward-time frame of reference. Since the process is thought of as occurring in reverse time, it is natural to use the same behavior, but such as it appears in the time-reversed frame of reference. Using equation $\tilde{h}_\alpha(\tilde{t}) = \varepsilon_\alpha h_\alpha(-\tilde{t})$ (see (1.2.41a)), we should write $\varepsilon h(-\tilde{\tau})$ instead of $h(\tau)$. Accordingly, instead of (1.2.79), it is more exact to write

$$m^* = \text{Tr}\{\hat{\rho}_0(b)\hat{L}[\varepsilon\tilde{B}(-\tilde{t})]\}_{\varepsilon h(-\tilde{\tau})}.$$
(1.2.80)

Let us now use the condition of time-reversal invariance, under which, for any operator expression $\hat{M}[\hat{B}(t)]$, the means in forward and reverse times are equal

$$\text{Tr}\{\tilde{\hat{\rho}}_0(b)\hat{M}[\tilde{\hat{B}}(\tilde{t})]\}_{\tilde{h}'(\tilde{\tau})} = \text{Tr}\{\hat{\rho}_0(a)\hat{M}[\hat{B}(t)]\}_{h''(\tau)}\,, \tag{1.2.81}$$

provided the behavior of the forces regarded in reverse time on the left-hand side of (1.2.81) coincides with that the forces in forward time on the right-hand side. In symbols, the condition is

$$\tilde{h}'(\tilde{\tau}) = f(\tilde{\tau})\,,$$
$$h''(\tau) = f(\tau)\,. \tag{1.2.82}$$

It is worth stressing that the behaviors of the forces on either side of (1.2.81) are actually different, although they are represented identically in various coordinate systems. Therefore we denote the forces differently: $h'(\tau)$ on the left-hand side and $h''(\tau)$ on the right-hand side.

We will choose the operator expression $\hat{M}[\hat{B}(t)]$ and the behavior of the forces $\tilde{h}'(\tilde{\tau})$ so that the left-hand side of (1.2.81) would coincide with the expression on the right of (1.2.80). We will need the following expressions

$$\hat{M}[\hat{B}(t)] = \hat{L}[\varepsilon\hat{B}(-t)]\,,$$
$$\tilde{h}'(\tilde{\tau}) = \varepsilon h(-\tilde{\tau})\,. \tag{1.2.83}$$

In accordance with (1.2.82) we have

$$f(\tilde{\tau}) = \varepsilon h(-\tilde{\tau})$$
$$h''(\tau) = \varepsilon h(-\tau)\,. \tag{2.84}$$

In the case of (1.2.83,84), applying (1.2.81) to (1.2.80), we obtain

$$m = \text{Tr}\{\hat{\rho}_0(a)\hat{L}[\varepsilon\hat{B}(-t)]\}^{*}_{\varepsilon h(-\tau)}\,. \tag{1.2.85}$$

Here, as in (1.2.74), the means are taken with the density matrix $\hat{\rho}_0(a)$. From (1.2.74,85), considering (1.2.72) we finally arrive at

$$\left\langle \overleftarrow{\exp}\left[-\int_0^\beta d\lambda(\exp(\lambda\hat{\mathcal{H}}_0)\hat{E}\exp(-\lambda\hat{\mathcal{H}}_0))_\lambda\right]\hat{L}[\hat{B}(t)]\right\rangle_{h(\tau)}$$
$$= \langle\hat{L}[\varepsilon\hat{B}(-t)]\rangle^{*}_{\varepsilon h(-\tau)}\,. \tag{1.2.86}$$

Using specific forms of the operator expression $\hat{L}[\hat{B}]$ we can obtain from this various special cases of quantum generating equations.

1.2.8 Several Specific Generating Equations

We will now consider some specific cases of $\hat{L}[\hat{B}(t)]$.

(a) We have time moments $t_1, t_2, \ldots, t_n$ and subscripts $\alpha_1, \ldots, \alpha_n$, and let

$$L[B] = B_{\alpha_1}(t_1)B_{\alpha_2}(t_2)\ldots B_{\alpha_n}(t_n)\,. \tag{1.2.87}$$

Substitution of this special form of the operator expression into (1.2.86) gives

$$\left\langle \overleftarrow{\exp}\left[-\int_0^\beta d\lambda(\exp(\lambda\mathcal{H}_0)E\exp(-\lambda\mathcal{H}_0))_\lambda\right]B_{\alpha_1}(t_1)\ldots B_{\alpha_n}(t_n)\right\rangle_{h(\tau)}$$
$$= \varepsilon_{\alpha_1}\ldots\varepsilon_{\alpha_n}\langle B_{\alpha_n}(-t_n)\ldots B_{\alpha_1}(-t)\rangle_{\varepsilon h(-\tau)}. \tag{1.2.88}$$

On the right-hand side of the above equation we have used (1.1.65). Equation (1.2.88) can be referred to as an incomplete generating equation since, unlike (1.2.37), it relates expressions that depend only on $h(t)$, and not on $v(t)$.

(b) We will now take the following operator expression:

$$L[B(t)] = \exp\left[\int_{-\infty}^{\infty}v(t)B(t)dt\right]. \tag{1.2.89}$$

If we then substitute it into (1.2.86) and note that

$$\int_{-\infty}^{\infty}v_\alpha(t)\varepsilon_\alpha B_\alpha(-t)dt = \int_{-\infty}^{\infty}\varepsilon_\alpha v_\alpha(-t')B_\alpha(t')dt', \tag{1.2.90}$$

we will have

$$\left\langle \overleftarrow{\exp}\left[-\int_0^\beta d\lambda(\exp(\lambda\mathcal{H}_0)E\exp(-\lambda\mathcal{H}_0))_\lambda\right]\exp\left[\int v(t)B(t)dt\right]\right\rangle_{h(\tau)}$$
$$= \left\langle \exp\left[\int \varepsilon v(-t)B(t)dt\right]\right\rangle_{\varepsilon h(-\tau)}. \tag{1.2.91}$$

The complex conjugate on the right has been discarded, since the operator (1.2.89) is Hermitian for the real-valued function $v(t)$, and so the mean on the right is also real-valued.

Instead of (1.2.89) we could take an exponential function with another operator ordering, say $\overrightarrow{\exp}$ ordered in time. In that case, on the right-hand side of (1.2.91) we will have $\overleftarrow{\exp}$ with the opposite ordering.

(c) If we take the operator expression

$$L[B(t)] = \exp\left[\int_{-\infty}^{\infty}v(t)\dot{B}(t)dt\right], \tag{1.2.92}$$

we will derive from (1.2.86) the generating equation

$$\left\langle \overleftarrow{\exp}\left[-\int_0^\beta d\lambda(\exp(\lambda\mathcal{H}_0)E\exp(-\lambda\mathcal{H}_0))_\lambda\right]\exp\left[\int v(t)J(t)dt\right]\right\rangle_{h(\tau)}$$
$$= \left\langle \exp\left[-\int \varepsilon v(-t)J(t)dt\right]\right\rangle_{\varepsilon h(-\tau)}, \tag{1.2.93}$$

which is a quantum generalization of (1.2.37).

1.2.9 Some Simple Uses of One of the Quantum Generating Equations

The generating equations (1.2.88) (taken at various n, $t_1, \ldots t_n$, $\alpha_1, \ldots, \alpha_n$) or (1.2.91) or (1.2.93), which we have derived above, can in principle be used to obtain all quantum linear and nonlinear FDRs. But this is a rather difficult task owing to the unwieldy calculations. That is why in Sects. 5.3,4 v.1 we have applied other techniques. We are not going to reproduce the derivation of FDRs here. We will confine ourselves to providing some simple consequences of the simplest of the equations derived, viz. of the equation (1.2.76), whose validity is not subject to the condition of time-reversal invariance.

Let us expand (1.2.76) as a Taylor series and take into account the operator ordering in λ. After applying the same technique as used for deriving (1.1.50) from (1.1.46), we have

$$
\begin{aligned}
0 = & -\int_0^\beta d\lambda \langle \exp(\lambda\mathcal{H}_0)(E)_{h(\tau)} \exp(-\lambda\mathcal{H}_0) \rangle_0 \\
& + \int_0^\beta d\lambda_1 \int_0^{\lambda_1} d\lambda_2 \langle \exp(\lambda_1\mathcal{H}_0)(E)_{h(\tau)} \exp[(\lambda_2 - \lambda_1)\mathcal{H}_0](E)_{h(\tau)} \\
& \times \exp(-\lambda_2\mathcal{H}_0) \rangle_0 \\
& - \int_0^\beta d\lambda_1 \int_0^{\lambda_1} d\lambda_2 \int_0^{\lambda_2} d\lambda_3 \langle \exp(\lambda_1\mathcal{H}_0)(E)_h \exp[(\lambda_2 - \lambda_1)\mathcal{H}_0](E)_h \\
& \times \exp[(\lambda_3 - \lambda_2)\mathcal{H}_0](E)_h \exp(-\lambda_3\mathcal{H}_0) \rangle_0 + \ldots .
\end{aligned}
\tag{1.2.94}
$$

Here E is supplied with the subscript $h = h(\tau)$ because the latter indicates how the operators $J_\alpha(t)$ vary and what form the $h_\alpha(t)$ have in

$$
E = \int_{-\infty}^{\infty} J_\alpha(t) h_\alpha(t) dt.
\tag{1.2.95}
$$

The averaging in (1.2.94) is carried out with the density matrix $\rho_0(a)$ and so the means are equilibrium ones.

We will next substitute (1.2.95) into (1.2.94), repeatedly take variational derivatives with respect to $h(t)$ and then let $h(t) \equiv 0$. In the process, we will use the equations

$$
\begin{aligned}
\frac{\delta(E)_h}{\delta h_1} &= \left(J_1 + \int \frac{\delta J_\alpha(t)}{\delta h_1} h_\alpha(t) dt \right)_h, \\
\frac{\delta^2(E)_h}{\delta h_2 \delta h_1} &= \left(\frac{\delta J_1}{\delta h_2} + \frac{\delta J_2}{\delta h_1} + \int \frac{\delta^2 J_\alpha(t)}{\delta h_1 \delta h_2} h_\alpha(t) dt \right)_h,
\end{aligned}
\tag{1.2.96}
$$

etc. At $h(t) = 0$ we will only have equilibrium means of the type $\langle C'D' \ldots Z' \rangle_0$, where $C', D', \ldots$ have the form

$$
D' = \exp(\lambda\mathcal{H}_0) D(t_1, \ldots t_k) \exp(-\lambda\mathcal{H}_0),
\tag{1.2.97}
$$

etc. In analogy with (1.1.36), we get

$$
\begin{aligned}
\exp(\lambda\mathcal{H}_0) D(t_1, \ldots, t_k) \exp(-\lambda\mathcal{H}_0) &= D(t_1 - i\hbar\lambda, \ldots, t_k - i\hbar\lambda) \\
&= \exp[-i\hbar\lambda(p_1 + \ldots + p_k)] \\
&\quad \times D(t_1, \ldots, t_k).
\end{aligned}
\tag{1.2.98}
$$

Differentiating (1.2.94) with respect to h_1 and h_2 at $h(t) \equiv 0$ and using (1.2.96) and (1.2.97) gives

$$-\langle \delta J_1/\delta h_2 \rangle_0 - \langle \delta J_2/\delta h_1 \rangle_0 + \beta \Phi(p_1, p_2)\langle J_1 J_2 \rangle_0 + \beta \Phi(p_2, p_1)\langle J_2 J_1 \rangle = 0 . \quad (1.2.99)$$

Here we have used the notation (1.1.51). Since $\langle J_1 \rangle = 0$ we may substitute in (1.2.99) the correlator $Y_{12} = \langle J_1, J_2 \rangle_0$ for the moment $\langle J_1 J_2 \rangle_0$. Using (1.1.54), it can be easily verified that

$$\Phi(p, -p) + \Phi(-p, p)\exp(-i\hbar\beta p) = \Phi(p) , \quad (1.2.100)$$

which is a special case of (A1.1). In analogy with (1.1.59), we will therefore have

$$\Phi(p_1, -p_1)Y_{12} + \Phi(-p_1, p_1)Y_{21} = \Phi(p_1)Y_{12} . \quad (1.2.101)$$

If we then use this equation and [(5.2.8) v.1] the equation (1.2.99) becomes

$$Y_{1,2} + Y_{2,1} = \beta \Phi(p_1)Y_{12} . \quad (1.2.102)$$

This coincides with [(5.3.92) v.1] because $\Phi(p) = [\Theta^+(p)]^{-1}$, which is easily verified.

Further, differentiating (1.2.94) with respect to h_1, h_2, h_3 and setting $h(t) \equiv 0$, we will have, using (1.2.96,98),

$$P_{(123)}\{-Y_{1,23} + \beta\Phi(p_1)\langle J_1, \delta J_2/\delta h_3 + \delta J_3/\delta h_2 \rangle\}$$
$$= \beta^2 \Phi(p_1, p_2)Y_{123} + \beta^2 \Phi(p_2, p_1)Y_{213} , \quad (1.2.103)$$

where we have also used (1.2.100) and an equation of the type (1.1.59) at $n = 3$. The relation (1.2.103) is a quantum generalization of the FDR (1.2.55). Using [(5.2.50) v.1] we easily find

$$\langle J_1 \delta J_2/\delta h_3 \rangle_0 = -(i\hbar p_3)^{-1}\langle J_1[J_2, J_3]\rangle_0 \eta(t_{23})$$
$$= -(i\hbar p_3)^{-1}(Y_{123} - Y_{132})\eta(t_{23}) . \quad (1.2.104)$$

From (1.2.103) we can therefore derive a relation that only involves $Y_{1,23}$ and Y_{123} and that does not follow from the condition of time reversibility.

1.2.10 Quasiclassical Generating Equation in the Quantum Case

Let us consider the integral

$$\hat{F} = \int_a^b v(t)\hat{J}(t)dt . \quad (1.2.105)$$

From (1.2.65) we have

$$\mathrm{Tr}\exp\left[-\beta\hat{\mathcal{H}}_0(a) - \beta\int_a^b h(t)\hat{J}(t)dt + \hat{F}\right] = \mathrm{Tr}\exp[-\beta\hat{\mathcal{H}}_0(b) + \hat{F}] . \quad (1.2.106)$$

Using (1.2.66,67), we can write this as

$$\text{Tr}\left\{\hat{\rho}_0(a)\exp[\beta\hat{\mathcal{H}}_0(a)]\exp\left[-\beta\hat{\mathcal{H}}_0(a)-\beta\int_a^b h(t)\hat{J}(t)dt+\hat{F}\right]\right\}$$
$$=\text{Tr}\{\hat{\rho}_0(b)\exp[\beta\hat{\mathcal{H}}_0(b)]\exp[-\beta\hat{\mathcal{H}}_0(b)+\hat{F}]\}. \tag{1.2.107}$$

We will assume that the Heisenberg operators $\hat{J}_\alpha(t)$ in (1.2.105,107) vary with time under the action of the external forces $h(t)$. The expression on the left-hand side of (1.2.107) will then be a mean of the type (1.2.72), i.e., it can be written as

$$\left\langle\exp[\beta\hat{\mathcal{H}}_0(a)]\exp\left[-\beta\hat{\mathcal{H}}_0(a)-\beta\int_a^b h(t)\hat{J}(t)dt+\int_a^b v(t)\hat{J}(t)dt\right]\right\rangle_{h(\tau)}. \tag{1.2.108}$$

If we now introduce the functional

$$\Psi[v(t),h(t)]=\left\langle\exp[\beta\hat{\mathcal{H}}_0(a)]\exp\left[-\beta\hat{\mathcal{H}}_0(a)+\int_a^b v(t)\hat{J}(t)dt\right]\right\rangle_{h(\tau)}, \tag{1.2.109}$$

then this expression will be nothing but

$$\Psi[v(t)-\beta h(t),h(t)]. \tag{1.2.110}$$

Turning to the right-hand side of (1.2.107), we will treat the matrix $\hat{\rho}_0^*(b)$ as a Heisenberg density matrix for processes evolving in reverse time. We can then transform the operator (1.2.105) that enters the right-hand side of (1.2.107) by substituting into it the equation $\hat{J}_\alpha(t)=-\varepsilon_\alpha\hat{\tilde{J}}_\alpha^*(-t)$ of the type (1.2.78). This gives

$$\hat{F}=-\int_a^b v_\alpha(t)\varepsilon_\alpha\hat{\tilde{J}}_\alpha^*(-t)dt=-\int_{-b}^{-a}\varepsilon_\alpha v_\alpha(-\tilde{t})\hat{\tilde{J}}_\alpha^*(\tilde{t})d\tilde{t}, \tag{1.2.111}$$

Substituting (1.2.111) into the right-hand side of (1.2.107) gives

$$\text{Tr}\left\{\hat{\rho}_0(b)\exp[\beta\hat{\mathcal{H}}_0(b)]\exp\left[-\beta\hat{\mathcal{H}}_0(b)-\int_{-b}^{-a}\varepsilon_\alpha v_\alpha(-\tilde{t})\hat{\tilde{J}}_\alpha^*(\tilde{t})d\tilde{t}\right]\right\}_{\varepsilon h(-\tilde{\tau})}. \tag{1.2.112}$$

The forces are represented here as they appear in the time-reversed frame of reference.

The counterpart of (1.2.109) in reverse time is

$$\Psi[\tilde{v}(\tilde{t}),\tilde{h}(\tilde{t})]$$
$$=\text{Tr}\left\{\hat{\rho}_0^*(b)\exp[\beta\hat{\mathcal{H}}_0^*(b)]\exp\left[-\beta\hat{\mathcal{H}}_0^*(b)+\int_{-b}^{-a}\tilde{v}(\tilde{t})\hat{\tilde{J}}(\tilde{t})d\tilde{t}\right]\right\}_{\tilde{h}(\tilde{\tau})}, \tag{1.2.113}$$

the functional $\Psi[\cdot,\cdot]$ in (1.2.109) and (1.2.113) being the same in the case of time-reversal invariance if $a=-b$. It is easily seen from (1.2.113) that the expression (1.2.112) can be written as

$$\Psi[-\varepsilon v(-t), \varepsilon h(-t)]^* . \tag{1.2.114}$$

The asterisk, which denotes complex conjugation, comes from the complex conjugation in (1.2.111). It can be removed because the functional (1.2.109) is real for real $v(t)$. In fact, by virtue of (1.2.72,66), it can be written as

$$\Psi[v(t), h(t)] = C_0 \text{Tr} \left\{ \exp\left[-\beta\hat{\mathcal{H}}_0(a) + \int v(t)\hat{J}(t)dt \right] \right\} . \tag{1.2.115}$$

The trace is real because the operators in the exponential function are Hermitian.

Since by virtue of (1.2.107) the expressions (1.2.110) and (1.2.114) are equal, we obtain

$$\Psi[v(t) - \beta h(t), h(t)] = \Psi[-\varepsilon v(-t), \varepsilon h(-t)] . \tag{1.2.116}$$

We can pass here to the limit as $b = -a \to \infty$.

This equation may be called the quasiclassical generating equation, since it is similar in form to the nonquantum equation (1.2.37). The functional (1.2.109) is a quasiclassical characteristic functional of the quantum theory. If the expansion of the nonquantum functional (1.2.11)

$$\Theta[v(t), h(t)] = \sum_{n=0}^{\infty} \frac{1}{n!} \langle J_1 \ldots J_n \rangle_{h(\tau)} v_1 \ldots v_n \tag{1.2.117}$$

yields the moments $\langle J_1 \ldots J_n \rangle_{h(\tau)}$, then the similar expansion

$$\Psi[v(t), h(t)] = \sum_{n=0}^{\infty} \frac{1}{n!} m_{1\ldots n}^{\text{q.c.}}[h(t)] v_1 \ldots v_n \tag{1.2.118}$$

defines functions that can be referred to as quasiclassical moments. Significantly, the derivatives with respect to $h(t)$ of these moments and also of the quasiclassical correlators, which are related to the former in the conventional way, obey, by (1.2.116), the same nonquantum relations derivable from (1.2.37).

1.2.11 Derivatives of Quasiclassical Moments and Correlators

Using (1.2.105) we can write (1.2.109) in the form

$$\Psi[v(t), h(t)] = \langle \exp[\beta\hat{\mathcal{H}}_0(a)] \exp[-\beta\hat{\mathcal{H}}_0(a) + \hat{F}[h]] \rangle_0 . \tag{1.2.119}$$

We have here allowed for the fact that the subscript $h(\tau)$ in (1.2.109) indicates how the Heisenberg operators evolve, whereas the mean is computed by using the equilibrium density matrix (1.2.66) and is an equilibrium mean. We have therefore assigned $h(\tau)$ to the operator $\hat{F}$. Using the equation

$$\exp(\beta\hat{\mathcal{H}}_0) \exp(-\beta\hat{\mathcal{H}}_0 + \hat{F}[h])$$
$$= \overleftarrow{\exp} \left\{ kT \int_0^\beta d\lambda [\exp(\lambda\hat{\mathcal{H}}_0)\hat{F}[h]\exp(-\lambda\hat{\mathcal{H}}_0)]_\lambda \right\} , \tag{1.2.120}$$

which is similar to (1.2.70), we will write (1.2.119) in the form

$$\Psi[v(t), h(t)] = \left\langle \overleftarrow{\exp}\left\{ kT \int_0^\beta d\lambda [\exp(\lambda\hat{\mathcal{H}}_0)\hat{F}[h]\exp(-\lambda\hat{\mathcal{H}}_0)]_\lambda \right\} \right\rangle_0 . \quad (1.2.121)$$

We will now expand the exponential function here into a Taylor series. The terms

$$\left\{ \int_0^\beta d\lambda [\exp(\lambda\hat{\mathcal{H}}_0)\hat{F}[h]\exp(-\lambda\hat{\mathcal{H}}_0)]_\lambda \right\}^n , \quad (1.2.122)$$

where the operators are arranged according to the value of λ (the larger λ the farther to the left is placed the operator marked by λ), will, similarly to (1.1.52), be represented in the form

$$P_{1\ldots n}\int_0^\beta d\lambda_1 \int_0^{\lambda_1} d\lambda_2 \ldots \int_0^{\lambda_{n-1}} d\lambda_n \langle [\exp(\lambda_1\hat{\mathcal{H}}_0)\hat{F}[h]\exp(-\lambda_1\hat{\mathcal{H}}_0)]$$
$$\times \ldots \times [\exp(\lambda_n\hat{\mathcal{H}}_0)\hat{F}[h]\exp(-\lambda_n\hat{\mathcal{H}}_0)]\rangle_0 . \quad (1.2.123)$$

Substituting (1.2.105) and carrying out the summation of similar expressions in accordance with the expansion of the exponential function, we will obtain, instead of (1.2.121),

$$\begin{aligned}
\Psi[v(t), h(t)] &= \sum_{n=0}^\infty (kT)^n (n!)^{-1} v_1 \ldots v_n \\
&\quad \times P_{1\ldots n}\int_0^\beta d\lambda_1 \int_0^{\lambda_1} d\lambda_2 \ldots \int_0^{\lambda_{n-1}} d\lambda_n \langle [\exp(\lambda_1\hat{\mathcal{H}}_0)\hat{J}_1[h] \\
&\quad \times \exp(-\lambda_1\hat{\mathcal{H}}_0)] \times \ldots \times [\exp(\lambda_n\hat{\mathcal{H}}_0)\hat{J}_n[h]\exp(-\lambda_n\hat{\mathcal{H}}_0)]\rangle_0 .
\end{aligned}$$
$$(1.2.124)$$

Here the summation convention is used. Comparing this expansion with (1.2.118), we find the quasiclassical moments

$$\begin{aligned}
m_{1\ldots n}^{\mathrm{q.c.}}[h(t)] &= (kT)^n P_{1\ldots n}\int_0^\beta d\lambda_1 \int_0^{\lambda_1} d\lambda_2 \\
&\quad \times \ldots \times \int_0^{\lambda_{n-1}} d\lambda_n \langle [\exp(\lambda_1\hat{\mathcal{H}}_0)\hat{J}_1[h]\exp(-\lambda_1\hat{\mathcal{H}}_0)] \\
&\quad \times \ldots \times [\exp(\lambda_n\hat{\mathcal{H}}_0)\hat{J}_n[h]\exp(-\lambda_n\hat{\mathcal{H}}_0)]\rangle_0 .
\end{aligned}$$
$$(1.2.125)$$

In deriving (1.2.125), we made use of the fact that its right-hand side is symmetric in the subscripts $1, 2, \ldots, n$. By using the identity (A1.12) we can reduce the number of terms on the right-hand side of (1.2.125), and so, instead of (1.2.125), we will have

$$\begin{aligned}
m_{1\ldots n}^{\mathrm{q.c.}}[h(t)] &= (kT)^{n-1} P_{1\ldots(n-1)}\int_0^\beta d\lambda_1 \int_0^{\lambda_1} d\lambda_2 \\
&\quad \times \ldots \times \int_0^{\lambda_{n-2}} d\lambda_{n-1} \langle [\exp(\lambda_1\hat{\mathcal{H}}_0)\hat{J}_1[h]\exp(-\lambda_1\hat{\mathcal{H}}_0)] \\
&\quad \times \ldots \times [\exp(\lambda_{n-1}\hat{\mathcal{H}}_0)\hat{J}_{n-1}[h]\exp(-\lambda_{n-1}\hat{\mathcal{H}}_0)]\hat{J}_n[h]\rangle_0 .
\end{aligned}$$
$$(1.2.126)$$

Besides (1.2.109), we can also look at the analogous quasiclassical characteristic functional

$$\Psi_B[u(t), h(t)] = \sum_{n=0}^{\infty} \left\langle \exp(\beta\hat{\mathcal{H}}_0) \exp\left[-\beta\hat{\mathcal{H}}_0 + \int_{-\infty}^{\infty} u(t)\hat{B}(t)\right] \right\rangle_{h(\tau)} , \quad (1.2.127)$$

where, instead of the fluxes $\hat{J}$, we have the internal parameters B and the integration limits are placed at infinity.

According to the formula

$$\Psi_B[u(t), h(t)] = \sum_{n=0}^{\infty} (n!)^{-1} M_{1\ldots n}^{\mathrm{q.c.}}[h(t)] u_1 \ldots u_n , \quad (1.2.128)$$

which is similar to (1.2.118), the functional defines the quasiclassical moments of B. Using

$$\int_{-\infty}^{\infty} v(t)\dot{\hat{B}}(t)dt = \int_{-\infty}^{\infty} v(t)d\hat{B}(t)$$
$$= -\int_{-\infty}^{\infty} \hat{B}(t)\dot{v}(t)dt , \quad (1.2.129)$$

we can easily derive the formula relating the two functionals:

$$\Psi[v(t), h(t)] = \Psi_B[-\dot{v}(t), h(t)] . \quad (1.2.130)$$

From this and (1.2.118,128) we find that the quasiclassical moments are related by

$$M_{1\ldots n}^{\mathrm{q.c.}}[h(t)] = (\partial/\partial t_1)^{-1} \ldots (\partial/\partial t_n)^{-1} m_{1\ldots n}^{\mathrm{q.c.}}[h(t)] . \quad (1.2.131)$$

These relationships have the same form as those for ordinary moments. By applying these to (1.2.126) we obtain a similar formula for the quasiclassical moments of $B(t)$:

$$\begin{aligned} M_{1\ldots n}^{\mathrm{q.c.}}[h(t)] = {} & (kT)^{n-1} P_{1\ldots(n-1)} \int_0^{\beta} d\lambda_1 \int_0^{\lambda_1} d\lambda_2 \ldots \int_0^{\lambda_{n-2}} d\lambda_{n-1} \\ & \times \langle [\exp(\lambda_1 \hat{\mathcal{H}}_0) \hat{B}_1[h] \exp(-\lambda_1 \hat{\mathcal{H}}_0)] \\ & \times \ldots \times [\exp(\lambda_{n-1} \hat{\mathcal{H}}_0) \hat{B}_{n-1}[h] \exp(-\lambda_{n-1} \hat{\mathcal{H}}_0)] \hat{B}_n[h] \rangle_0 . \end{aligned}$$
$$(1.2.132)$$

Letting $h(t) \equiv 0$ we obtain by the use of (1.1.36) formula (1.1.61). If we differentiate both sides of (1.2.132) with respect to h_{n+1} and put $h = 0$, we then find

$$\begin{aligned} [\delta M_{1\ldots n}^{\mathrm{q.c.}}/\delta h_{n+1}]_{h=0} = {} & (k_B T)^{n-1} (i/\hbar) P_{1\ldots(n-1)} \\ & \times \left\{ \sum_{k=1}^{n} \Phi(p_1, \ldots, p_{k-1}, p_k + p_{n+1}, p_{k+1}, \ldots, p_{n-1}) \right. \\ & \left. \times \langle \hat{B}_1 \ldots \hat{B}_{k-1}[\hat{B}_k, \hat{B}_{n+1}]\hat{B}_{k+1} \ldots \hat{B}_n \rangle_0 \eta(t_k - t_{n+1}) \right\} \end{aligned}$$
$$(1.2.133)$$

by [(5.2.50) v.1]. As in (1.1.61), the operators B here are the Heisenberg operators, which evolve with the unperturbed Hamiltonian $\mathcal{H}_0$. Likewise, from (1.2.132) we can find higher derivatives with respect to external forces at zero.

If we put $n = 1$ in (1.2.132), we will obtain the quantity $M_1^{\text{q.c.}}[h]$, which coincides with the conventional mean $A_\alpha[t, h(\tau)]$ appearing on the left-hand side of [(5.2.2) v. 1]. The derivatives of the latter with respect to forces at zero coincide with admittances [(5.2.3) v.1]. Further, letting $n = 2$ in (1.2.132), we will obtain the second quasiclassical moment $M_{12}^{\text{q.c.}}[h]$. Using this, we can construct the quasiclassical double correlator

$$
\begin{aligned}
K_{12}^{\text{q.c.}}[h] &= M_{12}^{\text{q.c.}}[h] - M_1^{\text{q.c.}}[h] M_2^{\text{q.c.}}[h] \\
&= M_2^{\text{q.c.}}[h] - (G_1 h_1 + \tfrac{1}{2} G_{1,34} h_3 h_4 + \dots) \\
&\quad \times (G_2 h_2 + \tfrac{1}{2} G_{2,56} h_5 h_6 + \dots).
\end{aligned}
\tag{1.2.134}
$$

In terms of the moments (1.2.132) we can also express quasiclassical correlators with more subscripts.

In analogy with the notation of Sects. 5.3,4 v.1, in which the expression

$$
\begin{aligned}
\langle \hat{B}_1[h], \dots, \hat{B}_m[h] \rangle_0 &= G_{1\dots m} + G_{1\dots m,(m+1)} h_{m+1} \\
&\quad + \tfrac{1}{2} G_{1\dots m,(m+1)(m+2)} h_{m+1} h_{m+2} + \dots
\end{aligned}
\tag{1.2.135}
$$

holds, we will introduce the following notation for the derivatives of quasiclassical correlators:

$$
\begin{aligned}
K_{1\dots m}^{\text{q.c.}}[h] &= G_{1\dots m}^{\text{q.c.}} + G_{1\dots m,(m+1)}^{\text{q.c.}} h_{m+1} \\
&\quad + \tfrac{1}{2} G_{1\dots m,(m+1)(m+2)}^{\text{q.c.}} h_{m+1} h_{m+2} \\
&\quad + \dots.
\end{aligned}
\tag{1.2.136}
$$

By virtue of the generating equation (1.2.116) analogous to (1.2.37), the functions $G_{\dots}^{\text{q.c.}}$, which enter into the above expression, must be related to the admittances $\bar{G}_{1,2\dots m}^{\text{q.c.}} = G_{1,2\dots m}$ and to one another by the same relations as are the functions $G_{\dots}$ (including admittance) in the nonquantum case. Using (1.2.132) and [(5.2.50) v.1], we can obtain from (1.2.134)

$$
\begin{aligned}
G_{12}^{\text{q.c.}} &= \Phi(p_1) \langle \hat{B}_1, \hat{B}_2, \rangle_0, \\
G_{12,3}^{\text{q.c.}} &= (i/\hbar)[\Phi(p_1 + p_3) \langle [\hat{B}_1, \hat{B}_3] \hat{B}_2 \rangle_0 \eta(t_{13}) \\
&\quad + \Phi(p_1) \langle \hat{B}_1 [\hat{B}_2, \hat{B}_3] \rangle_0 \eta(t_{23})], \\
G_{12,34}^{\text{q.c.}} &= (i/\hbar)^2 P_{34} \{ \Phi(p_1 + p_3 + p_4) \langle [[\hat{B}_1, \hat{B}_3], \hat{B}_4] \hat{B}_2 \rangle_0 \eta_{134} \\
&\quad + \Phi(p_1 + p_3) \langle [\hat{B}_1, \hat{B}_3][\hat{B}_2, \hat{B}_4] \rangle_0 \eta_{13} \eta_{24} \\
&\quad + \Phi(p_1) \langle \hat{B}_1 [[\hat{B}_2, \hat{B}_3], \hat{B}_4] \rangle_0 \eta_{234} \} - P_{34} G_{1,3} G_{2,}.
\end{aligned}
\tag{1.2.137}
$$

If in these and in

$$
G_{123}^{\text{q.c.}} = \Phi(p_1, p_2) \langle \hat{B}_1, \hat{B}_2, \hat{B}_3 \rangle_0 + \Phi(p_2, p_1) \langle \hat{B}_2, \hat{B}_1, \hat{B}_3 \rangle_0
\tag{1.2.138}
$$

we write the means in terms of [(5.2.59) v.1], using the equations of Sect. 5.2.6 v.1 and (1.1.54), we can readily verify that $G_{12}^{\text{q.c.}}, G_{123}^{\text{q.c.}}, G_{12,3}^{\text{q.c.}}, G_{12,34}^{\text{q.c.}}$ are represented through $V_{12}, V_{123}, V_{1234}$ in exactly the same manner as $G_{12}, G_{123}, G_{12,3}, G_{12,34}$ are represented through $V_{\dots}$ in the nonquantum case. The same is true of other $G_{\dots}^{\text{q.c.}}$.

1.2.12 Quantum Markov Processes

The generalization of the concept of the Markov process to the quantum case is complicated by the fact that the conventional definition of a Markov process is given in terms of conditional probabilities, and in a quantum treatment the concept of conditional probability becomes ambiguous because operators may be ordered now in a number of ways. Thus there are several ways of introducing the concept of the quantum Markov process. Of these we will only consider two. It would certainly be desirable to have a definition such that all the nonquantum Markov processes defined in physical terms would be quantum Markov process in the quantum case. However, here we leave steady ground.

Definition 1. *A process $\{\hat{B}_\alpha(t)\}$ is said to be a thermodynamic quantum Markov process at temperature T, if the quasiclassical characteristic functional (1.2.127), where $\beta = 1/kT$ and $\hat{\mathcal{H}}_0$ is a complete Hamiltonian (including a thermostat), coincides with the characteristic functional of some nonquantum Markov process.*

This definition means that all the equilibrium quasiclassical moments

$$\langle \hat{B}_{\alpha_1}(t_1) \ldots \hat{B}_{\alpha_n}(t_n) \rangle_0^{\text{q.c.}} = P_{1\ldots(n-1)}(\varPhi_{1\ldots(n-1)} \langle \hat{B}_{\alpha_1}(t_1) \ldots \hat{B}_{\alpha_n}(t_n) \rangle)_0 \ (1.2.139)$$

coincide with the moments for a conventional Markov process. It follows from the definition that the one-time probability density $w(B)$, which is defined by one-time quasiclassical moments $\langle \hat{B}_{\alpha_1}(t) \ldots \hat{B}_{\alpha_n}(t) \rangle^{\text{q.c.}}$, $n = 1, 2, \ldots$, obeys the master equation of the type $\dot{w} = Mw$.

The concept of the Markov processes in this definition is related to the concept of the quantum Langevin equations.

Definition 2. *The equations*

$$\dot{\hat{B}}_\alpha(t) = \chi_\alpha(\hat{B}(t)) + \psi_{\alpha\beta}(\hat{B}(t))\hat{\xi}_\beta(t) \tag{1.2.140}$$

which hold for any t are called the Langevin equations, if the correlators and commutators of the processes $\{\hat{\xi}_\beta(t)\}$ are known and if $\hat{\xi}_\beta(t)$ are such that the one-time commutators $[\hat{B}_\alpha(t), \hat{B}_\beta(t)]$ have the standard form and do not vary with time.

In (1.2.140) $\chi_\alpha(\hat{B}), \psi_{\alpha\beta}(\hat{B})$ are the inertia-free operator expressions, i.e. the order of operation of $\hat{B}_1, \ldots, \hat{B}_r$ in them must also be indicated. Initial conditions are here of no significance, since the initial time is assumed to be placed at minus infinity: $t_0 = -\infty$, and the functions $\chi_\alpha(\hat{B})$ are assumed to be such that the system has some dissipation. This makes the stationary process possible.

The following is a special case of the quantum Langevin equations

$$\begin{aligned}
\dot{\hat{q}} &= (i/\hbar)[\hat{\mathcal{H}}, \hat{q}]\,, \\
\dot{\hat{p}} &= (i/\hbar)[\hat{\mathcal{H}}, \hat{p}] - b(i/\hbar)[\hat{\mathcal{H}}, \hat{q}] + c\hat{\xi}(t)\,.
\end{aligned} \tag{1.2.141}$$

They describe a simple dynamic system with friction. In connection with the last equation we will formulate a theorem, which can be easily proved.

Theorem 1. *Let the following conditions hold: (1) A process $\hat{\xi}(t)$ is Gaussian; it has a zero mean and the moment*

$$\langle \hat{\xi}(t_1)\hat{\xi}(t_2) \rangle = kT\Theta_T^+(d/dt_1)\delta(t_{12}) \, ; \qquad (1.2.142)$$

(2) $c = (2b)^{1/2}$; *(3) The Hamiltonian $\hat{\mathcal{H}}$ is quadratic:* $\hat{\mathcal{H}} = (m^{-1}\hat{p}^2 + \kappa\hat{q}^2)/2$. *The process $\{\hat{q}(t), \hat{p}(t)\}$ will then be an equilibrium quantum Markov process at temperature T.*

Proof. According to condition (3), we obtain

$$[m(d/dt)^2 + bd/dt + \kappa]\hat{q} = c\hat{\xi} \quad (\hat{p} = m\dot{\hat{q}}) \, . \qquad (1.2.143)$$

via the equations (1.2.141). Solving this by taking the Fourier transforms and using (1.2.142), we find

$$\langle \hat{q}(\omega_1)\hat{q}(\omega_2) \rangle = [(m\omega_1^2 - \kappa)^2 + b^2\omega_1^2]^{-1}c^2kT\Theta_T^+(i\omega_1)\delta(\omega_1 + \omega_2) \, . \qquad (1.2.144)$$

We can also easily obtain

$$\begin{aligned}
\langle \hat{p}(\omega_1)\hat{q}(\omega_2) \rangle & \\
&= mi\omega_1[(m\omega_1^2 - \kappa)^2 + b^2\omega_1^2]^{-1}c^2kT\Theta_T^+(i\omega_1)\delta(\omega_1 + \omega_2) \, , & (1.2.145a) \\
\langle \hat{q}(\omega_1)\hat{p}(\omega_2) \rangle & \\
&= mi\omega_2[(m\omega_1^2 - \kappa)^2 + b^2\omega_1^2]^{-1}c^2kT\Theta_T^-(i\omega_2)\delta(\omega_1 + \omega_2) \, , & (1.2.145b)
\end{aligned}$$

and similarly for $\langle \hat{p}(\omega_1)\hat{p}(\omega_2) \rangle$. According to (1.2.139), to change to quasiclassical moments at $n = 2$ it is sufficient to multiply them by Φ_1 or, since $\Phi_1 = 1/\Theta_1^+$ (see (1.1.54a)), to divide by Θ_1^+. From (1.2.144), after the division we will have

$$\langle \hat{q}(\omega_1)\hat{q}(\omega_2) \rangle^{\text{q.c.}} = [(m\omega_1^2 - \kappa)^2 + b^2\omega_1^2]^{-1}c^2kT\delta(\omega_1 + \omega_2) \qquad (1.2.146)$$

and similarly for other moments. Thus the moments obtained in such a way coincide with the moments of a nonquantum Markov process, whose master equation has the form

$$\dot{w} = -(p/m)\partial w/\partial q + \partial[(bp/m + \kappa q)w]/\partial p + \tfrac{1}{2}c^2kT\partial^2 w/\partial p^2 \, . \qquad (1.2.147)$$

It can be shown that in the Gaussian case the quasiclassical characteristic functional is completely determined by twofold quasiclassical moments, just as the characteristic functional of the symmetrization ordering is determined by twofold symmetrized moments. Therefore, the moments of higher multiplicity may be overlooked. Consequently, the condition stated in Definition 1 is satisfied. It only remains to check the validity of the commutator, as mentioned in Definition 2. By using [(5.2.73) v.1], i.e. $\langle [\hat{D}(t), \hat{Q}] \rangle = \{\Gamma^+(d/dt)\}^{-1}\langle \hat{D}(t)\hat{Q} \rangle$, from (1.2.145) after exchanging subscripts $1 \rightleftarrows 2$ in (1.2.145b) we will have

$$\langle [\hat{p}(\omega_1), \hat{q}(\omega_2)] \rangle = mi\omega_1[(m\omega_1^2 - \kappa)^2 + b^2\omega_1^2]^{-1}c^2i\hbar(i\omega_1)\delta(\omega_1 + \omega_2) \qquad (1.2.148)$$

since $\Theta^+(p) - \Theta^-(p) = i\beta\hbar p[\Gamma^+(p) - \Gamma^-(p)] = i\beta\hbar p$. It follows in analogy with the known formula

$$\langle \hat{q}^2(t) \rangle = \frac{1}{2\pi} \int_{-\infty}^{\infty} S_q(\omega) d\omega \tag{1.2.149}$$

that

$$\frac{i}{\hbar} \langle [\hat{p}(t), \hat{q}(t)] \rangle = \frac{mc^2}{2\pi} \int_{-\infty}^{\infty} \frac{\omega^2 d\omega}{(m\omega^2 - \kappa)^2 + b^2\omega^2} = \frac{c^2}{2b}. \tag{1.2.150}$$

Thus, the commutation relations are standard ones if the condition (2) of the theorem is met. QED.

It can also be shown that the standard commutation relations are valid and their form is preserved not only in average. To prove the latter statement we will need an appropriate form of the commutator for $\hat{\xi}(t)$.

Theorem 2. *If near the thermodynamic equilibrium at temperature T the internal parameters obey the phenomenological equation of linear relaxation*

$$\dot{A}_\alpha = -d_{\alpha\beta} A_\beta \quad (A_\alpha = \langle \hat{B}_\alpha \rangle), \tag{1.2.151}$$

then the process $\{\hat{B}_\alpha(t)\}$ is a stationary Markov quantum process at temperature T.

Proof. Consider the Langevin equations

$$\dot{\hat{B}}_\alpha = -d_{\alpha\beta} \hat{B}_\beta + \hat{\xi}_\alpha(t) \tag{1.2.152}$$

that correspond to (1.2.151). The correlator of the random inputs in the above expression is given by the quantum linear FDR of the first kind [(5.5.27) v.1]. If we put $h(t) = 0$ in [(5.5.20) v.1], we will see that the matrix $\Phi_{1,2}$ in [(5.5.20) v.1] is related to the matrix $d_{\alpha\beta}$ in (1.2.151) by

$$\Phi_{\alpha,\gamma}(t_1, t_2) u_{\gamma,\beta} = d_{\alpha\beta} \delta(t_1 - t_2), \tag{1.2.153}$$

where

$$u_{\alpha\gamma} = \partial^2 F(B) / \partial B_\alpha \partial B_\gamma. \tag{1.2.154}$$

But according to (1.1.57a)

$$\partial^2 F(a) / \partial a_\alpha \partial a_\gamma = -\beta \langle \hat{B}_\alpha \hat{B}_\gamma \rangle^{\text{q.c.}}. \tag{1.2.155}$$

Considering that

$$\| u_{\alpha\gamma} \| = \| -\partial^2 F(a) / \partial a_\alpha \partial a_\gamma \|^{-1}, \tag{1.2.156}$$

we have

$$\Phi_{\alpha,\gamma}(t_1, t_2) = \beta d_{\alpha\delta} \langle \hat{B}_\delta(t) \hat{B}_\gamma(t) \rangle^{\text{q.c.}} \delta(t_1 - t_2). \tag{1.2.157}$$

Now from [(5.5.27) v.1] we obtain

$$\langle \hat{\xi}_\alpha(t_1) \hat{\xi}_\beta(t_2) \rangle = \Theta_1^+ (\partial/\partial t_1) [d_{\alpha\gamma} \langle \hat{B}_\gamma \hat{B}_\beta \rangle^{\text{q.c.}} + \langle \hat{B}_\alpha \hat{B}_\gamma \rangle^{\text{q.c.}} d_{\beta\gamma}] \delta(t_1 - t_2). \tag{1.2.158}$$

For quasiclassical moments the operator $\Theta_1^+(\partial/\partial t_1)$ is discarded and the fluctuational inputs become delta-correlated and the appropriate quasiclassical moment $\langle \hat{B}_\alpha(t_1)\hat{B}_\beta(t_2)\rangle^{\text{q.c.}}$ becomes

$$\| \langle \hat{B}_\alpha(t_1)\hat{B}_\beta(t_2)\rangle^{\text{q.c.}} \| = \exp[-\hat{D}(t_1 - t_2)] \| \langle \hat{B}_\alpha(t_2)\hat{B}_\beta(t_2)\rangle^{\text{q.c.}} \|, \quad t_1 \geq t_2$$

$$(1.2.159)$$

$(\hat{D} = \| d_{\alpha\beta} \|)$, which is common for the nonquantum Markov linear relaxation. The linear Markov process is Gaussian. Therefore, the quasiclassical characteristic functional is completely determined by the twofold quasiclassical moments and so higher moments can be ignored. QED.

These two theorems are concerned with linear approximation. It is of interest to know whether or not we may remove constraint (3) in Theorem 1. It would also be desirable for the following statement to be true: a process described by the Langevin equation (1.2.140) is Markovian if $\hat{\xi}_\beta(t)$ are quantum Gaussian white noises, i.e. if

$$\langle \hat{\xi}_\beta(t)\hat{\xi}_\gamma(t')\rangle = \Theta_T^+(\partial/\partial t)\delta(t - t')\delta_{\beta\gamma}, \quad \langle \hat{\xi}_\beta\rangle = 0.$$

$$(1.2.160)$$

Nonlinear generalizations have not been studied.

Definition 3. *We will refer to the quantum process $\{\hat{B}_1(t),\ldots,\hat{B}_r(t)\}$ as the specific quantum process if, in the Schrödinger representation, the derivative $\dot{\hat{\rho}}$ is linearly expressed through $\hat{\rho}$:*

$$\dot{\hat{\rho}}(t) = \hat{\mathbf{M}}\hat{\rho}(t).$$

$$(1.2.161)$$

Here $\hat{\rho}$ is the density matrix that operates in the same space as the operators $\hat{B}_1,\ldots,\hat{B}_r$ and which can be expressed in terms of them (in this case the set of operators must be complete enough, e.g. contain pairs of conjugate coordinates and momenta).

The operator $\hat{\mathbf{M}}$ in (1.2.161) acts on the Hilbert space $H \otimes H$ if $\hat{B}_\alpha$ and $\hat{\rho}$ act on H. It must of course meet some requirements, such as the preservation of normalization, of the Hermitian property, and of non-negative definiteness of a density matrix.

The mean of the arbitrary operator F on H is defined as usual:

$$\langle F\rangle_t = \text{Tr}\{\hat{F}\hat{\rho}(t)\}.$$

$$(1.2.162)$$

Differentiating both sides of this equation with respect to time and using (1.2.161), we obtain

$$\begin{aligned} d\langle F\rangle_t/dt &= \text{Tr}\{\hat{F}\dot{\hat{\rho}}(t)\} \\ &= \text{Tr}\{\hat{F}[\hat{M}\hat{\rho}(t)]\}. \end{aligned}$$

$$(1.2.163)$$

Let us introduce the transposed operator $\hat{M}^T$ defined by the formula

$$\text{Tr}\{\hat{A}[\hat{M}\hat{B}]\} = \text{Tr}\{[\hat{M}^T\hat{A}]\hat{B}\}$$

$$(1.2.164)$$

valid for arbitrary operators $\hat{A}, \hat{B}$ on H. Owing to this definition we obtain from (1.2.163)

$$d\langle \hat{F} \rangle_t / dt = \text{Tr}\{[\hat{M}^T \hat{F}]\hat{\rho}(t)\}$$

or

$$d\langle \hat{F} \rangle_t / dt = \langle \hat{M}^T \hat{F} \rangle_t. \tag{1.2.165}$$

It is easy to prove that the mean value (1.2.162) coincides with the mean

$$\langle \hat{F} \rangle_t = \text{Tr}_{\text{tot}}[\hat{F}_H(t) \otimes \hat{\rho}_{\text{tot}}] \tag{1.2.166}$$

of the Heisenberg operator acting on the Hilbert space $H_{\text{tot}} = H \otimes H_2$. Here H_2 is the Hilbert space corresponding to some ambient system, for example, to a thermal bath, and $\hat{\rho}_{\text{tot}}$ is the density matrix of the total system which is a combination of the original system with the ambient system. The trace Tr_{tot} is a total one.

In fact we have in the Schrödinger picture

$$\hat{\rho}(t) = \text{Tr}_2\{\hat{\rho}_{\text{tot}}(t)\}, \tag{1.2.167}$$

where Tr_2 is the partial trace, i.e. the trace with respect to variables of the ambient system. Substituting (1.2.167) into (1.2.162), we get

$$\langle \hat{F} \rangle_t = \text{Tr}_{\text{tot}}\{(\hat{F} \otimes \hat{I}_2)\hat{\rho}_{\text{tot}}(t)\}, \tag{1.2.168}$$

where $\hat{I}_2$ is the identity operator on H_2 and $\text{Tr}_{\text{tot}} = \text{Tr}\,\text{Tr}_2$. Passing to the Heisenberg picture, we obtain (1.2.166) from (1.2.168).

For the specific quantum Markov process the quantum regression theorem has been proved (see, e.g., [1.4]). The theorem states that from the equation

$$\dot{A}_\alpha = -d_{\alpha\gamma}(t)A_\gamma \quad (A_\alpha = \langle \hat{B}_\alpha \rangle) \tag{1.2.169}$$

it follows that

$$\partial\langle \hat{B}_\alpha(t_1)\hat{B}_\beta(t_2)\rangle / \partial t_1 = -d_{\alpha\gamma}(t_1)\langle \hat{B}_\gamma(t_1)\hat{B}_\beta(t_2)\rangle, \quad t_2 > t_1. \tag{1.2.170}$$

For $d_{\alpha\beta}$ independent of t, equations (1.2.169) coincide with (1.2.151), and equations (1.2.170) have the solution

$$\| \langle \hat{B}_\alpha(t_1)\hat{B}_\beta(t_2)\rangle \| = \exp[-\hat{D}(t_1 - t_2)] \| \langle \hat{B}_\alpha(t_2)\hat{B}_\beta(t_2)\rangle \|, \quad t_1 \geq t_2. \tag{1.2.171}$$

If, in addition to this equation, we also take account of a similar expression with $t_1 \rightleftarrows t_2$, we will have

$$\| \langle \hat{B}_\alpha(t_1)\hat{B}_\beta(t_2)\rangle \|$$
$$= \exp(-\hat{D}t_{12}\eta_{12}) \| \langle \hat{B}_\alpha(t)\hat{B}_\beta(t)\rangle \| \exp(-\hat{D}^T t_{21}\eta_{21}). \tag{1.2.172}$$

Owing to the formula $\langle \hat{B}_1\hat{B}_2 \rangle = \Theta_1^+ \langle \hat{B}_1, \hat{B}_2 \rangle^{\text{q.c.}}$, from equation (1.2.159) reduced to the similar form we have

$$\langle \hat{B}_\alpha(t_1)\hat{B}_\beta(t_2)\rangle = \Theta_1^+ (\partial/\partial t_1)[\exp(-\hat{D}t_{12}\eta_{12})]_{\alpha\gamma}\langle \hat{B}_\gamma(t)\hat{B}_\delta(t)\rangle$$
$$\times [\exp(-\hat{D}^T t_{21}\eta_{21})]_{\delta\beta}.$$

Comparing this with (1.2.172), we obtain

$$
\begin{aligned}
[\exp(-\hat{D}t_{12}\eta_{12})]_{\alpha\gamma}&\langle\hat{B}_\gamma(t)\hat{B}_\beta(t)\rangle^{\mathrm{q.c.}}[\exp(-\hat{D}^T t_{21}\eta_{12})]_{\beta\delta} \\
&= [\Theta_T^+(\partial/\partial t_1)]^{-1}[\exp(-\hat{D}t_{12}\eta_{12})]_{\alpha\gamma}\langle\hat{B}_\gamma(t)\hat{B}_\beta(t)\rangle[\exp(-\hat{D}^T t_{21}\eta_{21})]_{\beta\delta} \\
&\quad (\eta_{12} = \eta(t_1 - t_2))\,.
\end{aligned}
\tag{1.2.173}
$$

This equation, which must hold for equilibrium fluctuation processes at some temperature T, is rather difficult to explain for $\hbar \neq 0$, since the action of the operator $[\Theta_1^+]^{-1}$ must change the temporal factors $\exp[-\hat{D}t_{12}\eta_{12}], \exp[-\hat{D}^T t_{21}\eta_{21}]$. This casts a shadow of doubt on the applicability of specific Markov processes to thermodynamics.

In this connection it would be useful to consider an example of a specific Markov process [1.4], where the equation (1.2.161) has the form

$$
\begin{aligned}
\dot{\rho} &= -i\omega_0[\hat{a}^+\hat{a}, \hat{\rho}] + b(\bar{N} + 1)(2\hat{a}\hat{\rho}\hat{a}^+ - \hat{a}^+\hat{a}\hat{\rho} - \hat{\rho}\hat{a}^+\hat{a}) \\
&\quad + b\bar{N}(2\hat{a}^+\hat{\rho}\hat{a} - \hat{a}\hat{a}^+\hat{\rho} - \hat{\rho}\hat{a}\hat{a}^+)\,,
\end{aligned}
\tag{1.2.174}
$$

where $\omega_0, b, \bar{N}$ are constants, and $\hat{a} = (2\hbar)^{-1/2}(\hat{q} + i\hat{p})$, $\hat{a}^+ = (2\hbar)^{-1/2}(\hat{q} - i\hat{p})$. In this case, (1.2.165) reads

$$
\begin{aligned}
d\langle\hat{F}\rangle/dt &= i\omega_0\langle[\hat{a}\hat{a}^+, \hat{F}]\rangle + b(\bar{N} + 1)\langle[\hat{a}^+, \hat{F}]\hat{a} - \hat{a}^+[\hat{a}, \hat{F}]\rangle \\
&\quad + b\bar{N}\langle[\hat{a}, \hat{F}]\hat{a}^+ - \hat{a}[\hat{a}^+, \hat{F}]\rangle\,.
\end{aligned}
\tag{1.2.175}
$$

Putting $\hat{F} = \hat{a}$ and using the standard commutation relation $[\hat{a}, \hat{a}^+] = \hat{1}$ gives

$$
\langle\dot{\hat{a}}\rangle = -(b + i\omega_0)\langle\hat{a}\rangle\,.
\tag{1.2.176}
$$

The complex conjugates obey

$$
\langle\dot{\hat{a}}^+\rangle = -(b - i\omega_0)\langle\hat{a}^+\rangle\,.
\tag{1.2.177}
$$

Using the quantum regression theorem, these two equations yield the stationary correlators

$$
\begin{aligned}
\langle\hat{a}(t_1)\hat{a}^+(t_2)\rangle &= \langle\hat{a}(t)\hat{a}^+(t)\rangle\exp(-b\mid t_{12}\mid -i\omega_0 t_{12}) \\
&= \langle\hat{a}^+(t)\hat{a}(t) + 1\rangle\exp(-b\mid t_{12}\mid -i\omega_0 t_{12})\,, \\
\langle\hat{a}^+(t_2)\hat{a}(t_1)\rangle &= \langle\hat{a}^+(t)\hat{a}(t)\rangle\exp(-b\mid t_{12}\mid +i\omega_0 t_{21})\,.
\end{aligned}
\tag{1.2.178}
$$

We can easily find $\langle\hat{a}^+(t)\hat{a}(t)\rangle$ by substituting $\hat{F} = \hat{a}\hat{a}^+$ into (1.2.175). This gives

$$
d(\hat{a}^+\hat{a})/dt = -2b(\hat{a}^+\hat{a} - \bar{N})\,,
\tag{1.2.179}
$$

whence we find the stationary value $\langle\hat{a}^+\hat{a}\rangle = \bar{N}$. Substituting this into (1.2.178), we get

$$
\begin{aligned}
\langle\hat{a}(t_1)\hat{a}^+(t_2)\rangle &= (\bar{N} + 1)\exp(-b\mid t_{12}\mid -i\omega_0 t_{12})\,, \\
\langle[\hat{a}(t_1), \hat{a}^+(t_2)]\rangle &= \exp(-b\mid t_{12}\mid -i\omega_0 t_{12})\,.
\end{aligned}
\tag{1.2.180}
$$

Let us now apply to these expressions the formula [(5.2.72) v.1] for some test temperature $T = (k\beta)^{-1}$. Using the spectral language, we obtain

$$-2b[(\omega + \omega_0)^2 + b^2]^{-1} = [\exp(\beta\hbar\omega) - 1](\bar{N} + 1)2b[(\omega + \omega_0)^2 + b^2]^{-1}. \quad (1.2.181)$$

We will now consider possibilities for this equality to hold. To begin with, we assume that $b \neq 0$. Then the relation

$$[\exp(\beta\hbar\omega) - 1](\bar{N} + 1) = -1 \qquad (1.2.182)$$

must hold for all ω, which is impossible even in the limit as $\beta \longrightarrow 0, \bar{N} \longrightarrow \infty$ (in that case, the right-hand side of (1.2.182) will be $\hbar\omega \lim(\beta\bar{N})$, which depends on ω).

If $b = 0$, (1.2.181) will hold at all ω except at $\omega = -\omega_0$. It can be shown that $2b[(\omega + \omega_0)^2 + b^2]^{-1}$ becomes $2\pi\delta(\omega + \omega_0)$ as $b \longrightarrow 0$. Therefore, for (1.2.181) to hold throughout it is necessary that

$$\bar{N} + 1 = -(e^{\beta\hbar\omega_0} - 1)^{-1}. \qquad (1.2.183)$$

But at $b = 0, \bar{N} < \infty$, equation (1.2.175) becomes trivial, i.e. non-dissipative: $\dot{\hat{F}} = i\omega_0[\hat{a}^+\hat{a}, \hat{F}]$. For the nontrivial term in (1.2.175) not to vanish, we must have $\bar{N} = \infty, b\bar{N} > 0$. Still, by (1.2.183) the condition $\bar{N} = \infty$ means that $\exp(\beta\hbar\omega_0) = -1$, i.e. either $\omega_0 = 0$ or $T^{-1} = 0$, or else both parameters are zero. Thus, for $\hbar \neq 0$ and for nontrivial (1.2.175) we have two main possibilities: either $b = 0, \omega_0 = 0$ or $b = 0, T = \infty$. In either of these cases $\bar{N} = \langle \hat{a}^+\hat{a} \rangle = \infty$, i.e. no stationary distribution is possible.

The above demonstrates how difficult it is to apply specific quantum Markov processes to describe physical fluctuation-dissipational processes.

1.3 Notes on References to Chapter 1

The non-Markov generating equations were first derived in the works [1.2, 3] cited in text. The general generating equations were also considered in [1.5]. Formulas (1.2.47, 88) were obtained in [1.3].

2. Nonequilibrium Thermodynamics of Open Systems

When dealing with the general theory of open systems we do our best to bring out the parallelism and analogy between the nonequilibrium thermodynamics of open systems and the thermodynamics of conventional closed systems. Admittedly, the analogy should not be carried too far, since in the theory of open systems the important fundamental principle of time-reversal invariance does not exist. Therefore, in the case of open systems there are fewer FDRs than before, and they are not so effective. There is only one linear relation, one quadratic FDR and one cubic FDR.

There are two ways of extending the theory put forward above to cover the case of open systems: (1) to include an open system into a larger closed system and (2) to introduce and use quasi-frenergy (frenergy = free energy) related to the one-time stationary probability density of the internal parameters. For quasi-frenergy it is possible to prove the H-theorem of the Markov nonequilibrium thermodynamics. Much attention is paid to the nonequilibrium stationary state, which in the theory of open systems is the analogue of the equilibrium state.

This chapter deals with multistability and nonequilibrium kinetic phase transitions. Near these transitions it is convenient to use FDRs written in terms of derivatives of quasi-frenergy. It was found, using these FDRs, that the fluctuations of parameters near the kinetic phase transitions are not only anomalously large, i.e. much larger that away from these transitions, but also strongly non-Gaussian. For the one- and two-component case the order of magnitude was estimated of two-, three-, and fourfold correlators near the transitions of the first and second kind.

2.1 Open Systems. Examples

2.1.1 Open Systems and the Equations Describing Them

We say that a system is open if there exist inputs from the environment, i.e. nonzero fluxes $J_1^{ex}, \ldots, J_k^{ex}$ that can have either sign. A schematic representation of the open system is given in Fig. 2.1, where a large rectangle divided into small squares represents an open system S_0 which has a complex structure. Also shown in the figure are input and output external fluxes $J_1^{ex}, \ldots, J_k^{ex}$. The equations describing an open system can be reduced to first-order equations

$$\dot{A}_\alpha = f_\alpha^{(1)}(A, J^{ex}), \quad \alpha = 1, \ldots, r, \tag{2.1.1}$$

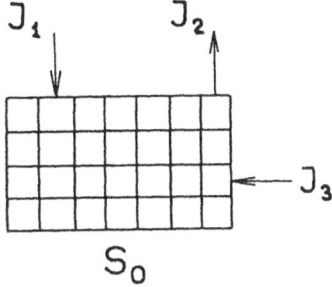

Fig. 2.1. A schematic picture of an open system to which external fluxes are directed

where A is a set of internal thermodynamic parameters $A_1, \ldots, A_r$, and J^{ex} is the set of fluxes mentioned above. The fluxes $J_1^{\text{ex}}, \ldots, J_k^{\text{ex}}$ can be time derivatives of some of the parameters A. Designating these parameters as Q_β, we will have $J_\beta^{\text{ex}} = \dot{Q}_\beta$, $\beta = 1, \ldots, k$. These equations can be regarded as belonging to (2.1.1). We say that S_0 is open in $Q_1, \ldots, Q_k$ and closed in other parameters that enter into the set $A_1, \ldots, A_r$.

If all the external fluxes in (2.1.1) are set to zero, then we will obtain the closed system described by

$$\dot{A}_\gamma = f_\gamma^{(1)}(A, 0), \quad \gamma = 1, \ldots, l \leq r . \tag{2.1.2}$$

An open system may be completely open, i.e. open in all its variables. Then $l = 0$ and there are no equations (2.1.2).

2.1.2 Inclusion of an Open System into a Closed System

Suppose that $\Delta Q_\beta = \int_t^{t+\Delta t} J_\beta^{\text{ex}} dt'$, $\beta = 1, \ldots, k$ are inputs to an open system S_0 from some storage systems, or "reservoirs", $P_1, \ldots, P_k$; or else they are outputs from S_0 to reservoirs (Fig. 2.2). The large system

$$S_L = S_0 + \sum_{\beta=1}^{k} P_\beta \tag{2.1.3}$$

(i.e., the open system plus the reservoirs) may now be treated as a closed system. The free energy F_L of the large system is the sum of the free energies of the open system and the reservoirs

$$F_L = F_0 + \sum_{\beta=1}^{k} F_\beta . \tag{2.1.4}$$

The same is true of the entropy. Let the internal parameters be $A_1, \ldots, A_l$, $Q_1, \ldots, Q_k$ (i.e. $r = l + k$). Using (2.1.4), we define the thermodynamic forces $x_\alpha = \partial F_L / \partial A_\alpha$, which are conjugates of the above internal parameters

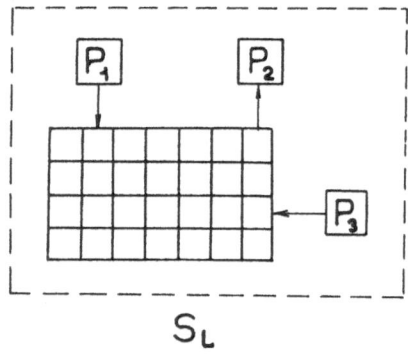

Fig. 2.2. A schematic picture of the "large" closed system consisting of the initial open system and of "reservoirs"

$$x_\gamma = \partial F_0 / \partial A_\gamma \quad \text{for} \quad \gamma \leq l \qquad (2.1.5a)$$

$$x_{l+\beta} = \partial F_0 / \partial Q_\beta + \partial F_\beta / \partial Q_\beta \quad \text{for} \quad \beta = 1, \ldots, k \qquad (2.1.5b)$$

assuming that F_β is independent of Q_γ at $\gamma \neq \beta$. Let Q'_β be the store of Q_β in the reservoir P_β. Increasing Q_β in S_0 reduces the store of it in the reservoir. Therefore,

$$dQ_\beta = -dQ'_\beta \,. \qquad (2.1.6)$$

The derivative

$$\partial F_\beta / \partial Q'_\beta = h_\beta \qquad (2.1.7)$$

has the sense of a thermodynamic force that is conjugate of Q'_β. From (2.1.6,7), we can write (2.1.5b) as

$$\begin{aligned} x_{l+\beta} &= \partial F_0 / \partial Q_\beta - \partial F_\beta / \partial Q'_\beta \\ &= \partial F_0 / \partial Q_\beta - h_\beta \,. \end{aligned} \qquad (2.1.8)$$

Now for the large closed system we can write equations [(4.3.1) v.1] in the standard form [(4.3.9) v.1]:

$$\dot{A}_\alpha = \kappa_\alpha (\partial F_0 / \partial A_\gamma, \partial F_0 / \partial Q_\beta - h), \quad \alpha = 1, \ldots, r \,. \qquad (2.1.9)$$

If now we take into account that F_0, and hence the derivatives $\partial F_0 / \partial A_\gamma$, $\partial F_0 / \partial Q_\beta$, are the functions of $A = (A_1, \ldots, A_l, Q_1, \ldots, Q_k)$, we may write (2.1.9) as

$$\dot{A}_\alpha = f_\alpha^{(2)}(A, h) \,. \qquad (2.1.10)$$

The forces h in (2.1.9,10) are external in relation to the open system S_0. Some of the equations (2.1.9,10) can, since $J_\beta^{\text{ex}} = \dot{Q}_\beta$, be written as

$$J_\beta^{\text{ex}} = \kappa_{l+\beta}(\partial F_0 / \partial A_\gamma, \partial F_0 / \partial A_\beta - h_\beta) \,. \qquad (2.1.11)$$

These equations relate the external fluxes J^{ex} to the external forces h. Using these relationships, we can transform (2.1.1) to (2.1.10), and vice versa.

We will assume that the forces h (or fluxes J) for the initial open system either vary very slowly with time or are constant. This condition can be formulated as follows: the time constant of their variation must be much longer than for the parameters A_γ, $\gamma \le 1$. For this to be the case, the reservoirs P_β must be fairly "capacious", i.e. must have fairly large generalized capacities given by

$$
\begin{aligned}
C_\beta &= \partial Q'_\beta / \partial h_\beta \\
&= (\partial^2 F_\beta(Q'_\beta)/\partial Q'^2_\beta)^{-1} \, .
\end{aligned}
\tag{2.1.12}
$$

Strictly speaking, an open system is obtained by passing to the limit $C_\beta \longrightarrow \infty$, $\beta = 1, \ldots, k$. If in the initial open system we want to maintain external fluxes J^{ex}, not forces at a constant or quasiconstant level, then we should include into P_β large resistances $R_\beta = h_\beta / J_\beta$ and make h sufficiently large.

If the free energy F_0 is independent of some of the parameters $Q_1, \ldots, Q_k$, then some of the $\partial F_0/\partial A_\beta$ on the right-hand side of (2.1.9) will be identically zero. Then, the open system may have stationary nonequilibrium states that are characterized by constant fluxes (i.e. some of the $J^{\text{ex}}_1, \ldots, J^{\text{ex}}_k$ will be nonzero and constant). Such nonequilibrium stationary states are interesting objects for study since they have no counterparts in the equilibrium theory. Let F_0 be dependent only on $A_1, \ldots, A_l$, $(l \le l' < r)$ and stationary values $A^0_1, \ldots, A^0_l$, of these parameters be possible. Putting $\dot{A}_\gamma = 0$ in (2.1.1) or (2.1.10) we get

$$
f^{(1)}_\gamma(A^0, J^{\text{ex}}) = 0
\tag{2.1.13a}
$$

or

$$
f^{(2)}_\gamma(A^0, h) = 0 \, , \quad \gamma = 1, \ldots, l' \, .
\tag{2.1.13b}
$$

These equations can be used to find the above stationary values A^0_γ, $\gamma = 1, \ldots, l'$ of those parameters that are fixed in a stationary state.

Note that, instead of (2.1.4,5,7–9,12), we may use appropriate formulas of the modified version, in which we deal with entropy instead of free energy.

2.1.3 Fluctuations in Open Systems with Relatively Small Nonlinearity

If an open system is linear or slightly nonlinear, i.e. not too far from equilibrium, then the statistical behavior of fluctuations of internal parameters, and also fluxes, can be calculated using the Markov and non-Markov FDRs (Sects. 4.3–5 and 6.1,2 v.1).

Let us take, for example, the formula [(6.2.11) v.1] derived with the help of the non-Markov FDRs, which determines the derivative of the correlator $\langle J(\omega_1), J(\omega_2) \rangle$ with respect to an external force. Knowing this derivative we can find the nonequilibrium correlator [(6.2.13) v.1], which consists of an equilibrium part and of a nonequilibrium part, the latter being due to an external force. However, if a system is exposed to an external force, it is open. Accordingly, the above formulas

[(6.2.11,13) v.1] enable us to work out the correlator of a flux (here an electric current) in the open system shown in Fig. 6.1 v.1. If the external e.m.f. is constant, the system comes to a stationary nonequilibrium state with the correlator [(6.2.15) v.1].

Another example may be provided by a body (say, a Brownian particle) moving in a medium with nonlinear friction (see Sect. 6.2.4 v.1). The system becomes open owing to the action of an external force f. For this case we have derived the expression [(6.2.38) v.1], which, to within two dissipationally undeterminable constants, determines the term added to the equilibrium correlator of velocities to take into account the external force. From [(6.2.38) v.1], we can also find the nonequilibrium correlator of velocities corresponding to a nonequilibrium stationary state under a constant force.

Likewise, fluctuations in the open system without aftereffect can be dealt with by the purely Markov techniques. Examples will follow below. In that case, it is essential that the dissipative nonlinearity be relatively slight. As for nondissipative nonlinearity, it may be strong, since it does not hinder the determination of fluctuation properties, although it adds difficulties to the computing of the correlators. For the fluctuations to be statistically determinable with arbitrary dissipative nonlinearity and degree of nonequilibrium, we will have to know the total nonequilibrium potential. Specifically, it can be found from the fluctuation-dissipative model of the process at hand.

2.1.4 Principle of Minimal Free Energy Decrease of Minimal Entropy Production

In equilibrium thermodynamics the second law, i.e. the law of free energy decrease and entropy production, enables stable states to be distinguished in a definite way: in a stable state the free energy has at least a local minimum, and the entropy, at least a local maximum. In the theory of open systems the second law of thermodynamics is no longer of help, since in nonequilibrium stationary states the free energy need not have a minimum, nor the entropy a maximum. This suggests that for open systems we have to seek another function that would help us to distinguish stable nonequilibrium stationary states from other states. One attempt at solving this problem involves the principle of minimal free energy decrease or minimal entropy production.

Consider the equation

$$\dot{A}_\gamma = f_\gamma(A) \quad \gamma = 1, \dots, l', \tag{2.1.14}$$

where $A = (A_1, \dots, A_{l'})$, describing the time variation of the parameters that tend to stationary values A_γ^0. In some cases these equations follow from (2.1.1) and f also depends on the constant fluxes J^{ex}; in other cases they follow from (2.1.10) and f depends on the constant forces h. These variables are omitted here for brevity. The stationary values $A_1^0, \dots, A_{l'}^0$ of the parameters are given by the equations

$$f_\gamma(A_1^0, \dots, A_{l'}^0) = 0, \quad \gamma = 1, \dots, l' \tag{2.1.15}$$

of the type (2.1.13).

If we now introduce the deviations $\delta A_\gamma = A_\gamma - A_\gamma^0$ from the stationary values, substitute $A_\gamma = A_\gamma^0 + \delta A_\gamma$ into (2.1.14) and linearize in δA, we obtain

$$\delta \dot{A}_\gamma = -D_{\gamma\sigma}\delta A_\sigma, \quad \gamma, \sigma = 1, \ldots, l', \tag{2.1.16}$$

where

$$D_{\gamma\sigma} = -\partial f_\gamma / \partial A_\sigma \quad \text{for} \quad A = A^0. \tag{2.1.17}$$

The free energy $F_0(A^0 + \delta A)$ can be expanded in terms of the deviations δA:

$$F_0(A^0 + \delta A) = F_0(A^0) + \sum_{n=1}^{\infty}(n!)^{-1}u_{\gamma_1 \ldots \gamma_n}\delta A_{\gamma_1} \ldots \delta A_{\gamma_n}. \tag{2.1.18}$$

Subtracting the linear terms gives

$$F_0^{(2)}(A^0 + \delta A) = F_0(A^0 + \delta A) - u_\gamma \delta A_\gamma. \tag{2.1.19}$$

We will now introduce the forces

$$\begin{aligned}\delta x_\gamma &= \partial F_0^{(2)}(A)/\partial A_\gamma \\ &= \partial F^{(2)}(A^0 + \delta A)/\partial(\delta A_\gamma)\end{aligned} \tag{2.1.20}$$

which are thermodynamically conjugates of δA_γ. It is easy to verify that δx_γ have the sense of differences $x_\gamma - x_\gamma^0$, i.e.

$$\delta x_\gamma = \partial F_0/\partial A_\gamma - (\partial F_0/\partial A_\gamma)^0, \tag{2.1.21}$$

for which one should substitute $F_0 = F_0^{(2)} + u_\gamma \delta A_\gamma$ into (2.1.21) and take into account that $\partial F_0/\partial A_\gamma = u_\gamma$ for $\delta A = 0$. If we then discard the nonlinear terms, then using (2.1.18,19), we obtain

$$\delta x_\gamma = u_{\gamma\sigma}\delta A_\sigma. \tag{2.1.22}$$

This formula resembles the analogous one of the linear theory of closed systems. Using (2.1.17), we transform the linearized equations (2.1.16) so that the right-hand side is expressed in terms of the forces

$$\delta \dot{A}_\gamma = l_{\gamma,\varepsilon}\delta x_\varepsilon, \tag{2.1.23}$$

where $l_{\gamma,\varepsilon} = -D_{\gamma\sigma}u_{\sigma\varepsilon}^{-1}$. Now we can formulate two theorems.

Theorem 1. *If the reciprocal relations*

$$l_{\gamma,\varepsilon} = l_{\varepsilon,\gamma} \tag{2.1.24}$$

hold and the condition

$$u_{\gamma\varepsilon} \equiv [\partial^2 F_0/\partial A_\gamma \partial A_\varepsilon]_{A=A^0} = \text{positive definite} \tag{2.1.25}$$

is met, then near a nonequilibrium stationary state the second derivative of $F_0^{(2)}$ with respect to time is nonnegative:

$$\ddot{F}_0^{(2)} \geq 0 \,. \tag{2.1.26}$$

Proof. From (2.1.20,23) we have

$$
\begin{aligned}
\dot{F}_0^{(2)} &= \delta x_\gamma \delta \dot{A}_\gamma \\
&= l_{\gamma,\varepsilon} \delta x_\gamma \delta x_\varepsilon \,.
\end{aligned} \tag{2.1.27}
$$

If now we differentiate this with respect to time using (2.1.24) and apply (2.1.23) once more, we get

$$
\begin{aligned}
\ddot{F}_0^{(2)} &= 2 l_{\gamma,\varepsilon} \delta x_\varepsilon (\partial x_\gamma / \partial A_\sigma) \delta \dot{A}_\sigma \\
&= 2 (\partial x_\gamma / \partial A_\sigma)(l_{\gamma,\varepsilon} \delta x_\varepsilon)(l_{\sigma,\rho} \delta x_\rho) \,.
\end{aligned} \tag{2.1.28}
$$

Due to (2.1.20), $\partial x_\gamma / \partial A_\sigma$ is simply $u_{\gamma\sigma}$. The expression $u_{\gamma\sigma} c_\gamma c_\sigma$ is nonnegative by (2.1.25), and we thus obtain (2.1.26). QED.

The proof just provided is absolutely analogous to that of the analogous theorem in Sect. 4.6 v.1.

It follows from (2.1.26) that in the nonstationary states passed by the system during relaxation, $\dot{F}_0^{(2)}$ is smaller than in a stationary state, i.e. it is nonpositive (since $\dot{F}_0^{(2)} = 0$ at $\delta A = 0$).

Theorem 2. *If for a matrix $l_{\gamma,\delta} + l_{\delta,\gamma}$ the condition of nonpositive definiteness is met, i.e.*

$$-l_{\gamma,\delta} - l_{\delta,\gamma} = \text{nonnegative definite}\,, \tag{2.1.29}$$

then near a stationary state

$$\dot{F}_0^{(2)} \leq 0 \,. \tag{2.1.30}$$

The inequality (2.1.30) clearly follows from (2.1.27,29).

In the modified version of the theory, instead of free energy we must take entropy with a minus sign. The equations $-\dot{F}_0^{(2)} \geq 0$, $\dot{S}_0^{(2)} \geq 0$ and $\dot{F}_0^{(2)} = 0$, $\dot{S}_0^{(2)} = 0$ (the last two refer to a stationary state) express the principle of minimal free energy decrease or minimal entropy production.

Note that the inequality (2.1.30) or the appropriate inequality $\dot{S}_0^{(2)} \geq 0$ for the entropy $S_0^{(2)}$ do not follow from the second law of thermodynamics, since the law describes the behavior of free energy or entropy in a large system S_L, i.e. the free energy (2.1.4) (and similarly for entropy), but not the free energy $F_0^{(2)}$.

The question arises of whether or not there exist stationary states for which the above theorems are valid. In Sect. 2.1.6 we will consider an example of a system for which the conditions of the theorems are fulfilled arbitrarily far away from equilibrium, i.e. in a markedly nonlinear region. But now it will be instructive to take a look at the linear region, where the linear approximation in deviations from equilibrium is applicable, and the stationary nonequilibrium state is fairly close to equilibrium.

Let the deviations from the equilibrium $A^{eq} = 0$ obey the linear equations of motion

$$\dot{A}_\gamma = \sum_{\alpha=1}^{r} l_{\gamma,\alpha} x_\alpha, \quad \gamma = 1, \ldots, l',$$

$$\dot{A}_{l'+\beta} = \sum_{\alpha=1}^{r} l_{l'+\beta,\alpha} x_\alpha, \quad \beta = 1, \ldots, r - l', \qquad (2.1.31)$$

where, as usual

$$x_\alpha = \partial F / \partial A_\alpha, \qquad (2.1.32)$$

so that

$$
\begin{aligned}
\dot{F} &= \sum_{\alpha=1}^{r} x_\alpha \dot{A}_\alpha \\
&= \sum_{\alpha,\mu=1}^{r} l_{\alpha,\mu} x_\alpha x_\mu \\
&= \tfrac{1}{2} \sum_{\alpha,\mu=1}^{r} (l_{\alpha,\mu} + l_{\mu,\alpha}) x_\alpha x_\mu .
\end{aligned}
\qquad (2.1.33)
$$

In order that the equations may be written in the form (2.1.31) the complete matrix $u_{\alpha\mu}$, $0 \leq \alpha, \mu \leq r$, must be nondegenerate. From this closed system we will obtain an open one by putting $x_{l+\beta} = -h_\beta$, $\beta = 1, \ldots, r - l'$, where h_β are constant external forces. Here the relations $x_\alpha = u_{\alpha\mu} A_\mu$ that follow from (2.1.32) are violated at $\alpha > l'$. As is seen from (2.1.31), the equations of motion for the open system now become

$$\dot{A}_\gamma = \sum_{\alpha=1}^{r} l_{\gamma,\alpha} x_\alpha = \sum_{\alpha=1}^{l'} l_{\gamma,\alpha} x_\alpha - \sum_{\beta=1}^{r-l'} l_{\gamma,l'+\beta} h_\beta . \qquad (2.1.34)$$

Next we will proceed to find an extremum of the time derivative (2.1.33) with respect to $x_1, \ldots, x_{l'}$, subject to the condition that $x_{l'+\beta} = -h_\beta$. We will thus obtain $\partial \dot{F} / \partial x_\gamma = 0$, $\gamma \leq l'$. Hence $\sum_{\alpha=1}^{r} l_{\gamma,\alpha} x_\alpha = 0$, or, by (2.1.34), $\dot{A}_\gamma = 0$, if the Onsager relations are valid, i.e. in particular if all the A_α, $\alpha \leq r$, have the same time parity. Moreover, from [(4.1.12) v.1] the matrix $l_{\alpha,\mu}$ is nonpositive definite, so that the extremum is a maximum. Consequently, in the situation under consideration at a nonequilibrium stationary point the function $\dot{F}$ has a maximum, i.e. the principle of minimal free energy decrease holds.

For an open system the free energy in (2.1.32,33) is the free energy (2.1.4) of a large system. For $x_{l'+\beta} = -h_\beta$ to hold, it is then necessary, from (2.1.8), that F_0 be independent of $A_{l'+\beta}$, $\beta \geq 1$. In this case we have

$$F_0(A) = \tfrac{1}{2} \sum_{\gamma,\sigma=1}^{l'} u_{\gamma\sigma} A_\gamma A_\sigma . \qquad (2.1.35)$$

Equation (2.1.34) at $\dot{A}_\gamma = 0$ yields stationary nonequilibrium values A_γ^0, which are proportional to h. For the deviations $\delta A_\gamma = A_\gamma - A_\gamma^0$ from (2.1.34) we have

$$\delta \dot{A}_\gamma = \sum_{\sigma,\rho=1}^{l'} l_{\gamma,\sigma} u_{\sigma\rho} \delta A_\rho \,, \tag{2.1.36}$$

i.e. equation (2.1.16). Substituting $A_\gamma = A_\gamma^0 + \delta A_\gamma$ into (2.1.35) and subtracting, by (2.1.19), the terms linear in δA, we get

$$F_0^{(2)}(A) = \tfrac{1}{2} \sum_{\gamma,\sigma=1}^{l'} u_{\gamma\sigma} \delta A_\gamma \delta A_\sigma + F_0(A^0) \,. \tag{2.1.37}$$

Applying (2.1.20) to (2.1.37) shows that (2.1.22) contains the submatrix $u_{\gamma\sigma}$, $\gamma, \sigma \leq l'$ of the earlier matrix $u_{\alpha\mu}$, $0 \leq \alpha, \mu \leq r$, which defined the transformation $x_\alpha = u_{\alpha\mu} A_\mu$ that followed from (2.1.32). This reduces (2.1.36) to the form (2.1.23), where $l_{\gamma,\sigma}$ is a submatrix of the earlier matrix $l_{\alpha,\mu}$ which enters (2.1.31). Just like the complete matrix, it is therefore symmetric and nonpositive definite, i.e. the conditions (2.1.24) and (2.1.29) are satisfied.

The state $A^{\mathrm{eq}} = 0$ being an equilibrium, the complete matrix $u_{\alpha\mu}$ is positive definite, and so is the submatrix $u_{\gamma\sigma}$. Hence, it is nondegenerate. Therefore, the condition (2.1.25) is also fulfilled, and hence theorems 1 and 2 are applicable. Consequently, apart from $\dot{F}_L \leq (\dot{F}_L)_{\mathrm{st}}$, the inequalities (2.1.26,30) will be obeyed. It is to be noted, however, that if in (2.1.31) we fix the fluxes $\dot{A}_{l'+\beta} = J_\beta^{\mathrm{ex}}$, not forces $x_{l'+\beta}$, it can be shown that the principle of minimal free energy decrease or minimal entropy production will not hold true.

That the entropy production is minimal in a nonequilibrium stationary state in a linear (relative to an equilibrium) region with the forces fixed was first noticed by *Prigogine* [2.1,2].

The function $P(A) = \dot{F}_0^{(2)}(A)$ is zero at $A = A^0$. If the matrix $l_{\gamma,\delta} + l_{\delta,\gamma}$ not only obeys (2.1.29) but is also nondegenerate, then in addition to (2.1.30) we will have

$$P(A) < 0 \quad \text{for} \quad \delta A \neq 0, \tag{2.1.38}$$

as is easily found from (2.1.27). This suggests that $P(A)$ has at least a local maximum at a stationary point. Then, if theorem 1 is applicable (and from (2.1.26) $P(A)$ is a nondecreasing function), the stationary point A^0 is stable.

In the above argument $P(A) = \dot{F}_0^{(2)}(A)$ plays the role of a Lyapunov function. The Lyapunov function is a function that has at least a local maximum or minimum at a point of interest and varies in a monotonic manner with time (nonincreasing in the case of a minimum and nondecreasing in the case of a maximum). This function can thus be used as a criterion of stability of motion.

It should be noted, however, that in this situation we can also prove that a stationary point is stable using the function $F_0^{(2)}(A)$ instead of $P(A)$. In effect, owing to (2.1.25) this function has a local minimum at A^0, and so the stability of that point can be judged from (2.1.30). Generally speaking, one cannot be confident that the conditions of the two theorems considered above are satisfied, especially far away from equilibrium. This lack of generality is a disadvantage of the principle of minimal free energy decrease or minimal entropy production. Another constraint is the fact that it is only valid in a linear (in δA) approximation, i.e. only if the equations are linearized.

Another comment is worth making: whereas entropy and free energy in thermo-dynamics are known a priori, i.e. before the nonequilibrium process is considered, we can only find the function $P = \dot{F}_0^{(2)}$, or $P = -\dot{S}_0^{(2)}$, by knowing the equations of motion (2.1.14) or (2.1.16). If then the evolution equations are known, we can handle the issue of stability of one or another stable state by using conventional techniques of the theory of dynamic stability, i.e. by analyzing the equation without turning to P.

We have thus seen that the principle of minimal free energy decrease or minimal entropy production in the theory of open systems can by no means play such an important role as the second law of thermodynamics in the theory of equilibrium states.

2.1.5 Stability Criteria for Stationary States.
Other Lyapunov Functions

As stated above, the stability of a nonequilibrium stationary state can be investi-gated by examining the linearized equations (2.1.16). Furthermore, we can also use various forms of the Lyapunov function and formulate respective stability criteria. It has been noted in the previous subsection that we may choose to have $\dot{F}_0^{(2)}$ for such a function. A simple example of the Lyapunov function is $\sum_\gamma (\delta A_\gamma)^2 \equiv |\delta A|^2$. In this connection we can formulate the following trivial stability criterion:

Criterion 1. If $|\delta A|^2$ does not grow in time, then a stationary state is stable.

A somewhat more complex example of the Lyapunov function is $F_0^{(2)}$, for which we have:

Criterion 2. If (1) $\dot{F}_0^{(2)} \leq 0$ in the neighborhood of A^0 and (2) $F_0^{(2)}$ has a local minimum at this point, i.e. if

$$F_0^{(2)}(A) > F_0^{(2)}(A^0) \tag{2.1.39}$$

at sufficiently small $|\delta A| > 0$, then the stationary state is stable.

This criterion has already been mentioned in the previous subsection. It was noted there that the condition (2.1.39) is equivalent to (2.1.25). In this criterion we have substituted $\delta^2 F_0 = \sum_{\gamma\sigma} u_{\gamma\sigma} \delta A_\gamma \delta A_\sigma$ for $\sum_\gamma (\delta A_\gamma)^2$. By virtue of (2.1.22), we may write the derivative

$$\partial \delta^2 F_0 / \partial t = 2 \sum_{\gamma\sigma} u_{\gamma\sigma} \delta A_\gamma \delta J_\sigma \quad (\delta J_\sigma = \delta \dot{A}_\sigma) \tag{2.1.40}$$

of the former sum as

$$\partial \delta^2 F_0 / \partial t = 2 \sum_{\gamma} \delta x_{\gamma} \delta J_{\gamma} .$$

Hence criterion 2 becomes

1) $\delta^2 F_0 > 0$ for $\delta A \neq 0$,

2) $\sum_{\gamma} \delta x_{\gamma} \delta J_{\gamma} \leq 0$.   (2.1.41)

In the entropy version, criterion 2 is given by

1) $\delta^2 S_0 < 0$ for $\delta A \neq 0$,

2) $\partial \delta^2 S_0 / \partial t = - \sum_{\gamma} \delta X_{\gamma} \delta J_{\gamma} \geq 0$.   (2.1.42)

This form of stability criterion was derived in [2.3], the only difference being that we have defined the forces with the opposite sign.

The next criterion uses another Lyapunov function.

Criterion 3. If

(1) $d_x P / dt \equiv \sum_{\gamma} J_{\gamma} \dot{x}_{\gamma} \geq 0$,   (2.1.43a)

(2) $\sum_{\gamma} J_{\gamma}(A) dx_{\gamma} = d\Phi(A)$,   (2.1.43b)

and

(3) $\Phi(A) < \Phi(A^0)$ (2.1.43c)

at sufficiently small $| \delta A | > 0$, then a stationary state is stable.

Just as in Sect. 4.6 v.1, here $P = \sum x_{\gamma} J_{\gamma}$. The specific "differential" d_x was introduced by *Glansdorff* and *Prigogine* [2.3]. As indicated in Sect. 4.5 v.1, the inequality $\dot{\Phi} \equiv \sum J_{\gamma} \dot{x}_{\gamma} \geq 0$ comes from $F(A)$ (in this case $F_0(A)$) being a convex function. Near a nonequilibrium stationary state the equality holds subject to (2.1.25).

The equation $\sum J_{\gamma} dx_{\gamma} = d\Phi$ means that $\sum J_{\gamma} dx_{\gamma}$ is a total differential. This condition obviously holds in a one-component case, whereas in a multicomponent case it does not always hold. As pointed out in [2.3], Φ can be sought using the integrating factor $\varepsilon^2(A)$ with conditon (2.1.43b) replaced by $\varepsilon^2(A) \sum J_{\gamma} dx_{\gamma} = d\Phi(A)$. For criterion 3 to be used, it is sufficient to require that condition (2.1.43b) be fulfilled only near a stationary point with $\sum J_{\gamma} dx_{\gamma}$ replaced by $\sum \delta J_{\gamma} d\delta x_{\gamma}$.

Criterion 3 can easily be written in terms of entropy as well.

Note that even if the Lyapunov function $\Phi(A)$ used in criterion 3 exists, this by no means implies that this criterion has an advantage over the other criteria. A trade-off of respective criteria and a discussion of their suitability (as compared with the examination of linearized equations) lie outside the scope of the book. We will confine ourselves here to illustrating the application of the above criteria to a specific example (Sect. 2.1.9) in which all three criteria are essentially equivalent.

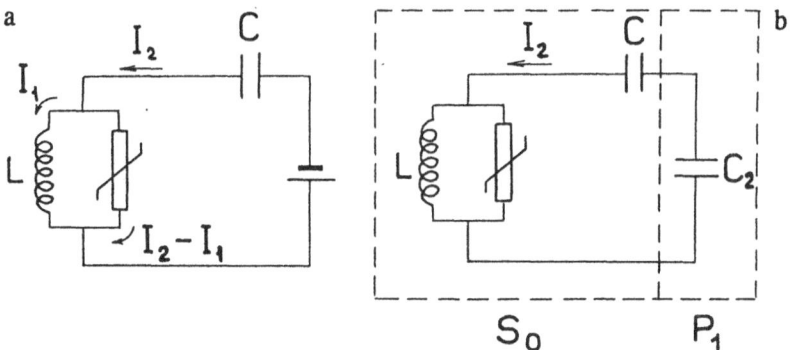

Fig. 2.3. Circuit with a nonlinear resistor and voltage source. (a) for the open system, (b) for the "large" closed system corresponding to the open system

2.1.6 Example of an Open Electrical System

Consider the electric circuit diagram of Fig. 2.3a, containing a nonlinear resistance, an inductance L, a capacitance C and a voltage source. If we replace the latter by a "reservoir" containing an electric charge Q', i.e. a capacitance C_2, we will have a large closed system (Fig. 2.3b). Let our internal parameters A_1 and A_2 be the "momentum" $p = LI_1$, where I_1 is the current flowing through L, and Q_2 be the charge on C. Instead of the free energy (2.1.4) we will deal here with the energy of the large system

$$W = \frac{p^2}{2L} + \frac{Q_2^2}{2C} + \frac{(Q')^2}{2C_2}. \tag{2.1.44}$$

Differentiating by (2.1.5) using the equation $dQ' = -dQ_2$, which corresponds to (2.1.8), we obtain

$$x_1 = \partial W/\partial p = p/L = I_1,$$
$$x_2 = \partial W/\partial Q_2 = Q_2/C - Q'/C_2 = Q_2/C - h_2. \tag{2.1.45}$$

It is easily seen that the large system is described by

$$L\dot{I}_1 + Q_2/C = Q'/C_2, \tag{2.1.46a}$$
$$g(I_2 - I_1) + Q_2/C = Q'/C_2, \tag{2.1.46b}$$

where $g(I) = V$ is the current-voltage characteristic of the nonlinear element. Solving (2.1.46b) for $I_2 = \dot{Q}_2$ and using (2.1.45), we obtain the standard form of the above equations

$$\dot{p} = -Q_2/C + h_2 \equiv -x_2, \tag{2.1.47a}$$
$$\dot{Q}_2 = p/L + f(h_2 - Q_2/C),$$
$$= x_1 + f(-x_2), \tag{2.1.47b}$$

where $f(V)$ is the inverse function of $g(I)$. These equations are examples of (2.1.9) or (2.1.10).

Furthermore (2.1.46b) becomes

$$h_2 = g(I^{\text{ex}} - I_1) + Q_2/C \tag{2.1.48}$$

if we treat I_2 as the external current I^{ex}. Substituting this into (2.1.47a) gives

$$\dot{p} = g(I^{\text{ex}} - p/L),$$
$$\dot{Q}_2 = I^{\text{ex}}. \tag{2.1.49}$$

These equations are a special case of (2.1.1). The total number of variables, r, here is two, and $l = 1$, i.e. the system is open in one variable and closed in the other. If in the system of Fig. 2.3b C_2 and Q' tend to infinity, then in (2.1.47) the force h_2 will be unchanged. In this case the use of (2.1.47) is preferable. If in the initial system we now remove C and instead connect a large resistance, and make the force h_2 large, then in (2.1.49) the external current I^{ex} will be almost timeindependent, and we should use precisely the last form of equations.

Notice that, in (2.1.47), however far away we are from equilibrium, the reciprocal relation

$$l_{1,2} = \varepsilon_1 \varepsilon_2 l_{2,1} = -l_{2,1}, \tag{2.1.50}$$

still holds, but not the Onsager relation. Therefore, we may not apply the theorems of the previous subsection here.

2.1.7 Nonlinear Open System with Two Capacitances

Let us take another example of an open system. This is a system with time-even parameters shown in Fig. 2.4. It has two capacitances C_1 and C_2, three nonlinear resistances $g_i(I)$, $i = 1, 2, 3$, and a current source with an e.m.f. h_3, the parameters A_1, A_2 being the charges Q_1 and Q_2 on the capacitances. Denoting the currents as in Fig. 2.4, we will have the equations

$$g_3(I_3) = h_3 + Q_2/C_2,$$
$$g_2(I_2 + I_3) = Q_1/C_1 - Q_2/C_2,$$
$$g_1(I_1 + I_2 + I_3) = -Q_1/C_1, \tag{2.1.51}$$

which determine the voltage balance. Solving these for currents gives

$$\dot{Q}_1 \equiv I_1 = f_1(-Q_1/C_1) - f_2(Q_1/C_1 - Q_2/C_2), \tag{2.1.52a}$$
$$\dot{Q}_2 = I_2 = f_2(Q_1/C_1 - Q_2/C_2) - f_3(h_3 + Q_2/C_2), \tag{2.1.52b}$$
$$I_3 = f_3(h_3 + Q_2/C_2), \tag{2.1.52c}$$

where f_i are the inverse functions of g_i. By equating to zero the left-hand sides of (2.1.52a,b), we find the equations for the stable values

$$Q_1^0/C_1 = x_1^0,$$
$$Q_2^0/C_2 = x_2^0. \tag{2.1.53}$$

Equation (2.1.52c) gives the stable value of I_3. Substituting

$$Q_i/C_i = x_i^0 + \delta Q_i/C_i, \quad i = 1, 2 \tag{2.1.54}$$

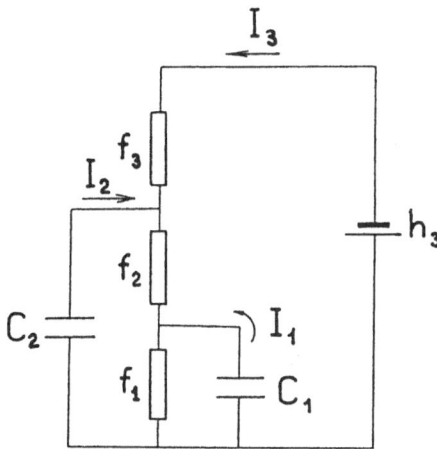

Fig. 2.4. Three-loop circuit with common voltage source

into (2.1.52a,b) and linearizing, we obtain equations for the deviations

$$\delta \dot{Q}_1 = -f_1'(-x_1^0)\delta Q_1/C_1 - f_2'(x_1^0 - x_2^0)(\delta Q_1/C_1 - \delta Q_2/C_2),$$
$$\delta \dot{Q}_2 = f_2'(x_1^0 - x_2^0)(\delta Q_1/C_1 - \delta Q_2/C_2) - f_3'(h_3 + x_2^0)\delta Q_2/C_2. \qquad (2.1.55)$$

The energy of the system is

$$F_0 = W = Q_1^2/(2C_1) + Q_2^2/(2C_2). \qquad (2.1.56)$$

Substituting (2.1.54) yields

$$F_0^{(2)} = F_0(Q^0) + \delta Q_1^2/(2C_1) + \delta Q_2^2/(2C_2) \qquad (2.1.57)$$

by (2.1.19). Hence, by virtue of (2.1.20),

$$\delta x_1 = \delta Q_1/C_1,$$
$$\delta x_2 = \delta Q_2/C_2. \qquad (2.1.58)$$

In terms of these forces, the expressions (2.1.55) become

$$\delta \dot{Q}_1 = -f_1'(-x_1^0)\delta x_1 - f_2'(x_1^0 - x_2^0)(\delta x_1 - \delta x_2) \equiv l_{1,1}\delta x_1 + l_{1,2}\delta x_2,$$
$$\delta \dot{Q}_2 = f_2'(x_1^0 - x_2^0)(\delta x_1 - \delta x_2) - f_3'(h_3 + x_2^0)\delta x_2 \equiv l_{2,1}\delta x_1 + l_{2,2}\delta x_2. \qquad (2.1.59)$$

These are special cases of (2.1.23). We see that the Onsager relations $l_{1,2} = l_{2,1}$ are obeyed at any h, i.e. arbitrarily far away from equilibrium. Furthermore, it can easily be verified that, according to (2.1.57), condition (2.1.25) is met. We can therefore apply here theorem 1 of Sect. 2.1.4, and so (2.1.26) holds. Moreover, for nonnegative derivatives $f_1'(-x_1^0), f_2'(x_1^0 - x_2^0), f_3'(h_3 + x_2^0)$ condition (2.1.29) is fulfilled, so that theorem 2 is applicable as well.

It is worth noting that, as shown in [2.4], for arbitrary electric circuits that contain linear inductances and capacitances and ordinary two-terminal nonlinear resistances the Onsager-Casimir relations hold for any distance from equilibrium.

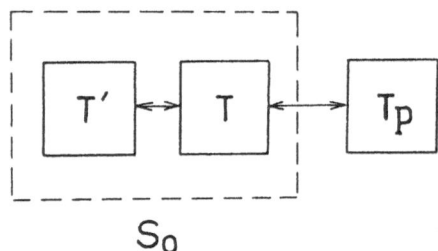

S_0

Fig. 2.5. Two bodies in thermal contact having heat exchange with a thermostat

If in the two examples just discussed a nonequilibrium stationary state lies not too far away from an equilibrium one, so that $f_j(V)$ in (2.1.47,52) can be represented as

$$f_j(V) = R_j^{-1}V + \tfrac{1}{2}\gamma_j V^2 \qquad\qquad (2.1.60)$$

or

$$f_j(V) = R_j^{-1}V + \tfrac{1}{2}\gamma_j V^2 + \tfrac{1}{6}\delta_j V^3 \,, \qquad\qquad (2.1.61)$$

then fluctuations in the system can be calculated using the Markov and non-Markov techniques laid down in Sects. 4.3–5 and 6.1,2 v.1.

2.1.8 Example of an Open System with Heat Exchange

Consider two bodies that exchange heat, their temperatures being T and T'. The heat input of the first body is $J^{ex} = dQ_{ex}/dt$, which makes the system open (Fig. 2.5). The law governing the heat exchange is assumed to be linear. The internal energies U, U' of the bodies will then be

$$\dot{U} = -\alpha(T - T') + dQ_{ex}/dt \,,$$
$$\dot{U}' = \alpha(T - T') \,. \qquad\qquad (2.1.62)$$

We will now include this open system into a large closed system by introducing the third body – a high-capacity thermal reservoir P (heat bath or thermostat). If we then assume that its heat exchange with the system is linear, i.e. $dQ_{ex}/dt = (T_P - T)$, we will have the following equations:

$$\dot{U} = \alpha(T' - T) + \gamma(T_P - T)$$
$$\dot{U}' = \alpha(T - T') \,. \qquad\qquad (2.1.63)$$

It is known that the entropies S, S', S_P of the three bodies of the combined system are related to their internal energies U, U', U_P by

$$dS = dU/T \,,$$
$$dS' = dU'/T' \,,$$
$$dS_P = dU_P/T_P \,. \qquad\qquad (2.1.64)$$

The heat exchange between the three bodies satisfies the energy conservation law $dU + dU' + dU_P = 0$ and so the total entropy obeys

$$dS_{sum} = (T^{-1} - T_P^{-1})dU + ((T')^{-1} - T_P^{-1})dU'. \tag{2.1.65}$$

Hence, using [(3.1.58) v.1] we can easily find the forces

$$X = -\partial S_{sum}/\partial U = T_P^{-1} - T^{-1},$$
$$X' = \partial S_{sum}/\partial U' = T_P^{-1} - (T')^{-1} \tag{2.1.66}$$

that are conjugates of U and U'. Here $-T_P^{-1}$ can be treated as the external force

$$H = -\partial S_P/\partial U_P = -T_P^{-1} \tag{2.1.67}$$

exerted by the thermal reservoir. Equations (2.1.66), i.e.

$$X = -T^{-1} - H,$$
$$X' = -1/T' - H \tag{2.1.68}$$

are an entropy version of (2.1.8) and equations (2.1.63) are an example of (2.1.10). In a linear approximation, (2.1.66) can be replaced by

$$X = T^{-2}(T - T_P),$$
$$X' = T^{-2}(T' - T_P). \tag{2.1.69}$$

Using these, we can write (2.1.63) as

$$\dot{U} = T^2\alpha(X' - X) - T^2\gamma X,$$
$$\dot{U}' = T^2\alpha(X - X'). \tag{2.1.70}$$

Comparing these equations with the general equations $\dot{A}_\alpha = L_{\alpha,\beta}X_\beta$ of the entropy version of linear nonequilibrium thermodynamics, we find the Onsager matrix

$$\| L_{\alpha,\beta} \| = T^2 \begin{pmatrix} -\alpha - \gamma & \alpha \\ \alpha & -\alpha \end{pmatrix}. \tag{2.1.71}$$

If now we add fluctuational forces to (2.1.71), we will arrive at the Langevin equations

$$\dot{U} = \alpha(T' - T) + \gamma(T_P - T) + \xi(t), \tag{2.1.72a}$$
$$\dot{U}' = \alpha(T - T') + \xi'(t). \tag{2.1.72b}$$

The random functions $\xi(t), \xi'(t)$ that enter into these expressions have zero means and, by virtue of the linear FDR

$$\langle\xi_\alpha(t_1)\xi_\beta(t_2)\rangle = -k(L_{\alpha,\beta} + L_{\beta,\alpha})\delta(t_{12}) \tag{2.1.73}$$

(see [(4.1.33a) v.1]) and (2.1.71), they have the correlators

$$\langle\xi(t_1)\xi(t_2)\rangle = 2kT^2(\alpha + \gamma)\delta(t_{12}),$$
$$\langle\xi'(t_1)\xi'(t_2)\rangle = -\langle\xi(t_1)\xi'(t_2)\rangle = 2kT^2\alpha\delta(t_{12}). \tag{2.1.74}$$

A fluctuational source whose intensity varies linearly with γ, but not with J^{ex}, appears in (2.1.72a) due to heat exchange with the heat bath. Using (2.1.72,74), we can readily find the statistical properties of energy fluctuations.

2.1.9 A Simple Autocatalytic Reaction.
Stability of the Stationary State

Let two substances Y and D be involved in the autocatalytic reaction

$$Y + D \xrightleftharpoons[k_-]{k_+} 2Y \tag{2.1.75}$$

in a volume V. As stated in Sect. 3.4.1 v.1, the molar concentrations $[Y], [D]$ in the ideal gas approximation obey

$$d[Y]/dt = k_+[D][Y] - k_-[Y]^2 . \tag{2.1.76}$$

We will assume that there is a constant output J^{ex} of Y to the environment. With enough mixing, the above equation becomes

$$d[Y]/dt = k_+[D][Y] - k_-[Y]^2 - J^{\text{ex}}/V . \tag{2.1.77}$$

In addition, we will suppose that the concentration $[D]$ is somehow maintained at a constant level. Since $[D]$ is constant and there is an output J^{ex}, the system in question is open and the stationary state is possible.

Introducing a new time $\tilde{t} = k_+[D]t/2$ and a new variable $y = 2k_-k_+^{-1}[Y]/[D]$, from (2.1.77) we have

$$dy/d\tilde{t} = 2y - y^2 - a^2 , \tag{2.1.78}$$

where

$$a^2 = 4k_-k_+^{-2}[D]^{-2}V^{-1}J^{\text{ex}} . \tag{2.1.79}$$

From the condition that y is constant, i.e. $dy/d\tilde{t} = 0$, we can, using (2.1.78), find the stationary values y^0, which are given by the equation $y^2 - 2y + a^2 = 0$. We arrive at the two stationary values

$$y_{1,2}^0 = 1 \pm \sqrt{1 - a^2} \tag{2.1.80}$$

assuming that $a^2 < 1$.

Linearizing (2.1.78), we find that the deviation $\delta y = y - y^0$ obeys

$$\begin{aligned} d\delta y/d\tilde{t} &= 2(1 - y^0)\delta y \\ &= \mp 2\sqrt{1 - a^2}\,\delta y . \end{aligned} \tag{2.1.81}$$

We see from these equations that the value $y_1^0 = 1 + (1 - a^2)^{1/2}$ is stable, and the value $y_2^0 = 1 - (1 - a^2)^{1/2}$ is unstable.

It is interesting to illustrate this application of stability criteria discussed in Sect. 2.1.5 to this case. Thus for criterion 1 we have

$$d\delta y^2/d\tilde{t} = 2\delta y\, d\delta y/d\tilde{t} . \tag{2.1.82}$$

Substituting (2.1.81) gives

$$d\delta y^2/d\tilde{t} = \mp 4\sqrt{1 - a^2}\,\delta y^2 . \tag{2.1.83}$$

Hence, by criterion 1, y_1^0 is stable and y_2^0 is unstable, which agrees with the earlier conclusion.

Turning to criterion 2, we will take it in the form (2.1.41). From [(3.4.25) v.1], we get

$$\delta x = RT\,\delta \ln[Y] = RT\,\delta y/y \approx RT\,\delta y/y^0\,. \tag{2.1.84}$$

As a consequence, the sum $\sum \delta x_\gamma \delta J_\gamma$, or in our case $\delta x \delta J = \delta x d\delta[Y]/d\tilde{t}$, will be const $\cdot \delta y d\delta y/d\tilde{t}$ with a positive constant. The inequality $\delta x \delta J \leq 1$ will therefore be equivalent to the inequality $d(\delta y^2)/d\tilde{t} \leq 0$, which enters into criterion 1, so that criterion 2 is equivalent to criterion 1.

Let us now consider criterion 3. We will ignore the difference between $\tilde{t}$ and t and notice that, from (2.1.84), $dx = RTy^{-1}dy$. Then

$$
\begin{aligned}
d\Phi(y) &= J dx \\
&= RT\dot{y}y^{-1}dy \\
&= RT(2y - y^2 - a^2)y^{-1}dy\,,
\end{aligned} \tag{2.1.85}
$$

hence

$$\Phi(y) = RT(2y - \tfrac{1}{2}y^2 - a^2 \ln y)\,. \tag{2.1.86}$$

We have thus found the function defined by (2.1.43b) in criterion 3. Condition (2.1.43a) is met because $\partial x/\partial y > 0$. It only remains to check (2.1.43c) near the stationary points. For (2.1.86) we obtain

$$d^2\Phi = RT(a^2 y_0^{-2} - 1)dy^2\,. \tag{2.1.87}$$

Near the points y_1^0 and y_2^0 we thus have, by (2.1.80),

$$
\begin{aligned}
d^2\Phi &= 2RT(y_{1,2}^0)^{-2}\left(\mp\sqrt{1-a^2} - 1 + a^2\right)dy^2 \\
&= 2RT(y_{1,2}^0)^{-2}\sqrt{1-a^2}\left[\mp 1 - \sqrt{1-a^2}\right]dy^2\,.
\end{aligned} \tag{2.1.88}
$$

The right-hand side is negative for y_1^0, so that $d^2\Phi < 0$ at $dy \neq 0$ and the conditions of validity of criterion 3 are satisfied, hence the point y_1^0 is stable. For y_2^0 we have $d^2\Phi > 0$ and therefore y_2^0 is unstable. We thus arrive again at the conclusions drawn earlier. Certainly, in more complex examples the different criteria may turn out to be nonequivalent.

2.1.10 Admixture Flux Through a Substance

Let us consider a three-dimensional medium through which some admixture diffuses. Its molar concentration $c(\mathbf{r}, t)$ obeys the diffusion equation

$$\dot{c} = D\Delta c\,. \tag{2.1.89}$$

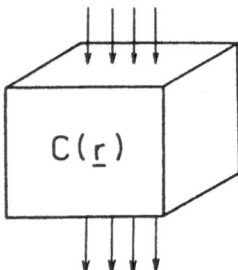

Fig. 2.6. A schematic picture of the three-dimensional space in which a constant flux of the admixture is maintained

Suppose further that the medium fills a rectangular parallelepiped and the admixture enters into it constantly from one side and emerges from the opposite side (Fig. 2.6). This gives rise to a stationary nonequilibrium state with a constant flux of the admixture through the medium, the flux being assumed to be homogeneous. We thus have some distribution of the concentration over the space, which, due to stationarity, is given by the equation $D\Delta c = 0$ that follows from (2.1.89), i.e., by the equation div $\mathbf{j} = 0$, where

$$\mathbf{j} = -D \ \text{grad} \ c. \tag{2.1.90}$$

If the x-axis is aligned with the flux, then (2.1.90) becomes

$$j_x = -D\partial c/\partial x,$$
$$j_y = -D\partial c/\partial y = 0,$$
$$j_z = -D\partial c/\partial z = 0. \tag{2.1.91}$$

It follows from the condition div $\mathbf{j} = 0$ that j_x is constant. This means that the concentration in the medium varies in a linear manner:

$$c(\mathbf{r}) = c_0 - (j_x/D)x. \tag{2.1.92}$$

Fluctuations here can be computed by using the linear FDR or the kinetic potential [(3.4.85) v.1] found in Sect. 3.4.6 v.1. According to [(3.1.8) or (3.1.9) v.1], with $\mathbf{r}$ substituted for α this potential enables us to work out for (2.1.89) the coefficient $K_{\mathbf{rr}'}[c]$ of the master equation. Since

$$\int K_{\mathbf{rr}'} y(\mathbf{r}) y(\mathbf{r}') d\mathbf{r} d\mathbf{r}' = 2\frac{kT}{RT} D \int c(\mathbf{r}) \mid \nabla y(\mathbf{r}) \mid^2 d\mathbf{r}, \tag{2.1.93}$$

we have

$$\begin{aligned} K_{\mathbf{rr}'}[c] &= -\frac{2D}{N_A}(\nabla c(r)\nabla)_{\mathbf{rr}'} \\ &= \frac{2D}{N_A}\frac{\partial}{\partial r_\alpha}\frac{\partial}{\partial r'_\alpha}[c(\mathbf{r})\delta(\mathbf{r}-\mathbf{r}')]. \end{aligned} \tag{2.1.94}$$

From this we find the stochastic equation

$$\dot{c}(\mathbf{r}) = D\Delta c(\mathbf{r}) + \text{div}[c^{1/2}(\mathbf{r})\zeta(\mathbf{r}, t)] \tag{2.1.95}$$

for the random concentration. Here $\zeta = (\zeta_1, \zeta_2, \zeta_3)$ is a vector whose components are independent random functions with zero mean and the correlator

$$\langle \zeta_i(\mathbf{r}, t)\zeta_j(\mathbf{r}', t')\rangle = (2D/N_A)^{1/2}\delta(\mathbf{r} - \mathbf{r}')\delta(t - t')\delta_{ij}. \tag{2.1.96}$$

It is easily verified that this corresponds to the coefficient (2.1.94). In a linear approximation we may in (2.1.95) substitute (2.1.92) for the random concentration $c(\mathbf{r})$.

Equation (2.1.95) holds both with or without a flux through the medium. The difference lies in the boundary conditions and the mean concentrations. Consequently, the flux exerts only an indirect influence owing to the mean concentration being inhomogeneous.

Now we would like to find the space – time spectral density. Note that the equation of the form (2.1.95), i.e.

$$\dot{c}(\mathbf{r}) - D\Delta c(\mathbf{r}) = \xi(\mathbf{r}, t), \tag{2.1.97}$$

can be conveniently solved by Fourier transformation in t and $\mathbf{r}$. For a random spectrum we have

$$(i\omega + Dk^2)c(\mathbf{k}, \omega) = \xi(\mathbf{k}, \omega), \tag{2.1.98}$$

hence

$$\langle c(\mathbf{k}_1, \omega_1), c(\mathbf{k}_2, \omega_2)\rangle$$
$$= (i\omega_1 + Dk_1^2)^{-1}(i\omega_2 + Dk_2^2)^{-1}\langle \xi(\mathbf{k}_1, \omega_1), \xi(\mathbf{k}_2, \omega_2)\rangle. \tag{2.1.99}$$

From (2.1.95) $\xi = \text{div}\,(c^{1/2}\zeta)$. The two-time correlator for ξ may be calculated using the mean concentration (2.1.92) as c. Therefore,

$$\langle \xi(r_1, t_1), \xi(r_2, t_2)\rangle = \frac{2D}{N_A}\frac{\partial^2}{\partial r_{1\alpha}\partial r_{2\alpha}}[(c_0 - \gamma x_1)\delta(r_{12})]\delta(t_{12}) \tag{2.1.100}$$

$(\gamma = j_x/D)$. Using the Fourier transforms [(A6.6) v.1] for each pair of variables, we will obtain the spectra

$$\langle \xi(\mathbf{k}_1, \omega_1), \xi(\mathbf{k}_2, \omega_2)\rangle$$
$$= -\frac{2D}{N_A}\mathbf{k}_1\mathbf{k}_2[c_0\delta(\mathbf{k}_1 + \mathbf{k}_2) + i\gamma\delta_x'(\mathbf{k}_1 + \mathbf{k}_2)]\delta(\omega_1 + \omega_2), \tag{2.1.101}$$

where

$$\delta_x'(\mathbf{k}) = \partial\delta(\mathbf{k})/\partial k_x = \delta'(k_x)\delta(k_y)\delta(k_z). \tag{2.1.102}$$

Substituting (2.1.101) into (2.1.99) gives

$$\langle c(\mathbf{k}_1, \omega_1), c(\mathbf{k}_2, \omega_2)\rangle$$
$$= -2DN_A^{-1}\mathbf{k}_1\mathbf{k}_2(i\omega_1 + Dk_1^2)^{-1}(i\omega_2 + Dk_2^2)^{-1}$$
$$\times [c_0\delta(\mathbf{k}_1 + \mathbf{k}_2) + i\gamma\delta_x'(\mathbf{k}_1 + \mathbf{k}_2)]\delta(\omega_1 + \omega_2). \tag{2.1.103}$$

This correlator corresponds to a stationary but non-homogeneous random process. The spectral density will then be given by [(A6.20) v.1]. It can be derived from (2.1.103) using [(A6.21) v.1], which here becomes

$$
S_c(\mathbf{k}, \omega, \mathbf{r}_0)
$$
$$
= \int \exp(-i k_0 r_0) \langle c(\tfrac{1}{2} \mathbf{k}_0 + \mathbf{k}, \omega), c(\tfrac{1}{2} \mathbf{k}_0 - \mathbf{k}, \omega_2) \rangle dk_0 d\omega_2 . \tag{2.1.104}
$$

Because the origin of coordinates $\mathbf{r} = 0$ is a typical point, we will find the spectral density at $r_0 = 0$. Substituting (2.1.103) into the above equation yields

$$
S_c(\mathbf{k}, \omega, 0) = 2 D N_A^{-1} k^2 (\omega^2 + D^2 k^4)^{-1} [c_0 + 2\gamma D \omega k_x (\omega^2 + D^2 k^4)^{-1}] . \tag{2.1.105}
$$

The second term in square brackets gives a correction for the inhomogeneity of average concentration which affects the noise level.

2.1.11 Admixture Flux with Nonlinear Diffusion

In the case of nonlinear diffusion, the flux of a diffusing substance through a medium will exert influence on fluctuations not only through the inhomogeneity of concentration. The equation will be

$$
\dot{c} = -\operatorname{div} \mathbf{j} , \tag{2.1.106}
$$

where

$$
j_l = -d_{lm} \frac{\partial c}{\partial r_m} - e_{lmi} \frac{\partial c}{\partial r_m} \frac{\partial c}{\partial r_i} \tag{2.1.107}
$$

in a linear–quadratic approximation, and

$$
j_l = -d_{lm} \frac{\partial c}{\partial r_m} - e_{lmn} \frac{\partial c}{\partial r_m} \frac{\partial c}{\partial r_n} - g_{lmni} \frac{\partial c}{\partial r_m} \frac{\partial c}{\partial r_n} \frac{\partial c}{\partial r_i} \tag{2.1.108}
$$

in a linear–quadratic–cubic approximation. The fact that the tensor e_{lmi} is nonzero is due to the medium being anisotropic. Let us dwell on this case assuming that the medium features preferable conduction along the direction indicated by the unit vector $\mathbf{n}$. In this connection, we will assume that e_{lmi} only has the longitudinal part $e_{lmi} = E n_l n_m n_i$, whereas we will assume that the tensor d_{lm} is the same: $d_{lm} = D \delta_{lm}$. For constant D and E in the linear–quadratic approximation we have

$$
\dot{c} = D \Delta c + E \mathbf{n} \cdot \nabla (\mathbf{n} \cdot \nabla c)^2 . \tag{2.1.109}
$$

When $\mathbf{n} = (1, 0, 0)$, i.e. $\mathbf{n}$ is aligned with the admixture flux, the above equation becomes

$$
\dot{c} = D \Delta c + E \frac{\partial}{\partial r_1} \left(\frac{\partial c}{\partial r_1} \right)^2 . \tag{2.1.110}
$$

We will compute the fluctuations using the Markov techniques, i.e. we will at first find $l_{\mathbf{r}, \mathbf{r}', \mathbf{r}''}$ and then the correction to the kinetic function $K_{\mathbf{r}\mathbf{r}'}[c]$. To this end, we will have to derive the equation in its standard form, using the equation

$$x(\mathbf{r}) = RT[c_0^{-1}(c(\mathbf{r}) - c_0) - \tfrac{1}{2}c_0^{-2}(c(\mathbf{r}) - c_0)^2]\,, \qquad (2.1.111)$$

that can be obtained from [(6.2.53) or (3.4.25) v.1]. Solving (2.1.111) for $c - c_0$ gives

$$c - c_0 = c_0[x/RT + \tfrac{1}{2}(x/RT)^2]\,. \qquad (2.1.112)$$

This expression must be substituted into the right-hand side of (2.1.110). The reasoning of the previous subsection is in essence equivalent to using the equation obtained by substituting (2.1.112) into the linear equation (2.1.89). We will here turn to the quadratic (in c) term in (2.1.110). Using the simple expression

$$c - c_0 = c_0 x/RT\,, \qquad (2.1.113)$$

the term becomes

$$\dot{c} = E\left(\frac{c_0}{RT}\right)^2 \frac{\partial}{\partial r_1}\left(\frac{\partial x}{\partial r_1}\right)^2 + \dots\,, \qquad (2.1.114)$$

where we have discarded the terms of no interest to us; r_1 is the longitudinal coordinate. Comparison of this with the standard equation $\dot{A}_\alpha = l_{\alpha,\beta}x_\beta + l_{\alpha,\beta\gamma}x_\beta x_\gamma/2$ gives

$$l_{\mathbf{r},\mathbf{r}'\mathbf{r}''} = 2E\left(\frac{c_0}{RT}\right)^2 \frac{\partial^3}{\partial r_1 \partial r_1' \partial r_1''}\delta(\mathbf{r} - \mathbf{r}')\delta(\mathbf{r} - \mathbf{r}'')\,. \qquad (2.1.115)$$

Using the FDR [(4.1.16) v.1] yields

$$l_{\mathbf{r}\mathbf{r}',\mathbf{r}''} = -kTl_{\mathbf{r},\mathbf{r}'\mathbf{r}''}$$
$$= -2EN_A^{-1}(RT)^{-1}c_0^2 \frac{\partial^3}{\partial r_1 \partial r_1' \partial r_1''}\delta(\mathbf{r} - \mathbf{r}')\delta(\mathbf{r} - \mathbf{r}'')\,. \qquad (2.1.116)$$

We thus obtain the following correction to the coefficient of the master equation:

$$K_{\mathbf{r}\mathbf{r}'}[c] = \dots + l_{\mathbf{r}\mathbf{r}',\mathbf{r}''}x_{\mathbf{r}''}[c]$$
$$= \dots + 2EN_A^{-1}(RT)^{-1}c_0^2 \frac{\partial^2}{\partial r_1 \partial r_1'}\left[\frac{\partial x(\mathbf{r})}{\partial r_1}\delta(\mathbf{r} - \mathbf{r}')\right]\,, \qquad (2.1.117)$$

or using (2.1.113) again,

$$K_{\mathbf{r}\mathbf{r}'}[c] = \dots + 2EN_A^{-1}c_0 \frac{\partial^2}{\partial r_1 \partial r_1'}\left[\frac{\partial c(\mathbf{r})}{\partial r_1}\delta(\mathbf{r} - \mathbf{r}')\right]\,. \qquad (2.1.118)$$

As in the previous subsection, we can compute the twofold correlator replacing $c(\mathbf{r})$ by the average concentration $\bar{c}(\mathbf{r}) = c_0 - \gamma r_1 \equiv c_0 - \gamma x$ of the type (2.1.92), but now γ is related to $j_x \equiv j_1$ by $j_x = D\gamma - E\gamma^2$. Then the above expression becomes

$$K_{\mathbf{r}\mathbf{r}'}[c] = \dots - 2N_A^{-1}\gamma c_0 E \frac{\partial^2}{\partial r_1 \partial r_1'}[\delta(\mathbf{r} - \mathbf{r}')]\,. \qquad (2.1.119)$$

We can now add to (2.1.110) a fluctuation source and obtain the Langevin equation

$$\dot{c} = D\Delta c + E\frac{\partial}{\partial r_1}\left(\frac{\partial c}{\partial r_1}\right)^2 + \xi(\mathbf{r}, t), \tag{2.1.120}$$

where $\xi(\mathbf{r}, t)$ is the fluctuation source having the correlator

$$\langle\xi(\mathbf{r}_1, t_1), \xi(\mathbf{r}_2, t_2)\rangle = \ldots - 2N_A^{-1}\gamma c_0 E\frac{\partial}{\partial x_1 \partial x_2}[\delta(\mathbf{r}_{12})]\delta(t_{12})$$

$$(x \equiv r_1). \tag{2.1.121}$$

The dots here stand for the expression on the right-hand side of (2.1.100). When calculating the twofold correlator we may discard the quadratic term in (2.1.120). Following along the same lines as in the previous subsection, we can easily obtain

$$S_c(\mathbf{k}, \omega, 0) = \ldots - 2N_A^{-1}\gamma c_0 Ek_x^2(\omega^2 + D^2 k^4)^{-1} \tag{2.1.122}$$

using the linear equation and (2.1.121). The dots here stand for the expression on the right-hand side of (2.1.105). The spectral density at the point $\mathbf{r}_0$ can be found replacing c_0 by $\bar{c}(\mathbf{r}_0)$ both here and on the right-hand side of (2.1.105).

We have thus determined how, due to nonlinearity of the diffusion equation, the impurity flux affects the spectral density, i.e. the correlator. Unlike the additional term considered in the previous subsection, the correction to the correlator is now proportional to k_x^2. Similar arguments can also be used in the case of a constant heat flux through a medium, of an electric current, and so on. What is more, we can work out higher correlators as well.

2.2 Some Generating Equations for Open Systems

2.2.1 The Markov Case

For a closed system S_L that includes an open system S_0, the generating equation [(3.2.50) v.1] of the Markov theory holds; this is known to be a consequence of the time-reversal symmetry. Naturally, from this equation we can derive a generating equation for an open system.

Let a system be closed in a subset B' of random internal thermodynamic parameters and open in B'', so that we can write $B = (B', B'')$. Parameters that are conjugates of B' will be denoted by x', and conjugates of B'', by x'', so that $x = (x', x'')$. Similarly, we can split into two subsets the variables $y = (y', y'')$ that enter [(3.2.50) v.1]. The latter equation can now be written as

$$R(y' + x', y'' + x'', x', x'') = R(-\varepsilon' y', -\varepsilon'' y'', \varepsilon' x', \varepsilon'' x'') \tag{2.2.1}$$

for a large closed system. The asymptotic form of this is

$$R(y' + \partial F_L/\partial B', y'' + \partial F_L/\partial B'', \partial F_L/\partial B', \partial F_L/\partial B'')$$
$$= R(-\varepsilon' y', -\varepsilon'' y'', \varepsilon' \partial F_L/\partial B', \varepsilon'' \partial F_L/\partial B''), \tag{2.2.2}$$

where F_L is the free energy of the large system. Or, from (2.1.5,8,9)

$$R(y' + \partial F_0/\partial B', y'' + \partial F_0/\partial Q_\beta + \partial F_\beta/\partial Q_\beta, \partial F_0/\partial B', \partial F_0/\partial Q_\beta + \partial F_\beta/\partial Q_\beta)$$
$$= R(-\varepsilon' y', -\varepsilon'' y'', \varepsilon' \partial F_0/\partial B', \varepsilon'' \delta F_0/\partial Q_\beta + \varepsilon'' \partial F_\beta/\partial Q_\beta). \tag{2.2.3}$$

We will now increase to infinity the capacities (2.1.12) of all reservoirs. In the limit, the subsystem S_0 – a part of the closed system S_L – will become open. As the capacities are increased, the time dependence of $\partial F_\beta(Q)/\partial Q_\beta$ will become ever weaker and in the limit it will become a constant, $-h_\beta$, which is independent of B''. Equation (2.2.3) will then become

$$R(y' + x', y'' + \partial F_0/\partial B'' - h'', x', \partial F_0/\partial B'' - h'')$$
$$= R(-\varepsilon'y', -\varepsilon''y'', \varepsilon'x', \varepsilon''\partial F_0/\partial B'' - \varepsilon''h'') \,. \tag{2.2.4}$$

This is the generating equation for the open system S_0.

We will confine ourselves to the discussion of the open systems that may have stationary nonequilibrium states. For such systems, at least, one of the $\partial F_0/\partial B''$ must vanish. If all $\partial F_0/\partial B''$ are zero, then from (2.2.4) we obtain the following generating equation:

$$R(y' + x', y'' - h'', x', -h'') = R(-\varepsilon'y', -\varepsilon''y'', \varepsilon'x', -\varepsilon''h'') \,. \tag{2.2.5}$$

Such open systems are described by the master equation

$$\dot{w}(B', B'') = N_{\partial,B}\beta V(-kT\partial/\partial B, B', h'')w(B', B'') \,. \tag{2.2.6}$$

Here $N_{\partial,B}$ is the operator ordering symbol, which has the same sense as in [(3.1.11) v.1]. The operator of this equation can be expressed in terms of the function R that enters into (2.2.5) using the asymptotic formula

$$N_{\partial,B}V(-kT\partial/\partial B, B', h'')$$
$$= N_{\partial,B}R(-kT\partial/\partial B', -kT\partial/\partial B'', \partial F_0(B')/\partial B', -h'') \,, \tag{2.2.7}$$

which is similar to [(3.1.37 v.1].

It sould be emphasized that the operator (2.2.7) does not contain B'', so that no stationary probability density for B'' can set in, but instead it spreads out in a diffusional manner.

If we integrate (2.2.7) with respect to B'', the probability density in B' will obey

$$\dot{w}(B') = N_{\partial,B'}\beta R(-kT\partial/\partial B', 0, \partial F_0(B')/\partial B', -h'') \,. \tag{2.2.8}$$

The generating equation (2.2.5) is equivalent to the following relationship for the operator (2.2.7):

$$N_{\partial,B'}V(-kT\partial/\partial B', y'' - h'', B', h'')w_0(B')$$
$$= w_0(B')\{N_{\partial,B'}V(-kT\varepsilon'\partial/\partial B', -\varepsilon''y'', \varepsilon'B', \varepsilon''h'')\}^T \,, \tag{2.2.9}$$

where

$$w_0(B') = \text{const} \cdot \exp(-F_0(B')/kT) \tag{2.2.10}$$

is the equilibrium probability density in B'.

Note that the right-hand side of (2.2.9) can be written as

$$w_0(B')N_{B',\partial}V(kT\varepsilon'\partial/\partial B', -\varepsilon''y'', \varepsilon'B', \varepsilon''h'') \,. \tag{2.2.11}$$

We find from [(3.2.36,37) v.1] that

$$\exp(-kT\partial^2/(\partial y'\partial B'))[V(y', y'' - h'', B', h'')w_0(B')]$$
$$= V(-\varepsilon'y', -\varepsilon''y'', \varepsilon'B', \varepsilon''h'')w_0(B') \tag{2.2.12}$$

is equivalent to (2.2.9). To see that (2.2.9) and (2.2.12) are valid, we will premultiply (2.2.12) by $\exp(\beta B'x')$ and integrate with respect to B' from $-\infty$ to ∞. Then, reasoning along the same lines as in the derivation of [(3.2.50) v.1] we will get

$$Q(y' + x', y'' - h'', x', h'') = Q(-\varepsilon'y', -\varepsilon''y'', \varepsilon'x', \varepsilon''h''), \tag{2.2.13}$$

where

$$Q(y, x', h'') = \int V(y, B', h'')w_{x'}(B')dB', \tag{2.2.14}$$

$$w_{x'}(B') = \exp(\beta B'x')w_0(B') \Big/ \int \exp(\beta B'x')w_0(B')dB'. \tag{2.2.15}$$

Since $kT = \beta^{-1}$ is small from a macroscopic point of view, the probability density (2.2.15) is fairly sharp, being concentrated near the point found from the equation $\partial F_0(B')/\partial B' = x'$. Therefore, we obtain from (2.2.14) the asymptotic equation

$$Q(y, \partial F_0(B')/\partial B', h'') = V(y, B', h''). \tag{2.2.16}$$

If, in addition, we take into account (2.2.7), we get

$$Q(y, x', h'') = R(y, x', -h''). \tag{2.2.17}$$

Consequently, (2.2.12,13), and hence (2.2.9), are equivalent to (2.2.5).

If the system in question is open in all variables, i.e., if $B = B''$, then (2.2.5) will be

$$R(y'' - h'', -h'') = R(-\varepsilon''y'', -\varepsilon''h'') \tag{2.2.18a}$$

or

$$R(y - h, -h) = R(-\varepsilon y, -\varepsilon h). \tag{2.2.18b}$$

According to (2.2.17), we have

$$Q(y - h, h) = Q(-\varepsilon y, \varepsilon h), \tag{2.2.19}$$

which is similar to [(3.2.50) v.1].

2.2.2 Generating Equation of the Non-Markov Theory for Systems Open in All Variables

The generating equation for the non-Markov nonquantum case is derived using (1.2.39). Recall that, by (1.2.11,38), the functional in (1.2.39) is of the form

$$\Pi[y(t), h(t)] = \beta^{-1}\ln\left\{\left\langle \exp[\beta \int y_\alpha(t)J_\alpha(t)dt]\right\rangle_{h(\tau)}\right\}. \tag{2.2.20}$$

Let the forces $h(t) = h^0$ be independent of time for $-\tau/2 < t < \tau/2$ and be zero otherwise. Likewise,

$$y_\alpha(t) = \begin{cases} y_\alpha^0 & \text{for} \quad -\tau/2 < t < \tau/2, \\ 0 & \text{for} \quad t < -\tau/2 \text{ or } t > \tau/2, \end{cases} \tag{2.2.21}$$

where y^0 are constants. Then,

$$\int y_\alpha(t) J_\alpha(t) dt = y_\alpha^0 [B_\alpha(\tau/2) - B_\alpha(-\tau/2)] \equiv y_\alpha^0 \Delta B_\alpha. \tag{2.2.22}$$

It is convenient to introduce the function

$$G(y^0, h^0) = \tau^{-1}\beta^{-1} \ln\{\langle \exp(\beta y_\alpha^0 \Delta B_\alpha) \rangle_{h^0}\}, \tag{2.2.23}$$

which, by [(2.1.10) v.1], can be represented by the series

$$G(y^0, h^0) = \tau^{-1} \sum_{m=1}^{\infty} \beta^{m-1}(m!)^{-1}\langle \Delta B_{\alpha_1}, \ldots, \Delta B_{\alpha_m}\rangle_{h^0} y_{\alpha_1}^0 \cdots y_{\alpha_m}^0, \tag{2.2.24}$$

where

$$\langle \Delta B_{\alpha_1}, \ldots, \Delta B_{\alpha_m}\rangle_{h^0} = \int_{-\tau/2}^{\tau/2} \cdots \int_{-\tau/2}^{\tau/2} \langle J_1, \ldots, J_m\rangle dt_1 \ldots dt_m. \tag{2.2.25}$$

Comparing (2.2.20) and (2.2.23) at the above $y(t)$ and $h(t)$ gives

$$\Pi[y(t), h(t)] = \tau G(y^0, h^0). \tag{2.2.26}$$

Hence from (1.2.39) we have

$$G(y^0 - h^0, h^0) = G(-\varepsilon y^0, \varepsilon h^0). \tag{2.2.27}$$

Let τ_{cor} be the correlation time for the random process $J(t)$. If the various components have different correlation times, then we should take for τ_{cor} the maximal time. If then we denote by τ_{rel} the depletion time for the reserves Q_β in the reservoirs P_β, for sufficiently large reservoirs we will have

$$\tau_{\text{cor}} \ll \tau_{\text{rel}}. \tag{2.2.28}$$

Now we will take τ in (2.2.24,25) such that

$$\tau_{\text{cor}} \ll \tau \ll \tau_{\text{rel}}. \tag{2.2.29}$$

As the reservoir capacities grow infinite, the value of τ in (2.2.29) may tend to infinity. It follows from (2.2.25) then that the quantities

$$\langle \Delta B_{\alpha_1}, \ldots, \Delta B_{\alpha_m}\rangle_{h^0}/\tau \tag{2.2.30}$$

will tend to the generalized diffusion coefficients

$$D_{\alpha_1 \ldots \alpha_m}(h^0) = \int_{-\infty}^{\infty} \cdots \int_{-\infty}^{\infty} \langle J_1 \ldots J_m\rangle_{h^0} dt_2 \ldots dt_m \tag{2.2.31}$$

and (2.2.24) will be

$$G(y^0, h^0) = \sum_{m=1} \beta^{m-1}(m!)^{-1} D_{\alpha_1 \ldots \alpha_m}(h^0) y^0_{\alpha_1} \ldots y^0_{\alpha_m} \,. \qquad (2.2.32)$$

This limiting function will obey the same equation (2.2.27).

The generating equation (2.2.27) is similar in form to (2.2.19). This is quite natural since h in (2.2.19) and h^0 in (2.2.27) have the same meaning of constant forces and since $G(y, h)$ has a meaning similar to $Q(y, h)$. In effect, at $\tau \gg \tau_{\text{cor}}$ the probability density $w(\Delta B)$ for increments of B obeys

$$\dot{w}(\Delta B) = \sum_{m=1}^{\infty} \frac{(-1)^m}{m!} \frac{\partial^m}{\partial(\Delta B_{\alpha_1}) \ldots \partial(\Delta B_{\alpha_m})} [D_{\alpha_1 \ldots \alpha_m}(h) w]$$

$$= \beta G\left(-kT \frac{\partial}{\partial(\Delta B)}, h\right) w \qquad (2.2.33)$$

by (2.2.32). This equation describes the diffusion spreading of the above probability density. Differentiation here may be carried out with respect to either ΔB or B. This equation is a special case of (2.2.6) for the system open in all variables. Comparison of (2.2.6) and (2.2.33) yields the equations $G(y, h) = V(y, h) = Q(y, h)$ that we sought.

The generating equations (2.2.5) and (2.2.27) can be used both to check the validity of a model chosen for a fluctuation-dissipation process, and to derive specific FDRs.

Since the above-mentioned generating equations are similar in form to the standard generating equation [(3.2.50)) v.1] in the theory of closed systems, individual FDRs obtained from them will be essentially the same as those derived in Sect. 4.1. v.1. By way of illustration, we will provide, from (2.2.27), some simple FDRs

$$D_{\alpha\beta} = 2kT \vartheta^+_{\alpha\beta} \partial D_\alpha / \partial h_\beta \,,$$

$$\partial D_\alpha / \partial h_\beta = \varepsilon_\alpha \varepsilon_\beta \partial D_\beta / \partial h_\alpha \,,$$

$$\partial D_{\alpha\beta} / \partial h_\gamma = kT(-\varepsilon_\alpha \varepsilon_\beta \varepsilon_\gamma \partial^2 D_\gamma / \partial h_\alpha \partial h_\beta + \partial^2 D_\alpha / \partial h_\beta \partial h_\gamma + \partial^2 D_\beta / \partial h_\alpha \partial h_\gamma) \,,$$

$$D_{\alpha\beta\gamma} = 2(kT)^2 \vartheta^-_{\alpha\beta\gamma}(\partial^2 D_\alpha / \partial h_\beta \partial h_\gamma + \partial^2 D_\beta / \partial h_\alpha \partial h_\gamma + \partial^2 D_\gamma / \partial h_\alpha \partial h_\beta) \qquad (2.2.34)$$

that hold at $h = 0$. These relations only differ from [(4.1.11,13,16,18) v.1] in that some of them have the opposite sign. Since these FDRs are only valid at $h = 0$ and since higher derivatives cannot be determined uniquely, (2.2.34) are of help in finding the diffusion coefficients $D_{\alpha\beta}(h), D_{\alpha\beta\gamma}(h)$ at small h only, i.e. at relatively small deviations from equilibrium.

2.3 H-Theorems and Relations
for Nonequilibrium Stationary States

2.3.1 Quasi-Frenergy or Quasi-Entropy

We will here deal with open systems in which nonequilibrium stationary states are possible. The FDRs suitable for strongly nonequilibrium stationary states cannot be derived from the generating equation (2.2.5). It is expedient to construct the theory for strongly nonequilibrium states on another basis, namely taking as a basis state a nonequilibrium stationary state, not an equilibrium one as was done in Chap. 3 v.1. It is true that with this approach we cannot use the symmetry in time, proceeding solely from the dynamic equilibrium principle. This thus sets narrow bounds to the theory in comparison with that put forward in Chaps. 3 and 4 of volume 1.

Integrating (2.2.6) with respect to the variables B'', (the variables in which the system is open) gives

$$\dot{w}(B) = \beta N_{\partial,B} V(-kT\partial/\partial B, B)w(B), \tag{2.3.1}$$

where

$$V(y', B') \equiv V(y', B', h'') = V(y', 0, B', h''). \tag{2.3.2}$$

For simplicity we do not write her the external forces h'' or inputs $J^{\mathrm{ex}} = \dot{B}''$, on which the kinetic potential $V(y', B')$ is dependent. Furthermore, we denote the ensemble of internal parameters in which the system is closed by B, not by B'.

We will write the stationary probability density $w_{\mathrm{st}}(B)$ as

$$w_{\mathrm{st}}(B) = \mathrm{const} \cdot \exp(-\Psi(B)/\kappa), \tag{2.3.3}$$

where $\kappa > 0$ is a parameter characterizing the noise intensity, which is equal, say, to kT or k, if the noise intensity is of the same order of magnitude as in an equilibrium state; $\Psi(B)$ is a certain function. Equation (2.3.3) is similar to the equation [(2.2.61) v.1] which gives the equilibrium distribution. Therefore, $\Psi(B)$ can be called quasi-frenergy. (Here "frenergy" stands for "free energy"). If now instead of $\Psi(B)$ we in (2.3.3) introduce the function $\Sigma(B) = -\Psi(B)$, then we will obtain a formula similar to [(2.2.65) v.1]. For this reason we can refer to $-\Sigma(B)$ as quasi-entropy.

With the help of the parameter κ appearing in (2.3.3), equation (2.3.1) can be written as

$$\dot{w}(B) = \kappa^{-1} N_{\partial,B} V(-\kappa\partial/\partial B, B)w(B), \tag{2.3.4}$$

where $V(y, B)$ is the kinetic potential.

The stationary probability density will then obey

$$N_{\partial,B} V(-\kappa\partial/\partial B, B)w_{\mathrm{st}}(B) = 0. \tag{2.3.5}$$

2.3.2 Generating Equation

By analogy with [(3.1.31) v.1], we introduce the biased probability density

$$\begin{aligned} w_x(B) &= \mathrm{const} \cdot \exp(xB/\kappa)w_{\mathrm{st}}(B) \\ &= \mathrm{const} \cdot \exp[-(\Psi(B) - xB)/\kappa]. \end{aligned} \tag{2.3.6}$$

If now we multiply (2.3.5) by $\exp(xB/\kappa)$ and integrate with respect to B in infinite terms, then using equations of the type [(3.1.33,34) v.1] we obtain

$$\int V(x,B)w_x(B)dB = 0 \qquad (2.3.7)$$

or

$$R(x,x) = 0 \qquad (2.3.8)$$

in terms of the kinetic potential image

$$R(y,x) = \int V(y,B)w_x(B)dB. \qquad (2.3.9)$$

In the case of the small parameter κ the probability density (2.3.6) is concentrated near the point B_x, where this density, and hence the function $Bx - \Psi(B)$, have the maximum. This point is found from

$$\partial\Psi(B)/\partial B = x. \qquad (2.3.10)$$

So, for the fairly narrow probability density (2.3.5) from (2.3.9) we have

$$R(y,x) \approx V(y,B_x). \qquad (2.3.11)$$

Therefore, (2.3.8) can be written in the asymptotic form $V(x,B_x) = 0$ or, using (2.3.10), in the form

$$V(\partial\Psi(B)/\partial B, B) = 0. \qquad (2.3.12)$$

This equation and also (2.3.8) can be applied in two ways. If the operator of the master equation (2.3.1) is specified, then (2.3.12) enables the function $\Psi(B)$ to be determined, and hence the stationary probability density (2.3.3). Further, if the one-time stationary probability density, and hence $\Psi(B)$, are known (e.g., found experimentally), then (2.3.12) imposes constraints on the operator of the master equation. And from this we can then obtain various FDRs, i.e. it is a generating equation.

2.3.3 Conjugate of $\Psi(B)$.

Let us introduce the function

$$\Phi(x) = -\kappa \ln\left\{\int \exp[(Bx - \Psi(B))/\kappa]\right\}dB \qquad (2.3.13)$$

which is the conjugate of $\Psi(B)$. It follows from (2.3.13) that the normalization constant in (2.3.3) can be expressed through $\Phi(x)$, and then (2.3.3) will become

$$w_{\mathrm{st}}(B) = \exp[(\Phi(0) - \Psi(B))/\kappa]. \qquad (2.3.14)$$

We now wish to find the characteristic function

$$\Theta(iu) = \int \exp(iuB)w_{\mathrm{st}}(B)dB \qquad (2.3.15)$$

corresponding to this probability density. Substituting (2.3.14) into (2.3.15) and using (2.3.13) we easily find

$$\kappa \ln \Theta(v) = \Phi(0) - \Phi(\kappa v) \,. \tag{2.3.16}$$

Consequently, one-time stationary correlators can be rather simply expressed in terms of $\Phi(x)$. In fact, using [(2.1.10) v.1], from (2.3.16) we obtain

$$\langle B_{\alpha_1}, \ldots, B_{\alpha_m} \rangle = -\kappa^{m-1} \partial^m \Phi(x)/\partial x_{\alpha_1}, \ldots, \partial x_{\alpha_m} \tag{2.3.17}$$

for $x = 0$. Specifically,

$$\langle B_\alpha \rangle = -\partial \Phi/\partial x_\alpha \quad \text{for} \quad x = 0 \,. \tag{2.3.18}$$

These exact formulas are a nonequilibrium generalization of [(2.2.30) v.1].

If we want to compute the correlators corresponding to the biased probability density (2.3.6), then in (2.3.17,18) we should not let $x = 0$:

$$\langle B_\alpha \rangle_x = -\partial \Phi(x)/\partial x_\alpha \,, \tag{2.3.19}$$

$$\langle B_{\alpha_1}, \ldots, B_{\alpha_m} \rangle_x = -\kappa^{m-1} \partial^m \Phi(x)/\partial x_{\alpha_1} \ldots x_{\alpha_m} \,. \tag{2.3.20}$$

When the fluctuation intensity parameter κ is small and when the function $\Psi(B) - Bx$ has the only minimum, the means (2.3.19) are close to B_x obtained from the condition (2.3.10) of an extremum of the probability density (2.3.6). It follows that under the above conditions the functions $\partial \Psi(B)/\partial B$ and $-\partial \Phi(x)/\partial x$ are asymptotically mutually inverse and that the functions $\Psi(B)$ and $\Phi(x)$ are approximately the Legendre transforms of one another:

$$\Phi(x) = \Psi(B_x) - xB_x \quad (\partial \Psi(B_x)/\partial B = x) \,. \tag{2.3.21}$$

This equation is proved in Appendix A1 v.1.

In this case, it holds that

$$\| \partial^2 \Psi/\partial B_\alpha \partial B_\beta \|^{-1} = - \| \partial^2 \Phi(x)/\partial x_\alpha \partial x_\beta \| \,. \tag{2.3.22}$$

From (2.3.22), the equation $\langle B_\alpha, B_\beta \rangle_x = -\kappa \partial^2 \Phi/\partial x_\alpha \partial x_\beta$ derivable from (2.3.20) at $m = 2$, can be written as

$$\| \langle B_\alpha, B_\beta \rangle_x \| = \kappa \| \partial^2 \Psi(B)/\partial B_\alpha \partial B_\beta \|^{-1} \tag{2.3.23}$$

at $B = B_x$. This equation can also be obtained by Gaussian approximation of the probability density (2.3.6). Here we assume that at the point B_x the matrix $\partial^2 \Psi/\partial B_\alpha \partial B_\beta$ is positive definite. Unlike (2.3.17–20), the formulas (2.3.21–23) are asymptotic.

2.3.4 The H-Theorem

Let the motion of a system under consideration be characterized by the phenomenological (macroscopic) equation

$$\dot{A} = f(A) \,. \tag{2.3.24}$$

This equation, especially when fluctuations are not too small, needs some clarification. Suppose that the right-hand side of (2.3.24) has the following exact meaning:

$$f_\alpha(A) = \int K_\alpha(B) w_{x(A)}(B) dB \,, \tag{2.3.25}$$

where the dependence $x(A)$ is the reverse of the dependence $A = -\partial\Phi(x)/\partial x$, i.e., of the dependence (2.3.19), and

$$K_\alpha(B) = \lim_{\Delta t \to 0} \langle \Delta B_\alpha / \Delta t \rangle_B \tag{2.3.26}$$

is the drift coefficient. This means that the derivative $\dot{A}$ in (2.3.24) is understood as a result of the averaging of $\dot{B}$, the averaging being carried out not at fixed A, but at fixed forces $x(A)$.

We will introduce the function

$$\bar{\Psi}(A) = \Phi(x(A)) + A x(A), \quad (A(x) = -\partial\Phi(x)/\partial x) \,, \tag{2.3.27}$$

where $x(A)$ has the same sense as in (2.3.25), i.e. as the Legendre transform of $\Phi(x)$. If (2.3.21) is valid, then $\bar{\Psi}(A)$ coincides with $\Psi(A)$, and if (2.3.21) is invalid, then $\bar{\Psi}(A)$ differs from $\Psi(A)$.

From the definition of (2.3.27) and from the equation $A = -\partial\Phi(x)/\partial x$, we obtain

$$\partial\bar{\Psi}(A)/\partial A_\alpha = x_\alpha \,. \tag{2.3.28}$$

Differentiating (2.3.27) with respect to time and using (2.3.24), we obtain

$$\frac{d\bar{\Psi}(A)}{dt} = \frac{\partial\bar{\Psi}}{\partial A}(A)\dot{A} = \frac{\partial\bar{\Psi}}{\partial A_\alpha}(A) f_\alpha(A) \tag{2.3.29}$$

or, by virtue of (2.3.28),

$$d\bar{\Psi}(A)/dt = x_\alpha f_\alpha(A(x)) \tag{2.3.30}$$

for $x = x(A)$. Substituting (2.3.25) gives

$$d\bar{\Psi}(A)/dt = x_\alpha \int K_\alpha(B) w_x(B) dB \quad \text{for} \quad x = x(A) \tag{2.3.31}$$

or, in terms of [(3.1.42) v.1],

$$d\bar{\Psi}(A)/dt = x_\alpha \kappa_\alpha(x) \quad \text{for} \quad x = x(A) \,. \tag{2.3.32}$$

Considering [(3.1.41) v.1], where instead of β we have to take κ^{-1}, and also [(3.1.42) v.1], the equation (2.3.8) can be written as the expansion

$$\sum_{m=1}^{\infty} (m!)^{-1} \kappa^{1-m} \langle K_{\alpha_1 \dots \alpha_m}(B) \rangle_x x_{\alpha_1} \dots x_{\alpha_m} = 0 \tag{2.3.33}$$

or, from [(2.3.25) v.1],

$$x_\alpha \kappa_\alpha(x) + \sum_{m=2}^{\infty} (m!)^{-1} \kappa^{1-m} \lim_{\Delta t \to 0} (\Delta t)^{-1} \langle \Delta B_{\alpha_1} \dots \Delta B_{\alpha_m} \rangle_x x_{\alpha_1} \dots x_{\alpha_m} = 0 \,. \tag{2.3.34}$$

Hence,

$$
\begin{aligned}
x_\alpha \kappa_\alpha(x) &= -\lim_{\Delta t \to 0} (\Delta t)^{-1} \left\langle \sum_{m=2}^{\infty} (m!)^{-1} \kappa^{1-m} (x_\alpha \Delta B_\alpha)^m \right\rangle_x \\
&= -\lim_{\Delta t \to 0} \kappa (\Delta t)^{-1} \langle g(x \Delta B / \kappa) \rangle_x ,
\end{aligned}
\tag{2.3.35}
$$

where

$$
g(z) = \exp z - 1 - z
\tag{2.3.36}
$$

is a nonnegative function, which is strictly positive everywhere except at the point $z = 0$. Taking into account this nonnegativeness, from (2.3.35) we obtain

$$
x_\alpha \kappa_\alpha(x) \le 0
\tag{2.3.37}
$$

and consequently for (2.3.32) we have

$$
d\bar{\Psi}(A)/dt \le 0 .
\tag{2.3.38}
$$

The equality sign here is only valid in the trivial case of the absence of fluctuations.

This H-theorem is similar to the appropriate theorem of Sect. 4.6.3 v.1.

Thus, in the case of open systems the function $\bar{\Psi}(A)$, varying according to the phenomenological equations, decreases monotonically like the free energy of closed systems.

2.3.5 Another Form of the H-Theorem

Consider a theorem that to a certain degree is akin to the theorem of the previous subsection. Let us introduce the quantity

$$
H = \kappa \int w(B) \ln[w(B)/w_{\text{st}}(B)] dB
\tag{2.3.39}
$$

that depends on the stationary probability density and on arbitrary probability density $w(B)$. Quantities of this kind were first introduced by *Kullback* [2.5]. Therefore, H can be referred to as the Kullback entropy.

It can be readily proved that it is nonnegative. In fact, from the inequatlity

$$
\ln x \le x - 1
\tag{2.3.40}
$$

(the equality sign is valid only for $x = 1$), we have

$$
w \ln(w_{\text{st}}/w) \le w(w_{\text{st}}/w - 1) = w_{\text{st}} - w .
\tag{2.3.41}
$$

Hence,

$$
\int w(B) \ln[w_{\text{st}}(B)/w(B)] dB \le \int w_{\text{st}}(B) dB - \int w(B) dB = 0 .
\tag{2.3.42}
$$

From (2.3.39) and (2.3.42) we get

$$
H \ge 0 .
\tag{2.3.43}
$$

Since, as is easily seen, $H = 0$ at $w = w_{st}$, the entropy H as a functional of $w(B)$ has a minimum at the point $w(B) = w_{st}(B)$.

The Kullback entropy (2.3.39) varies with time, since $w(B)$ varies according to the master equation (2.3.4). Let us prove that H decreases monotonically

$$dH/dt \leq 0 .\tag{2.3.44}$$

Proof. Consider the variation of the probability density $w(B)$ within a time interval $\Delta = t_2 - t_1 > 0$. If at t_1 the probability density is $w(B)$, then at t_2 it will be

$$\tilde{w}(B) = \int w_\Delta(B \mid B')w(B')dB' ,\tag{2.3.45}$$

where $w_\Delta(B \mid B')$ are the transition probabilities in [(2.3.10) v.1]. They naturally meet the normalization condition

$$\int w_\Delta(B \mid B')dB = 1 .\tag{2.3.46}$$

The stationary probability density does not vary with time, so that

$$\int w_\Delta(B \mid B')w_{st}(B')dB' = w_{st}(B) .\tag{2.3.47}$$

If at time t_1 the probability density $w(B)$ has the entropy (2.3.39), then at time t_2 we will have the entropy

$$\tilde{H} = \kappa \int \tilde{w}(B) \ln[\tilde{w}(B)/w_{st}(B)]dB .\tag{2.3.48}$$

By introducing the function

$$\varphi(z) = z \ln z, \quad z \geq 0 ,\tag{2.3.49}$$

we can write the above entropies (2.3.39,48) as

$$\tilde{H} = \kappa \int w_{st}(B)\varphi[\tilde{w}(B)/w_{st}(B)]dB ,\tag{2.3.50}$$

Consider the difference $H - \tilde{H}$. By (2.3.50) and (2.3.46,47) this difference can be written as

$$H - \tilde{H} = \kappa \int w_\Delta(B \mid B')w_{st}(B')$$
$$\times \{\varphi[w(B')/w_{st}(B')] - \varphi[\tilde{w}(B)/w_{st}(B)]\}dBdB' .\tag{2.3.51}$$

The function (2.3.49) is convex for $z \geq 0$, since $\varphi''(z) = z^{-1} \geq 0$. Hence, we have

$$\varphi(z_1) - \varphi(z_2) \geq (z_1 - z_2)\varphi'(z_2) ,$$
$$z_1, z_2 \geq 0 ,\tag{2.3.52}$$

where $\varphi'(z) = d\varphi(z)/dz$. Using it at $z_1 = w(B')/w_{st}(B')$ and $z_2 = \tilde{w}(B)/w_{st}(B)$, from (2.3.51) we obtain

$$H - \tilde{H} \;\geq\; \kappa \int w_\Delta(B \mid B') w_{\text{st}}(B')[w(B')/w_{\text{st}}(B') - \tilde{w}(B)/w_{\text{st}}(B)]$$
$$\times \varphi'[\tilde{w}(B)/w_{\text{st}}(B)]dBdB' \equiv J_1 - J_2 \,. \tag{2.3.53}$$

Here

$$J_1 \;=\; \kappa \int w_\Delta(B \mid B') w_{\text{st}}(B')[w(B')/w_{\text{st}}(B')]\varphi'[\tilde{w}(B)/w_{\text{st}}(B)]dBdB'$$
$$=\; \kappa \int \tilde{w}(B)\varphi'[\tilde{w}(B)/w_{\text{st}}(B)]dB \tag{2.3.54}$$

by (2.3.45).

Furthermore,

$$J_2 \;=\; \kappa \int w_\Delta(B \mid B') w_{\text{st}}(B')[\tilde{w}(B)/w_{\text{st}}(B)]\varphi'[\tilde{w}(B)/w_{\text{st}}(B)]dBdB'$$
$$=\; \kappa \int \tilde{w}(B)\varphi'[\tilde{w}(B)/w_{\text{st}}(B)]dB \tag{2.3.55}$$

by (2.3.47). Since the integrals (2.3.54) and (2.3.55) coincide, the inequality (2.3.53) gives

$$H - \tilde{H} \geq 0 \,. \tag{2.3.56}$$

Dividing this by Δ and going to the limit as $\Delta \longrightarrow 0$, we obtain (2.3.44). QED.

Note that the entropy (2.3.39) decreases monotonically regardless of whether or not the space of values of the parameters is continuous or discrete. This H-theorem is equally valid for open and closed systems with the Markov processes.

2.3.6 One Consequence of the Last Theorem

Before we come to using this theorem, we will ask the question of what probability density $w(B)$ at fixed means

$$\int B_\alpha w(B)dB = A_\alpha = \text{fix} \tag{2.3.57}$$

and fixed normalization conditions will minimize the entropy (2.3.39). To answer this question we should use the Lagrange factor technique to form the functional

$$K[w(B)] \;=\; \kappa \int w(B)\ln[w(B)/w_{\text{st}}(B)]dB$$
$$+ \sum_\alpha \lambda_\alpha \int B_\alpha w(B)dB + \lambda_0 \int w(B)dB \tag{2.3.58}$$

and write the condition for its extremum $\delta K/\delta w(B) = 0$. The latter gives

$$\kappa \ln[w(B)/w_{\text{st}}(B)] + \kappa + \lambda_\alpha B_\alpha + \lambda_0 = 0 \,, \tag{2.3.59}$$

i.e.

$$w(B) = \exp[-1 - (\lambda_\alpha B_\alpha + \lambda_0)/\kappa]w_{\text{st}}(B) \,. \tag{2.3.60}$$

The probability density obtained coincides with (2.3.6) if we put $\lambda_\alpha = -x_\alpha$. Using (2.3.13), we can find the normalization constant for the biased probability density (2.3.60):

$$w(B) = \exp\{\kappa^{-1}[\Phi(x) - \Psi(B) + Bx]\}.\tag{2.3.61}$$

The parameters x_α are obtained from the condition (2.3.57), which by (2.3.19) can be written as

$$-\partial\Phi(x)/\partial x_\alpha = A_\alpha.\tag{2.3.62}$$

It remains to make sure that the external probability density (2.3.61) corresponds exactly to the conditional minimum of the entropy (2.3.39). For this purpose, we will consider the second derivative of the functional (2.3.58)

$$\frac{\delta^2 K}{\delta w(B)\delta w(B')} = \frac{\kappa}{w(B)}\delta[w(B) - w(B')].\tag{2.3.63}$$

Since this matrix is nonnegative definite at nonnegative $w(B)$, at the above-mentioned extremal point the functional (2.3.58) has a minimum, and hence the entropy H has a conditional minimum.

Distributions of the type (2.3.61) form the hypersurface Γ in the space whose points are probability densities. Suppose that at time t_1 the probability density $w(B)$ belongs to the above-mentioned hypersurface, i.e. has the form (2.3.61) at certain x. In a time $\Delta = t_2 - t_1$ the above probability density is transformed according to the master equation and yields the probability density (2.3.45), for which the inequality (2.3.56) is valid. The new probability density $\tilde{w}(B)$ may not belong to the hypersurface Γ. Let us bring it back to this hypersurface, i.e. replace it by the probability density $w_\Delta(B)$ of the type (2.3.61) by requiring that $w_\Delta(B)$ would give the same means $\langle B_\alpha \rangle$, as $\tilde{w}(B)$:

$$\int Bw_\Delta(B)dB = \int B\tilde{w}(B)dB.\tag{2.3.64}$$

Since the probability densities from Γ have the minimal entropy H among all the probability densities with the same $\langle B_\alpha \rangle$, it is clearly seen that the entropy H_Δ of the probability density w_Δ will obey

$$H_\Delta \leq \tilde{H}.\tag{2.3.65}$$

Combining (2.3.56) and (2.3.65) gives

$$H_\Delta \leq H.\tag{2.3.66}$$

Thus the transition $w(B) \longrightarrow w_\Delta(B)$ is accompanied by a decrease in H.

We will carry out the Δ-division of the time axis and regard the sequence of changes of the probability density in accordance with the master equation and of returns (of the above type) to the hypersurface Γ as a prolonged process. In the limit as $\Delta \longrightarrow 0$ we obtain continuous motion in Γ. By (2.3.65) this motion will obey

$$dH/dt \leq 0.\tag{2.3.67}$$

In the case of the motion in Γ just described, the mean $A_\alpha = \langle B_\alpha \rangle$ will vary exactly according to (2.3.24) for the functions (2.3.25).

Substituting (2.3.61,14) into (2.3.39) gives

$$
\begin{aligned}
H &= \Phi(x) - \Phi(0) + x_\alpha \int B_\alpha w(B) dB \\
&= \Phi(x) - x \frac{\partial \Phi}{\partial x}(x) - \Phi(0)
\end{aligned}
\tag{2.3.68}
$$

by (2.3.19).

Comparing this with (2.3.27) yields

$$
H = \bar{\Psi}(A) - \Phi(0) \quad \text{for} \quad A = A(x) .
\tag{2.3.69}
$$

Consequently, (2.3.67) is just the inequality (2.3.38) obtained earlier. Thus, the use of (2.3.56), which (if the derivative exists) is equivalent to (2.3.44), gives another proof of (2.3.38).

As can be seen from the foregoing, (2.3.38) is exact and is true at any fluctuation level, provided the motion in the space of distributions proceeds in Γ as just described, for which the continual returns to Γ are required. One may well ask: is the variation of $H(t)$ with time in this motion similar to the variation of $H(t)$ in real motion when there are no returns to Γ? The difference between these variations is small, if the functions

$$
\begin{aligned}
f_\alpha^{(1)}(A) &= \int K_\alpha(B) \tilde{w}(B) dB , \quad \tilde{w}(B) \notin \Gamma , \\
f_\alpha^{(2)}(A) &= \int K_\alpha(B) w_\Delta(B) dB , \quad w_\Delta(B) \in \Gamma
\end{aligned}
\tag{2.3.70}
$$

(which are simply the right-hand sides of (2.3.24) for the various forms of treatment of this equation) are approximately equal. It is supposed in (2.3.70) that mean values $A = \int B \tilde{w}(B) dB$ and $A = \int B w_\Delta(B) dB$ for both distributions coincide. The functions (2.3.70) differ slightly if the fluctuation level is small. But if the fluctuations in a system are high — for example, it is exposed to intense external fluctuations — then the derivative in (2.3.38) may differ markedly from the derivative $\partial \Psi / \partial t$. Then the phenomenological equation (2.3.24) loses its practical value. Instead of A we should now take $w(B)$, and instead of (2.3.24), the equation $\dot{w} = F[w]$, i.e., the master equation, and instead of $\bar{\Psi}(A)$ the entropy $H[w]$, which obeys the inequality (2.3.44).

2.3.7 Entropy as a Measure of Uncertainty of the Parameters B

The Kullback entropy (2.3.39), unlike the ordinary entropy, is not a measure of uncertainty (it is a measure of the closeness of the probability densities w and w_{st}). We will now introduce the entropy

$$
S_B = -k \int w(B) \ln[w(B)] dB
\tag{2.3.71}
$$

which characterizes the uncertainty of B, or, in other words, their statistical spread.

Substituting $w(B) = \exp\{[-S + S(B)]/k\}$ into (2.3.71) gives

$$S = S_B + \langle S(B) \rangle \tag{2.3.72}$$

for the equilibrium state of closed systems. The same result can be obtained by substituting $w(B) = \exp\{[F - F(B)]/kT\}$. Equation (2.3.72) should be understood as follows: the total uncertainty in a system is equal to the sum of uncertainty of the parameters B and the mean uncertainty of the dynamic variables which remains after the B are fixed.

Returning to the stationary state of open systems, we substitute (2.3.14) into (2.3.71), to obtain

$$S_B = k\kappa^{-1}[\langle \Psi(B) \rangle - \Phi(0)] . \tag{2.3.73}$$

Suppose that the entropy (2.3.71) changes slightly owing to the variation δw of the stationary distribution caused, say, by a change in fluctuation level. The corresponding differential is

$$
\begin{aligned}
dS_B &= -k \int [\ln w(B) + 1]\delta w(B)dB \\
&= -k \int \ln[w(B)]\delta w(B)dB
\end{aligned}
\tag{2.3.74}
$$

(the integral $\int \delta w dB$ is zero because the normalization condition remains unchanged). Substituting (2.3.14) gives

$$
\begin{aligned}
dS_B &= k\kappa^{-1} \int [\Psi(B) - \Phi(0)]\delta w(B)dB \\
&= k\kappa^{-1} \int \Psi(B)\delta w(B)dB .
\end{aligned}
\tag{2.3.75}
$$

If $\Psi(B)$ is independent of the fluctuation level (which is exactly the case in the asymptotic region by virtue of (2.3.12)), then

$$
\begin{aligned}
\int \Psi(B)\delta w(B)dB &= d \int \Psi(B)w(B)dB \\
&= d\langle \Psi(B) \rangle ,
\end{aligned}
\tag{2.3.76}
$$

and, by (2.3.75),

$$dS_B = (\kappa/k)^{-1}d\langle \Psi(B) \rangle . \tag{2.3.77}$$

This equality is an analog of the well-known formula $dS = dQ/T$ in equilibrium thermodynamics, where instead of $dQ = dU$ we have $d\langle \Psi \rangle$, and instead of the temperature T we have κ/k. Moreover, (2.3.73) is analogous to $TS = U - F$.

To be sure, one should attach no great importance to this analogy. In fact, $\Psi(B)$ need not have the nature and dimensions of energy, and the entropy S_B accounts for a negligible part of the total physical entropy, because the number of parameters B_α is very small compared with the number of molecular dynamic variables q_i, p_i. The main entropy in the case of stationary nonequilibrium states moves as follows: it is produced continuously in an open system due to the nonequilibrium nature of processes going in it, and it is transferred to a heat bath (say, the environment). Concurrently, energy is transferred from external sources to the heat bath.

2.3.8 Fluctuation-Dissipation Relations

Let us represent the image (2.3.9) by a Taylor series

$$R(y,x) = \sum_{m=1}^{\infty} \sum_{n=0}^{\infty} (m!n!)^{-1} \kappa^{1-m} l_{\alpha_1 \ldots \alpha_m, \beta_1 \ldots \beta_n} y_{\alpha_1} \cdots y_{\alpha_m} x_{\beta_1} \cdots x_{\beta_n}, \quad (2.3.78)$$

where

$$l_{\alpha_1 \ldots \alpha_m, \beta_1 \ldots \beta_n} = \frac{\partial^n \kappa_{\alpha_1 \ldots \alpha_m}(x)}{\partial x_{\beta_1} \ldots \partial x_{\beta_n}} \quad \text{for} \quad x = 0, \quad (2.3.79)$$

$$\kappa_{\alpha_1 \ldots \alpha_m}(x) = \int K_{\alpha_1 \ldots \alpha_m}(B) w_x(B) dB. \quad (2.3.80)$$

Equations (2.3.78) and (2.3.79,80) are analogous to [(3.1.41) v.1] and [(4.1.6), (3.1.42) v.1].

Substituting (2.3.78) into (2.3.8) and collecting the terms of various orders in x we can readily obtain FDRs of various orders. So, isolating the terms linear in x gives

$$l_\alpha = 0. \quad (2.3.81)$$

Equating the sum of terms quadratic in x to zero, we get

$$l_{\alpha,\beta} + l_{\beta,\alpha} + \kappa^{-1} l_{\alpha\beta} = 0. \quad (2.3.82)$$

Considering the cubic terms, we obtain

$$\kappa(l_{\alpha,\beta\gamma} + l_{\beta,\alpha\gamma} + l_{\gamma,\alpha\beta}) + l_{\alpha\beta,\gamma} + l_{\beta\gamma,\alpha} l_{\gamma\alpha,\beta} + \kappa^{-1} l_{\alpha\beta\gamma} = 0. \quad (2.3.83)$$

The fourth-order terms give

$$\kappa^2 (l_{\alpha,\beta\gamma\delta} + l_{\beta,\alpha\gamma\delta} + l_{\gamma,\alpha\beta\delta} + l_{\delta,\alpha\beta\gamma})$$
$$+\kappa(l_{\alpha\beta,\gamma\delta} + l_{\alpha\gamma,\beta\delta} + l_{\alpha\delta,\beta\gamma} + l_{\beta\gamma,\alpha\delta} + l_{\beta\delta,\alpha\gamma} + l_{\gamma\delta,\alpha\beta})$$
$$+l_{\alpha\beta\gamma,\delta} + l_{\alpha\beta\delta,\gamma} + l_{\alpha\gamma\delta,\beta} + l_{\beta\gamma\delta,\alpha} + \kappa^{-1} l_{\alpha\beta\gamma\delta} = 0. \quad (2.3.84)$$

We thus end up with a set of FDRs, which obviously is not so powerful as that of Sect. 4.1.2 v.1, since now we cannot use the time-reversal symmetry.

These relations can be presented another way. We will use the expansion

$$K_{\alpha_1 \ldots \alpha_m}(B) = \sum_{s=0}^{\infty} (s!)^{-1} k_{\alpha_1 \ldots \alpha_m, \gamma_1 \ldots \gamma_s} B_{\gamma_1} \cdots B_{\gamma_s} \quad (2.3.85)$$

of kinetic coefficients at the point where the probability density (2.3.3) is maximal, i.e., at the point where the function $\Psi(B)$ is minimal (this point is assumed to be the origin of coordinates). If now we average (2.3.85) with the probability density $w_x(B)$ using (2.3.80) and take into account (2.3.20), we can express $\kappa_{\alpha_1 \ldots \alpha_m}(x)$ in terms of derivatives of $\Phi(x)$. For example

$$\kappa_\alpha(x) = k_\alpha - k_{\alpha,\gamma}\frac{\partial\Phi(x)}{\partial x_\gamma} + \tfrac{1}{2}k_{\alpha,\gamma\delta}\left(\frac{\partial\Phi}{\partial x_\gamma}\frac{\partial\Phi}{\partial x_\delta} - \kappa\frac{\partial^2\Phi}{\partial x_\gamma\partial x_\delta}\right) + \dots, \qquad (2.3.86a)$$

$$\kappa_{\alpha\beta}(x) = k_{\alpha\beta} - k_{\alpha\beta,\gamma}\frac{\partial\Phi(x)}{\partial x_\gamma} + \dots. \qquad (2.3.86b)$$

Differentiating these expressions at zero in accordance with (2.3.79) gives

$$
\begin{aligned}
l_{\alpha,\beta} &= -k_{\alpha,\gamma}\frac{\partial^2\Phi}{\partial x_\gamma\partial x_\beta} \\
&\quad + \tfrac{1}{2}k_{\alpha,\gamma\delta}\left(2\frac{\partial\Phi}{\partial x_\gamma}\frac{\partial^2\Phi}{\partial x_\beta\partial x_\delta} - \kappa\frac{\partial^3\Phi}{\partial x_\beta\partial x_\gamma\partial x_\delta}\right) + \dots, \\
l_{\alpha,\beta_1\beta_2} &= -k_{\alpha,\gamma}\frac{\partial^3\Phi}{\partial x_{\beta_1}\partial x_{\beta_2}\partial x_\gamma} \\
&\quad + \tfrac{1}{2}k_{\alpha,\gamma\delta}\left(2\frac{\partial^2\Phi}{\partial x_{\beta_1}\partial x_\gamma}\frac{\partial^2\Phi}{\partial x_{\beta_2}\partial x_\delta} + \dots\right) + \dots \qquad (2.3.87)
\end{aligned}
$$

and so on. The derivatives of $\Phi(x)$ at zero that enter these expressions can, by (2.3.17), be expressed through one-time stationary correlators

$$\langle B_{\alpha_1}(t), \dots, B_{\alpha_m}(t)\rangle \equiv \mu_{\alpha_1\dots\alpha_m}. \qquad (2.3.88)$$

The value $\langle B_\alpha\rangle = \mu_\alpha$ is of the order of κ if the origin of coordinates is placed at the maximum of probability, as is seen from [(A4.9) v.1]. Keeping only the main terms, we can obtain from (2.3.86,87)

$$
\begin{aligned}
l_{\alpha,\beta} &= \kappa^{-1}k_{\alpha,\gamma}\mu_{\gamma\beta}, \\
l_{\alpha\beta} &= k_{\alpha\beta}, \\
l_{\alpha,\beta_1\beta_2} &= \kappa^{-2}(k_{\alpha,\gamma}\mu_{\gamma\beta_1\beta_2} + k_{\alpha,\gamma\delta}\mu_{\gamma\beta_1}\mu_{\delta\beta_2}) \\
l_{\alpha\beta,\delta} &= \kappa^{-1}k_{\alpha\beta,\gamma}\mu_{\gamma\delta}, \\
l_{\alpha\beta\delta} &= k_{\alpha\beta\delta}. \qquad (2.3.89)
\end{aligned}
$$

Therefore, relations (2.3.82,83) now become

$$k_{\alpha,\gamma}\mu_{\gamma\beta} + k_{\beta,\gamma}\mu_{\gamma\alpha} = -k_{\alpha\beta}, \qquad (2.3.90)$$

$$
\begin{aligned}
&k_{\alpha,\delta}\mu_{\delta\beta\gamma} + k_{\beta,\delta}\mu_{\delta\alpha\gamma} + k_{\gamma,\delta}\mu_{\delta\alpha\beta} + k_{\alpha,\rho\sigma}\mu_{\rho\beta}\mu_{\sigma\gamma} \\
&\quad + k_{\beta,\rho\sigma}\mu_{\rho\alpha}\mu_{\sigma\gamma} + k_{\gamma,\rho\sigma}\mu_{\rho\alpha}\mu_{\sigma\beta} + k_{\alpha\beta,\delta}\mu_{\delta\gamma} \\
&\quad + k_{\beta\gamma,\delta}\mu_{\delta\alpha} + k_{\alpha\gamma,\delta}\mu_{\delta\beta} + k_{\alpha\beta\delta} = 0. \qquad (2.3.91)
\end{aligned}
$$

It is remarkable that these do not include the parameter κ that indicates the level of fluctuations. It is to be noted, however, that these relations are approximate, whereas (2.3.82,83) are exact.

The relations obtained above relate three objects:

1. the coefficients of expansion of the function that enters into the right-hand side of the phenomenological equation (2.3.24) and which, in the first approximation, coincides with $K_1(A)$, so that

$$f_\alpha(A) \approx k_{\alpha,\beta} A_\beta + \frac{1}{2} k_{\alpha,\beta\gamma} A_\beta A_\gamma + \dots \tag{2.3.92}$$

2. the correlators (2.3.88) that correspond to the one-time stationary probability density, and

3. nonstationary fluctuational characteristics $k_{\alpha\beta}, k_{\alpha\beta,\gamma}, k_{\alpha\beta\gamma}, \dots$ that describe the properties of the random inputs ξ in the Langevin equation

$$\dot{B} = K_1(B) + \xi(B,t) \,. \tag{2.3.93}$$

As follows from (2.3.90,91), knowing $k_{\alpha,\gamma}$ and $\mu_{\gamma\beta}$, we can find $k_{\alpha\beta}$, but unlike the case of closed systems, knowing $k_{\alpha,\gamma}, k_{\alpha,\beta\gamma}, \mu_{\alpha\beta}, \mu_{\alpha\beta\gamma}$, we cannot determine $k_{\alpha\beta,\gamma}, k_{\alpha\beta\gamma}$ in a unique manner.

2.3.9 FDRs in the Non-Markov Case

Instead of (2.3.24), for the non-Markov process $B(t)$ we will have the equation with delay

$$\begin{aligned}
\dot{B}_\alpha(t) &= \chi_\alpha[B(t)] \\
&= \int \chi_{\alpha,\beta}(t;t_1) B_\beta(t_1) dt_1 \\
&\quad + \frac{1}{2} \int \chi_{\alpha,\beta\gamma}(t;t_1,t_2) B_\beta(t_1) B_\gamma(t_2) dt_1 dt_2 + \dots
\end{aligned} \tag{2.3.94}$$

and the corresponding Langevin equation

$$\dot{B}_\alpha(t) = \chi_\alpha[B(t)] + \zeta_\alpha(t) \,, \tag{2.3.95}$$

where ζ_α are random functions with zero mean. In this case, the linear relationship that is a generalization of (2.3.90) will have the form

$$\chi_{\alpha,\gamma}(t,t') \mu_{\gamma\beta} + \chi_{\beta,\gamma}(t,t') \mu_{\gamma\alpha} = -\langle \zeta_\alpha(t), \zeta_\beta(t') \rangle_0 \tag{2.3.96}$$

or

$$\Phi_{1,2} + \Phi_{2,1} = \Phi_{12} \,, \tag{2.3.97}$$

where

$$\begin{aligned}
\Phi_{\alpha,\beta}(t,t') &= -\chi_{\alpha,\gamma}(t,t') \mu_{\gamma\beta} \,, \\
\langle \zeta_1, \zeta_2 \rangle_0 &= \Phi_{12} \,.
\end{aligned} \tag{2.3.98}$$

When B_α is a component of the combined Markov process (B,C) and when the stationary probability density $w(B,C)$ can be represented as the product

$$w(B,C) = w(B)w(C) \,, \tag{2.3.99}$$

equation (2.3.97) can be proved by the method used in Sect. 5.1.3 v.1. In fact, here [(5.1.8) v.1] hold at $u_{ij} = \kappa^{-1}\mu_{ij}^{-1}$ and also $\mu_{\alpha\sigma} = \mu_{\sigma\alpha} = 0$, by (2.3.99). Therefore, we can here apply the reasoning of Sect. 5.1.3 v.1, but now, instead of [(5.1.31) v.1], we should use the formula (2.3.90) written for the combined process.

Further, for the above case, we can use the method applied in Sect. 5.1.5 v.1 (of course by more unwieldy calculations) to prove for the combined process, using (2.3.91), the relation

$$\Phi_{123} - \Phi_{12,3} - \Phi_{13,2} - \Phi_{23,1} + \Phi_{1,23} + \Phi_{2,13} + \Phi_{3,12} = 0\,, \qquad (2.3.100)$$

which is of the type [(5.1.108) v.1]. Here

$$\Phi_{1,23} = -\chi_{\alpha_1,\gamma}(t_1,t_2)\delta(t_{23})\mu_{\gamma\alpha_1\alpha_3} - \chi_{\alpha_1,\gamma\delta}(t_1,t_2,t_3)\mu_{\gamma\alpha_2}\mu_{\delta\alpha_3}\,, \qquad (2.3.101a)$$

$$\Phi_{123} = -\langle\zeta_1,\zeta_2,\zeta_3\rangle_0\,, \qquad (2.3.101b)$$

$$\Phi_{12,3} = \frac{\delta\langle\zeta_1,\zeta_2\rangle_B}{\delta B_\gamma(t_3)}\mu_{\gamma\alpha_3} \qquad (2.3.101c)$$

for $B = 0$. The functional derivative that enters the last equation is understood in the same sense as in Sect. 5.1.6 v.1, i.e. the function $\zeta_\alpha(B,t)$ is taken as some functional expression $F_\alpha[t,\xi(t),B(t)]$, say,

$$\zeta_\alpha(t) = \sum_\sigma \left[\int S_{\alpha\beta}^{(\sigma)}(t;t')\xi_\beta^{(\sigma)}(t')dt' \right.$$
$$\left. + \int S_{\alpha\beta\gamma}^{(\sigma)}(t;t',t'')\xi_\beta^{(\sigma)}(t')B_\gamma(t'')dt'dt'' \right]\,, \qquad (2.3.102)$$

where $\xi^{(\sigma)}$ are random functions with zero mean, and $B_\gamma(t)$ are independently specified argument functions. The expression (2.3.102) has not yet been substituted into (2.3.95), and so the correlations between $B(t)$ and $\xi^{(\sigma)}(t)$ have not yet been established. To the expression (2.3.102) corresponds the correlator

$$\langle\zeta_{\alpha_1}(t_1),\zeta_{\alpha_2}(t_2)\rangle_B = \langle F_{\alpha_1}[t_1,\xi,B]F_{\alpha_2}[t_2,\xi,B]\rangle_B \qquad (2.3.103)$$

(the averaging is over ξ at fixed $B(t)$). The functional derivative in (2.3.101c) is the derivative of this mean

$$\frac{\delta\langle\zeta_1,\zeta_2\rangle_B}{\delta B_3} = \frac{\delta}{\delta B_3}\langle F_{\alpha_1}[t_1,\xi,B]F_{\alpha_2}[t_2,\xi,B]\rangle_B\,. \qquad (2.3.104)$$

Relation (2.3.100) is a non-Markov generalization of (2.3.91).

2.4 Methods of Calculating Correlators Near Nonequilibrium Kinetic Phase Transitions in the Markov Case

2.4.1 Equations for Derivatives of Quasi-Frenergy

If the coefficients of the master equation for the Markov process $B(t)$ are known, then the relations (2.3.90,91) and others enable one-time stationary correlators

(2.3.88) to be determined. So, by solving (2.3.90) we can find $\mu_{\alpha\beta}$. Using the obtained values of $\mu_{\alpha\beta}$ we can then, from (2.3.91), find $\mu_{\alpha\beta\gamma}$, and so on. In so doing, we will at all times have to solve a linear set of equations. We should take into account, however, that the above-mentioned equations are approximate and their accuracy diminishes quickly as points of nonequilibrium phase transitions are approached. These points will be considered later in the book. At such critical points and close to them equations (2.3.90,91) and others are no longer valid. To arrive at equations suitable near the phase transitions it would be advisable to go over to equations for derivatives of the quasi-frenergy $\Psi(B)$.

To introduce these, we will first derive the equations for the derivatives

$$\varphi_{\alpha_1...\alpha_s} = \partial^s \Phi(x)/\partial x_{\alpha_1} \ldots \partial x_{\alpha_s} \tag{2.4.1}$$

of the function (2.3.13) at $x = 0$. Keeping the main terms in (2.3.87), we have

$$l_{\alpha,\beta} = -k_{\alpha,\gamma}\varphi_{\gamma\beta}\,,$$
$$l_{\alpha,\beta\gamma} = -k_{\alpha,\sigma}\varphi_{\sigma\beta\gamma} + k_{\alpha,\sigma\tau}\varphi_{\sigma\beta}\varphi_{\tau\gamma} \tag{2.4.2}$$

and also

$$l_{\alpha\beta} = k_{\alpha\beta}\,,$$
$$l_{\alpha\beta,\gamma} = -k_{\alpha\beta,\sigma}\varphi_{\sigma\gamma}\,,$$
$$l_{\alpha\beta\gamma} = k_{\alpha\beta\gamma} \tag{2.4.3}$$

by equations of the type (2.3.86). Substitution of these expressions into (2.3.82,83) gives

$$k_{\alpha,\sigma}\varphi_{\sigma\beta} + k_{\beta,\sigma}\varphi_{\sigma\alpha} = \kappa^{-1}k_{\alpha\beta}\,, \tag{2.4.4}$$

$$-3\{k_{\alpha,\sigma}\varphi_{\sigma\beta\gamma}\}_{\mathrm{sym}} + 3\{k_{\alpha,\sigma\tau}\varphi_{\sigma\beta}\varphi_{\tau\gamma}\}_{\mathrm{sym}}$$
$$-3\kappa^{-1}\{k_{\alpha\beta,\sigma}\varphi_{\sigma\gamma}\}_{\mathrm{sym}} + \kappa^{-2}k_{\alpha\beta\gamma} = 0\,. \tag{2.4.5}$$

The subscript 'sym' here denotes symmetrization over those subscripts over which summation is not carried out, say

$$\{k_{\alpha,\sigma}\varphi_{\sigma\beta\gamma}\}_{\mathrm{sym}} = \tfrac{1}{3}(k_{\alpha,\sigma}\varphi_{\sigma\beta\gamma} + k_{\beta,\sigma}\varphi_{\sigma\alpha\gamma} + k_{\gamma,\sigma}\varphi_{\sigma\alpha\beta})\,. \tag{2.4.6}$$

Let us now introduce the derivatives

$$\psi_{\alpha_1...\alpha_s} = \partial^s \Psi(B)/\partial B_{\alpha_1} \ldots \partial B_{\alpha_s} \tag{2.4.7}$$

of the quasi-frenergy at $B = 0$, where the function has a minimum. Subsequently, we will consider only several lower derivatives from the large number of derivatives (2.4.7) ($s \leq 4$).

At small κ we can assume that $\Psi(B)$ and $\Phi(B)$ are related by the Legendre transformation

$$\Phi(x) = \Psi(B(x)) - B(x)x \qquad (x(B) = \partial\Psi(B)/\partial B)\,. \tag{2.4.8}$$

Asymptotically this relation differs slightly from the exact relation (2.3.13). It follows from the Legendre transformation (2.4.8) that the functions

$$\partial \Psi(B)/\partial B_\alpha = x_\alpha \,,$$
$$-\partial \Phi(x)/\partial x_\alpha = B_\alpha \qquad (2.4.9)$$

are mutually inverse. We will represent the functions in (2.4.9) as a Taylor expansion using (2.4.1,7). We thus obtain the following mutually inverse expansions:

$$\psi_{\alpha\beta}B_\beta + \tfrac{1}{2}\psi_{\alpha\beta\gamma}B_\beta B_\gamma + \ldots = x_\alpha \,, \qquad (2.4.10)$$

$$\psi_{\alpha\beta}x_\beta + \tfrac{1}{2}\varphi_{\alpha\beta\gamma}x_\beta x_\gamma + \ldots = -B_\alpha \,. \qquad (2.4.11)$$

We write (2.4.10) as

$$B_\beta = \psi_{\beta\alpha}^{-1}(x_\alpha - \tfrac{1}{2}\psi_{\alpha\sigma\tau}B_\sigma B_\tau - \ldots) \,. \qquad (2.4.12)$$

After iterations, we get

$$B_\alpha = \psi_{\alpha\rho}^{-1}(x_\rho - \tfrac{1}{2}\psi_{\rho\sigma\tau}\psi_{\sigma\beta}^{-1}\psi_{\tau\gamma}^{-1}x_\beta x_\gamma - \ldots) \,. \qquad (2.4.13)$$

Comparing (2.4.13) with (2.4.11) gives

$$\| \varphi_{\alpha\beta} \| = - \| \psi_{\alpha\beta} \|^{-1} \,,$$
$$\varphi_{\alpha\beta\gamma} = \psi_{\alpha\rho}^{-1}\psi_{\rho\sigma\tau}\psi_{\sigma\beta}^{-1}\psi_{\tau\gamma}^{-1} \,. \qquad (2.4.14)$$

Now we substitute (2.4.14) into (2.4.4,5) to obtain

$$\psi_{\alpha\gamma}k_{\gamma,\beta} + \psi_{\beta\gamma}k_{\gamma,\alpha} = -\psi_{\alpha\gamma}k_{\gamma\delta}^0\psi_{\delta\beta} \,, \qquad (2.4.15)$$

$$-3\{k_{\alpha,\sigma}\psi_{\sigma\rho}^{-1}\psi_{\rho\tau\pi}\psi_{\tau\beta}^{-1}\psi_{\pi\gamma}^{-1}\}_{\text{sym}} \;+\; 3\{k_{\alpha,\sigma\tau}\psi_{\sigma\beta}^{-1}\psi_{\tau\gamma}^{-1}\}_{\text{sym}}$$
$$+\; 3\{k_{\alpha\beta,\sigma}^0\psi_{\sigma\gamma}^{-1}\}_{\text{sym}} + k_{\alpha\beta\gamma}^0 = 0 \,, \qquad (2.4.16)$$

where

$$k_{\alpha_1 \ldots \alpha_m, \beta_1 \ldots \beta_n}^0 = \kappa^{1-m}k_{\alpha_1 \ldots \alpha_m, \beta_1 \ldots \beta_n} \,. \qquad (2.4.17)$$

Multiplying (2.4.16) from the left by $\| \psi_{\alpha\beta} \|$, we find

$$3\{\psi_{\alpha\mu}k_{\mu,\nu}\psi_{\nu\lambda}^{-1}\psi_{\lambda\beta\gamma}\}_{\text{sym}} = 3\{\psi_{\alpha\mu}k_{\mu,\beta\gamma}\}_{\text{sym}}$$
$$+3\{\psi_{\alpha\mu}\psi_{\beta\nu}k_{\mu\nu,\gamma}^0\}_{\text{sym}} + \psi_{\alpha\mu}\psi_{\beta\nu}\psi_{\gamma\lambda}k_{\mu\nu\lambda}^0 \,. \qquad (2.4.18)$$

Using the above technique, we can obtain, from (2.3.84), the third equation

$$4\{\psi_{\alpha\mu}k_{\mu,\nu}\psi_{\nu\lambda}^{-1}\psi_{\lambda\beta\gamma\delta}\}_{\text{sym}} = 4\{\psi_{\alpha\mu}k_{\mu,\beta\gamma\delta}\}_{\text{sym}}$$
$$+6\{k_{\mu,\alpha\beta}\psi_{\mu\gamma\delta}\}_{\text{sym}} + 3\{\psi_{\alpha\beta\mu}k_{\mu\nu}^0\psi_{\nu\gamma\delta}\}_{\text{sym}} + 12\{\psi_{\alpha\mu}k_{\mu\nu,\beta}^0\psi_{\nu\gamma\delta}\}_{\text{sym}}$$
$$+6\{\psi_{\alpha\mu}\psi_{\beta\nu}k_{\mu\nu,\gamma\delta}^0\}_{\text{sym}} + 6\{\psi_{\alpha\mu}\psi_{\beta\nu}k_{\mu\nu\lambda}^0\psi_{\lambda\gamma\delta}\}_{\text{sym}}$$
$$+4\{\psi_{\alpha\mu}\psi_{\beta\nu}\psi_{\gamma\lambda}k_{\mu\nu\lambda,\delta}^0\}_{\text{sym}} + \psi_{\alpha\kappa}\psi_{\beta\lambda}\psi_{\gamma\mu}\psi_{\delta\nu}k_{\kappa\lambda\mu\nu}^0 \qquad (2.4.19)$$

where (2.4.18) has been used among other formulas. Here in each term the number before the braces coincides with the minimum number of terms needed to symmetrize the expression within the braces in the subscripts $\alpha, \beta, \gamma, \delta$.

It is to be noted that the equations obtained can also be derived from (2.3.12).

Note that in the case of the Fokker-Planck equation and when the potentiality conditions

$$\partial(K_{\alpha\gamma}^{-1}K_\gamma)/\partial B_\beta = \partial(K_{\beta\gamma}^{-1}K_\gamma)/\partial B_\alpha \qquad (2.4.20)$$

are satisfied, it is expedient to seek the function $\Psi(B)$ and derivatives (2.4.7) not from (2.4.15,18,19), but from equation (2.3.12), which assumes the form

$$\left(K_\alpha + \frac{1}{2\kappa}K_{\alpha\beta}\frac{\partial\Psi}{\partial B_\beta}\right)\frac{\partial\Psi}{\partial B_\alpha} \doteq 0. \qquad (2.4.21)$$

If the potentiality condition is fulfilled, the expression in parentheses vanishes, and we get

$$\partial\Psi(B)/\partial B_\alpha = -2\kappa K_{\alpha\beta}^{-1}K_\beta(B). \qquad (2.4.22)$$

From this we can readily find the derivatives (2.4.7) and, by contour integration, the function $\Psi(B)$ as well. If, in addition, $K_{\alpha\beta}$ is independent of B, then we see that in the nonlinear region the Onsager equation $\langle\dot{B}_\alpha\rangle \equiv K_\alpha = -L_{\alpha\beta}\partial\Psi(B)/\partial B_\beta$ is valid at $L_{\alpha\beta} = (2\kappa)^{-1}K_{\alpha\beta}$. In the general case the simplification just described does not apply.

2.4.2 Multistability. Nonequilibrium Kinetic Phase Transitions

Suppose that the operator of the master equation, and hence the kinetic potential as well, are dependent on the external parameter, or parameters, Θ. The function $\Psi(B)$, which obeys (2.3.12), i.e. the equation

$$V(\partial\Psi/\partial B, B, \Theta) = 0, \qquad (2.4.23)$$

will then be dependent on Θ. Here Θ can be, say, the external forces h or external fluxes J^{ex}.

Suppose, at a fixed Θ, that the function $\Psi(B, \Theta)$ has isolated minima (perhaps local) at the points $B_1^0, \ldots, B_s^0$, which naturally vary with Θ and which are called stable points. Then, if the minima do not lie at the boundary of possible values B, the following equations must hold

$$\partial\Psi(B, \Theta)/\partial B = 0 \quad \text{for} \quad B = B_i^0(\Theta) \quad i = 1, \ldots, s. \qquad (2.4.24)$$

Generally, the minimality condition coincides with the positive definiteness condition for the second-derivative matrix

$$\partial^2\Psi(B, \Theta)/\partial B_\alpha\partial B_\beta = \text{positive definite}. \qquad (2.4.25)$$

If there are several stable points, this situation is referred to as multistability.

When there is one minimum point, according to the results of Appendix A1, v.1 $\Psi(B, \Theta)$ is asymptotically close to (2.3.27). By the H-theorem considered in Sect. 2.3.4 it cannot increase; moreover, it decreases in nontrivial cases. Consequently, the deviations of $B(t) - B^0(\Theta)$ from the minimum decrease with time. Therefore, $\Psi(B, \Theta)$ plays the role of the Lyapunov function, and its minimum is stable. This accounts for the term "stable point" for the minimum point $B^0(\Theta)$.

In the case of several minimum points these arguments do not hold good, since $\Psi(B)$ now differs markedly from (2.3.27) even at small κ. In order that the above reasoning may be extended to include this case, we should draw demarcation lines or "watersheds" between the minimum points and break up the space of values of B into regions $E_1, \ldots, E_s$, each of which would only contain one minimum point. Further, we should introduce the conditional functions

$$\Phi_i(x) = -\kappa \ln \int_{E_i} \exp[(Bx - \Psi(B))/\kappa]dB, \quad i - 1, \ldots, s. \tag{2.4.26}$$

It is easily seen that the unconditional function (2.3.13) is related to the conditional ones by

$$\exp[-\Phi(x)/\kappa] = \sum_i \exp[-\Phi_i(x)/\kappa]. \tag{2.4.27}$$

Let us introduce the conditional stationary probability density $w_{\mathrm{st}}(B \mid E_i)$, which corresponds to the condition that a random point B would appear in E_i. It is given by the relationship

$$w_{\mathrm{st}}(B \mid E_i) = \exp\{[\Phi_i(0) - \Psi(B)]/\kappa\} \quad \text{for } B \in E_i, \tag{2.4.28}$$

(and $w_{\mathrm{st}}(B \mid E_i) = 0$ outside E_i) which is similar to (2.3.14). In analogy with (2.3.27) we can introduce the conditional functions

$$\bar{\Psi}_i(B) = \Phi_i(x(B)) + Bx(B), \tag{2.4.29}$$

where

$$B(x) = -\partial \Phi_i(x)/\partial x.$$

The process of relaxation of the nonstationary probability density $w(B)$ to a stationary one is characterized by two time constants. The first constant τ_1 characterises the time during which the nonstationary conditional probability density $w(B \mid E_i)$ goes over into the conditional stationary distribution (2.4.28). This transition occurs much faster than the second process, characterized by τ_2, i.e. the process by which the nonstationary probabilities

$$P_i = \int_{E_i} w(B)dB \tag{2.4.30}$$

corresponding to the event $B \in E_i$ tend to the appropriate stationary probabilities

$$(P_i)_{\mathrm{st}} = \int_{E_i} w_{\mathrm{st}}(B)dB = \exp[(\Phi(0) - \Phi_i(0))/\kappa]. \tag{2.4.31}$$

We have used here (2.3.14) and (2.4.26). The different duration of these processes is expressed by the inequality $\tau_1 \ll \tau_2$; the smaller κ, the greater is the ratio τ_2/τ_1 since at small κ jumps between various minima are very rare.

The fast process of changing $w(B \mid E_i)$ into (2.4.28) proceeds as if there were no crossing the demarcation lines, i.e. the boundaries of E_i. Almost nothing would be changed if we install impermeable walls. With these walls we can apply to the conditional function (2.4.29) the H-theorem of Sect. 2.3.4, according to which $d\bar{\Psi}_i(A)/dt \leq 0$. Further, for the conditional function (2.4.26) we can go through the procedure given in Appendix A1, v.1, and so we can prove that difference between $\bar{\Psi}_i(A)$ and $\Psi(A)$ is small for $A \in E_i$. It follows that $\Psi(A)$, $A \in E_i$, is a Lyapunov function for processes occurring in E_i and having the time-scale τ_1, so that the point B_i^0 of the minimum of $\Psi(A)$ for $A \in E_i$, is stable.

We will now change the parameter(s) Θ. In the process, stable points may shift. But other, anomalous changes of stable points are also possible: stable points may vanish, appear, or merge; isolated points may become nonisolated, and so forth. Such anomalous changes are said to be nonequilibrium kinetic phase transitions, and the values of the parameters at which such a phase transition occurs are called critical and denoted by Θ_c. In a phase transition the condition of positive definiteness (2.4.25) may be violated, i.e. for at least one point B_i^0

$$\frac{\partial \Psi}{\partial B_\alpha}(B^0(\Theta_c), \Theta_c) = 0, \tag{2.4.32a}$$

$$\det \left\| \frac{\partial^2 \Psi}{\partial B_\alpha \partial B_\beta}(B^0(\Theta_c), \Theta_c) \right\| = 0 \tag{2.4.32b}$$

must hold simultaneously, whereas in an arbitrarily small neighborhood $\mid \Theta - \Theta_c \mid < \varepsilon$ of the critical value Θ_c the formulas (2.4.24,25) hold. Equation (2.4.32b) means that at least one eigenvalue of the matrix of second derivatives becomes zero.

Of course, the above definition (2.4.32) of the phase transition is not the most general one because the presence, emergence or disappearance of a minimum may be determined by the behavior of higher derivatives when the second derivative matrix is zero. This definition corresponds to a simple phase transition. Within the concept of a simple transition we will also include the requirement that the diffusion matrix $k_{\alpha\beta}$ is nondegenerate.

We will say that a phase transition is of the first kind if the minimal relative height of a "watershed" determined by $\Psi(B, \Theta)$ and measured from the lowest (in E_i) point B_i^0 tends to zero as $\Theta \longrightarrow \Theta_c$. The minimal relative height of the "watershed" measured from the other side must not tend to zero as $\Theta \longrightarrow \Theta_c$. Otherwise, the phase transition is of the second kind.

It can be easily verified that if all $\psi_{\alpha\beta\gamma}^c = 0$ and simultaneously $\psi_{\alpha\beta\gamma\delta}^c a_\alpha a_\beta a_\gamma a_\delta > 0$ for any nonzero vectors $\mathbf{a}$, where $\psi_{\alpha\beta\gamma} = \psi_{\alpha\beta\gamma}(\Theta_c)$, $\psi_{\alpha\beta\gamma\delta}^c = \psi_{\alpha\beta\gamma\delta}(\Theta_c)$, we have a second-order transition. If some of $\psi_{\alpha\beta\gamma}^c \neq 0$ or $\psi_{\alpha\beta\gamma\delta}^c a_\alpha a_\beta a_\gamma a_\delta < 0$ at least for one nonzero vector $\mathbf{a}$, then we have a phase transition of the first kind. If all $\psi_{\alpha\beta\gamma}^c = 0$ and all $\psi_{\alpha\beta\gamma\delta}^c = 0$, then the type of a phase transition is determined by the higher derivatives.

2.4.3 Solution of the Equations for $\varphi_{\alpha\beta}, \psi_{\alpha\beta}$

In the case of multistability, equations (2.4.4,15,18,19) are valid separately for each stable point B_i^0. In that case, $\psi_{\alpha_1 \dots \alpha_m}$ are derivatives (2.4.7) at $B = B_i^0$, and $\parallel \varphi_{\alpha\beta} \parallel$ $= - \parallel \psi_{\alpha\beta} \parallel^{-1}$. We are only interested in those stable points that correspond to a phase transition, i.e. ones where, at critical values of the parameters, equations (2.4.32) hold.

Consider first the equations (2.4.4,15) of the linear-Gaussian approximation. We refer to it in this way because the matrix $k_{\alpha,\beta}$ specifies the linear equation

$$\dot{A}_\alpha = k_{\alpha,\beta}(A_\beta - B_\beta^0) \tag{2.4.33}$$

(see (2.3.24,92)) and because in the approximation $\Psi(B) = \psi_{\alpha\beta}(B_\alpha - B_\alpha^0)(B_\beta - B_\beta^0)/2$ the probability density (2.3.3) becomes Gaussian. Instead of (2.4.15), it is convenient to deal with equation (2.4.4) since it is linear. Denoting

$$- \parallel k_{\alpha,\beta} \parallel \; = \hat{A}, \; \parallel k_{\alpha\beta}^0 \parallel \; = \hat{N}, \; - \parallel \varphi_{\alpha\beta} \parallel \; = \hat{H}, \tag{2.4.34}$$

we can represent (2.4.4) in the matrix form:

$$\hat{A}\hat{H} + \hat{H}\hat{A}^T = \hat{N}. \tag{2.4.35}$$

Note that if we carry out the transformation

$$\hat{A}' = \hat{N}^{-1/2}\hat{A}\hat{N}^{1/2}$$
$$\hat{H}' = \hat{N}^{-1/2}\hat{H}\hat{N}^{-1/2} \tag{2.4.36}$$

then (2.4.35) becomes

$$\hat{A}'\hat{H}' + \hat{H}'(\hat{A}')^T = \hat{I}, \tag{2.4.37}$$

where $\hat{I}$ is the identity matrix. Such a transformation can be performed when the matrix $\hat{N}$ is nondegenerate, and, because it is positive definite, we can select $\hat{N}^{1/2}$ so that it is real and positive definite.

Let α_i be the eigenvalues of the matrix $\hat{A}$, and hence of $\hat{A}'$, and let λ_i be the eigenvalues of the matrix $\hat{H}$. If λ_i' are the eigenvalues of $\hat{H}'$, then by the inertia law of linear algebra, the $\{\lambda_i\}$ and $\{\lambda_i'\}$ each comprise exactly the same number of positive and negative values, and zeros (these numbers are real because $\hat{H}$ and $\hat{H}'$ are symmetric).

By the *Taussky* theorem [2.6] (see also [2.7]), subject to

$$\alpha_i + \alpha_k^* \neq 0 \quad \text{for any } i,k = 1,\dots,r \tag{2.4.38}$$

equation (2.4.37), and hence (2.4.35), has a unique solution, and among λ_i' and Re α_i, and hence among λ_i and Re α_i, the number of positive values is the same. Since the matrix $\hat{A}$ is real, the number α_k^* is an eigenvalue of that matrix provided α_k is also one of eigenvalues. Therefore, the conditions (2.4.38) of the theorem can be replaced by

$$\alpha_i + \alpha_k \neq 0 \quad \text{for any } i,k = 1,\dots,r. \tag{2.4.39}$$

It follows that if the point B^0 is stable for the motion (2.4.33), so that all Re $\alpha_i > 0$ and if condition (2.4.38) is fulfilled, the matrices H and $\| \psi_{\alpha\beta} \| = H^{-1}$ are positive definite. Consequently, the condition (2.4.25) is met. We see that the earlier definition of stability does not contradict a direct definition of the stability of motion described by (2.4.33).

Let us clarify the need for condition (2.4.38). Suppose that the matrix A can be reduced to diagonal form by the unitary transformation U:

$$U^+ A U = \| \alpha_i \delta_{ik} \| \tag{2.4.40}$$

(moreover $U^+ A^+ U = \| \alpha_i^* \delta_{ik} \|$). Then, after the transformation to the new representation, (2.4.35) becomes

$$\alpha_i h_{ik} + h_{ik} \alpha_k^* = n_{ik} \quad (\| h_{ik} \| = U^+ H U). \tag{2.4.41}$$

Solving this equation gives

$$h_{ik} = n_{ik}/(\alpha_i + \alpha_k^*). \tag{2.4.42}$$

If the condition (2.4.38) is not satisfied, then some elements of (2.4.42) are equal to infinity for the general matrix n_{ik}, thus suggesting the absence of a solution of (2.4.35). If A is not reducible to diagonal form by unitary transformation, then it is undoubtedly reducible by it to upper triangular form. Then the equation for the elements h_{ik} of H assumes the form $D\mathbf{x} = \mathbf{e}$, where $\mathbf{x}$ is an unknown vector consisting of h_{ik}, $\mathbf{e}$ is a known vector consisting of n_{ik} and D is the upper triangular matrix with the determinant $\prod_{i,k=1}^r (\alpha_i + \alpha_k^*)$. It is clear from this that the condition (2.4.38) is necessary.

At the critical point $\Theta = \Theta_c$ the condition (2.4.38) (or (2.4.39)) is not met, so that some eigenvalues become zero or purely imaginary. Then there is no solution of (2.4.35); however, a solution of (2.4.15) must exist. This solution should be looked for in the form

$$\| \psi_{\alpha,\beta}(\Theta_c) \| = \lim_{\Theta \to \Theta_c} H^{-1}(\Theta) \tag{2.4.43}$$

where $H(\Theta)$ is the solution of (2.4.35).

To take a closer look at this limit we will assume that A is reducible to diagonal form, i.e. for some (not necessarily unitary) matrix S we have

$$SAS^{-1} = A_d \equiv \| \alpha_i \delta_{ik} \| \tag{2.4.44}$$

and, besides,

$$(S^T)^{-1} A^T S^T = A_d^T. \tag{2.4.45}$$

If now we operate on (2.4.35) with S from the left, and with S^T from the right, from the above equations, we will have

$$A_d SHS^T + SHS^T A_d = SNS^T. \tag{2.4.46}$$

The solution of this equation has the form

$$b_{ik} = c_{ik}/(\alpha_i + \alpha_k),$$ (2.4.47)

where

$$\| b_{ik} \| = SHS^T,$$
$$\| c_{ik} \| = SNS^T.$$ (2.4.48)

It exists, by (2.4.39), off the critical point. Hence,

$$H = S^{-1} \| c_{ik}/(\alpha_i + \alpha_k) \| (S^T)^{-1}.$$ (2.4.49)

From this we find the inverse matrix

$$\psi_{\alpha\beta} = \det^{-1} \| c_{ik}(\alpha_i + \alpha_k) \| A_{ik} s_{i\alpha} s_{k\beta},$$ (2.4.50)

where A_{ik} are algebraic adjuncts of matrix $\| c_{ik}/(\alpha_i + \alpha_k) \|$; $\| s_{ik} \| = S$. We see here that the limiting process (2.4.43) can readily be effected to yield a formula of the same type. In other words, (2.4.50) is also valid at $\Theta = \Theta_c$.

2.4.4 Solution of Equations Determining Higher Derivatives

Equations (2.4.18,19) have the following structure:

$$v_{\alpha\sigma}\psi_{\sigma\beta\gamma} + v_{\beta\sigma}\psi_{\sigma\alpha\gamma} + v_{\gamma\sigma}\psi_{\sigma\alpha\beta} = z_{\alpha\beta\gamma},$$ (2.4.51)

$$v_{\alpha\sigma}\psi_{\sigma\beta\gamma\delta} + v_{\beta\sigma}\psi_{\sigma\alpha\gamma\delta} + v_{\gamma\sigma}\psi_{\sigma\alpha\beta\delta} + v_{\delta\sigma}\psi_{\sigma\alpha\beta\gamma} = z_{\alpha\beta\gamma\delta},$$ (2.4.52)

where

$$v_{\alpha\sigma} = \psi_{\alpha\beta} k_{\beta,\gamma} \psi_{\gamma\sigma}^{-1},$$ (2.4.53)

and $z_{\alpha\beta...}$ are some expressions that are completely known before the equations in which they enter are solved. Note that, using (2.4.15), we can represent the matrix $v_{\alpha\sigma}$ in the form

$$v_{\alpha\sigma} = -k_{\sigma,\alpha} - \psi_{\alpha\gamma} k_{\gamma\sigma}^0.$$ (2.4.54)

Equations (2.4.51,52) form a set of linear equations, and they are to be solved in a conventional manner. In addition, equations (2.4.51,52) can be solved and examined using the method of reducing $v_{\alpha\sigma}$ to diagonal form.

Assuming that $v_{\alpha\sigma}$ is reducible to diagonal form by some transformation $r_{\gamma\alpha}$, we have

$$r_{\gamma\alpha} v_{\alpha\sigma} r_{\sigma\rho}^{-1} = v_\gamma \delta_{\gamma\rho},$$ (2.4.55)

where v_γ are eigenvalues of $v_{\alpha\sigma}$. Multiplying (2.4.51) by $r_{\lambda\alpha}, r_{\mu\beta}, r_{\nu\gamma}$ and summing over α, β, γ, by (2.4.55) we obtain

$$(v_\lambda + v_\mu + v_\nu)\psi'_{\lambda\mu\nu} = z'_{\lambda\mu\nu},$$ (2.4.56)

where

$$\psi'_{\lambda\mu\nu} = r_{\lambda\alpha}r_{\mu\beta}r_{\nu\gamma}\psi_{\alpha\beta\gamma},$$
$$z'_{\lambda\mu\nu} = r_{\lambda\alpha}r_{\mu\beta}r_{\nu\gamma}z_{\alpha\beta\gamma}. \tag{2.4.57}$$

Solving (2.4.56) gives

$$\psi'_{\lambda\mu\nu} = z'_{\lambda\mu\nu}/(v_\lambda + v_\mu + v_\nu),$$
$$\psi_{\alpha\beta\gamma} = r^{-1}_{\alpha\lambda}r^{-1}_{\beta\mu}r^{-1}_{\gamma\nu}z'_{\lambda\mu\nu}/(v_\lambda + v_\mu + v_\nu). \tag{2.4.58}$$

This solution can be used when at the critical point $v_\lambda + v_\mu + v_\nu \neq 0$ for any $\lambda, \mu, \nu = 1, \ldots, r$ or when the ratio $z'_{\lambda\mu\nu}/(v_\lambda + v_\mu + v_\nu)$ remains finite at $\Theta \longrightarrow \Theta_c$.

Reasoning along the same lines we can solve (2.4.52) and obtain

$$\psi_{\alpha\beta\gamma\delta} = r^{-1}_{\alpha\lambda}r^{-1}_{\beta\mu}r^{-1}_{\gamma\nu}r^{-1}_{\delta\sigma}z'_{\lambda\mu\nu\sigma}/(v_\lambda + v_\mu + v_\nu + v_\sigma), \tag{2.4.59}$$

where

$$z'_{\lambda\mu\nu\sigma} = r_{\lambda\alpha}r_{\mu\beta}r_{\nu\gamma}r_{\sigma\delta}z_{\alpha\beta\gamma\delta}. \tag{2.4.60}$$

For (2.4.59) to be meaningful in the critical region, we must assume either $v_\lambda + v_\mu + v_\nu + v_\sigma \neq 0$ at $\Theta = \Theta_c$ and at any $\lambda, \mu, \nu, \sigma$ or for various $\lambda, \mu, \nu, \sigma$ the ratio $z'_{\lambda\mu\nu\sigma}/(v_\lambda + v_\mu + v_\nu + v_\sigma)$ does not tend to infinity as $\Theta \longrightarrow \Theta_c$.

Consider the special case when all $\psi_{\alpha\beta}(\Theta_c) = 0$. Then the expression (2.4.54) for the critical point yields $\| v_{\alpha\sigma} \| = - \| k_{\sigma,\alpha} \| = A^T$. Hence, for $\Theta = \Theta_c$,

$$v_\alpha(\Theta_c) = \alpha_k(\Theta_c); \tag{2.4.61a}$$
$$v_\lambda + v_\mu + v_\nu = \alpha_\lambda + \alpha_\mu + \alpha_\nu$$
$$v_\lambda + v_\mu + v_\nu + v_\sigma = \alpha_\lambda + \alpha_\mu + \alpha_\nu + \alpha_\sigma \quad \text{for} \quad \Theta = \Theta_c. \tag{2.4.61b}$$

At a critical point for some eigenvalues α_i we have Re $\alpha_i = 0$. Accordingly, at certain i and k we have $\alpha_i + \alpha_k = 0$. It follows that

$$v_\lambda + v_\mu + v_\nu + v_\sigma = \alpha_\lambda + \alpha_\mu + \alpha_\nu + \alpha_\sigma = 0 \tag{2.4.62}$$

at $\Theta = \Theta_c$ and for certain values of $\lambda, \mu, \nu, \sigma$. This, however, does not necessarily mean that $z'_{\lambda\mu\nu\sigma}/(v_\lambda + v_\mu + v_\nu + v_\sigma)$ tends to infinity as $\Theta \longrightarrow \Theta_c$ since the matrix $z'_{\lambda\mu\nu\sigma}$ can also vanish at $\Theta = \Theta_c$. In many cases, the matrix $\psi_{\alpha\beta\gamma\delta}$, which is determined by (2.4.19), is finite at the critical point. Then the quantities $v_\lambda + v_\mu + v_\nu + v_\sigma$ and $z'_{\lambda\mu\nu\sigma}$ vanish simultaneously. For an example of this the reader is referred to Sect. 2.7.4 where the sum $\alpha_1 + \alpha_2 = a + b$, which becomes zero in the case of a vibrational phase transition, cancels out since it enters both the numerator and the denominator in (2.7.18).

In all likelihood the matrix $\psi_{\alpha\beta\gamma\delta}$ may also have infinite values at a critical point if the matrix is given by (2.4.19). This infinity means that one should compute $\psi_{\alpha\beta\gamma\delta}(\kappa)$, and not $[\psi_{\alpha\beta\gamma\delta}]_{\kappa=0}$ and deal with the quasi-frenergy $\Psi(B, \kappa)$, which is a function of κ as well (the dependence of Ψ on κ here is nonanalytical). Now for finding $\Psi(B, \kappa)$, and hence $\psi_{\alpha\beta\gamma\delta}(\kappa)$, we should use the exact equation

$$V\left(-\kappa\frac{\partial}{\partial B}, B\right)\exp[-\Psi(B)/\kappa] = 0 \tag{2.4.63}$$

instead of the approximate equation (2.3.12). When there are none of the above-mentioned infinities, it is expedient to apply the techniques just described, which are simpler that solving (2.4.63). After the derivatives $\psi_{\alpha\beta}, \psi_{\alpha\beta\gamma}, \psi_{\alpha\beta\gamma\delta}$ have been found, we can use the approximate probability density

$$
\begin{aligned}
w(B) \;=\; & \text{const} \cdot \exp[-\kappa^{-1}(\tfrac{1}{2}\psi_{\alpha\beta}\Delta B_\alpha \Delta B_\beta + \tfrac{1}{6}\psi^c_{\alpha\beta\gamma}\Delta B_\alpha \Delta B_\beta \Delta B_\gamma \\
& + \tfrac{1}{24}\psi^c_{\alpha\beta\gamma\delta}\Delta B_\alpha \Delta B_\beta \Delta B_\gamma \Delta B_\delta)], \quad \Delta B_\alpha = B_\alpha - B^0_\alpha \qquad (2.4.64)
\end{aligned}
$$

to work out the correlators $\langle B_{\alpha_1}, \dots, B_{\alpha_m} \rangle$ in the vicinity of a critical point. Some results derived in this way will be presented in Sects. 2.6,7.

2.4.5 On the Accuracy of the Equations (2.3.90,91)

At the beginning of Section 2.4.1 it was stated that equations (2.3.90,91) are not applicable at a critical point, and so in the critical region we should apply the equations for the derivatives of $\Psi(B)$. We will now explain this in more detail.

As follows from their derivation, equations (2.4.15,18) are equivalent to equations (2.4.4,5) if $\varphi_{\alpha\beta}, \varphi_{\alpha\beta\gamma}$ and $\psi_{\alpha\beta}, \psi_{\alpha\beta\gamma}$ are related by (2.4.14). However, formulas (2.4.14) hold when $\varphi_{\alpha\beta}, \varphi_{\alpha\beta\gamma}$ are derivatives (at the zero point) of (2.4.8). To distinguish the latter function (the Legendre transformation of $\Psi(B)$) from the exact function (2.3.13), we will denote the function (2.4.8) by $\Phi_0(x)$. Equations (2.4.4,5), on the one hand, and (2.4.15,18), on the other, are equivalent if

$$
\varphi_{\alpha\beta} = \frac{\partial^2 \Phi_0(x)}{\partial x_\alpha \partial x_\beta},
$$

$$
\varphi_{\alpha\beta\gamma} = \frac{\partial^3 \Phi_0(x)}{\partial x_\alpha \partial x_\beta \partial x_\gamma}, \quad \text{for} \quad x = 0. \qquad (2.4.65)
$$

Further, (2.3.90,91) and (2.4.4,5) are equivalent if

$$
\begin{aligned}
\mu_{\alpha_1 \dots \alpha_m} &\equiv \langle B_{\alpha_1}, \dots, B_{\alpha_m} \rangle \\
&= -\kappa^{m-1}_{\;\cdot}\varphi_{\alpha_1 \dots \alpha_m} \\
&\equiv -\kappa^{m-1}\partial^m \Phi_0(x)/\partial x_{\alpha_1} \dots \partial x_{\alpha_m}. \qquad (2.4.66)
\end{aligned}
$$

These expressions are similar to (2.3.17), but contain another function $\Phi_0(x)$. At $x = 0$ the exact correlators are given by

$$
\mu_{\alpha_1 \dots \alpha_m} = -\kappa^{m-1}\partial^m \Phi(x)/\partial x_{\alpha_1} \dots \partial x_{\alpha_m}, \qquad (2.4.67)
$$

We thus see that (2.3.90,91) are nearly equivalent to equations (2.4.15,18) if (2.4.66) holds approximately, i.e. if the difference

$$
\partial^m \Phi(x)/\partial x_{\alpha_1} \dots \partial x_{\alpha_m} - \partial^m \Phi_0(x)/\partial x_{\alpha_1} \dots \partial x_{\alpha_m} \qquad (2.4.68)
$$

at $x = 0, m = 2, 3$ is negligible. To examine this difference we draw on the results of Appendix A1, v.1 in which the correction to the function $\Phi_0(x)$ is computed:

$$
\Phi(x) - \Phi_0(x) = \frac{\kappa}{2}\ln \det \left\| \frac{\psi_{\alpha\beta}}{2\pi\kappa} \right\| + \frac{\kappa^2}{8}\psi_{\alpha\beta\gamma\delta}\psi^{-1}_{\alpha\beta}\psi^{-1}_{\gamma\delta} + O(\kappa^4). \qquad (2.4.69)
$$

This correction can be viewed as a measure of the difference $\Phi(x) - \Phi_0(x)$. Discarding in (2.4.69) the term of the order of κ^2 and differentiating yields

$$\frac{\partial \Phi(x)}{\partial x_\alpha} - \frac{\partial \Phi_0(x)}{\partial x_\alpha} = \frac{\kappa}{2} \psi_{\alpha\lambda}^{-1} \psi_{\lambda\mu\nu} \psi_{\mu\nu}^{-1} + O(\kappa^2) \tag{2.4.70}$$

[see (A4.6, v.1)]. A second differentiation yields

$$\frac{\partial^2 \Phi(x)}{\partial x_\alpha \partial x_\beta} - \frac{\partial^2 \Phi_0(x)}{\partial x_\alpha \partial x_\beta}$$
$$= \frac{\kappa}{2} \psi_{\alpha\gamma}^{-1} \psi_{\beta\delta}^{-1} (\psi_{\gamma\delta\mu\nu} - \psi_{\gamma\delta\rho} \psi_{\rho\sigma}^{-1} \psi_{\sigma\mu\nu} - \psi_{\gamma\mu\rho} \psi_{\rho\sigma}^{-1} \psi_{\sigma\nu\delta}) \psi_{\mu\nu}^{-1} + O(\kappa^2) . \tag{2.4.71}$$

As a critical point is approached and $\Theta \longrightarrow \Theta_c$, the matrix $\psi_{\alpha\beta}^{-1}$ tends to infinity. Hence the difference (2.4.71) quickly tends to infinity. The difference of higher derivatives grows even faster. This suggests that the dependence of $\Phi(x, \kappa)$ on κ is not analytical at $\kappa = 0$ and $\Theta = \Theta_c$ when $\Psi(B, \kappa)$ is analytical. Accordingly, we cannot make use of (2.4.66) in the critical region, and the equations (2.3.90,91) are not applicable.

2.4.6 Two-Subscript Relations for Complex Variables

In some cases the real and imaginary parts $z_1', \ldots, z_n', z_1'', \ldots, z_n''$ of complex quantities $z_j = z_j' + iz_j''$, $j = 1, \ldots, n$ are taken as the internal parameters B_α. In those cases, the matrix $H = - \| \varphi_{\alpha\beta} \| = \kappa^{-1} \langle B_\alpha, B_\beta \rangle$, which we will denote by H_0, will have block form

$$H_0 = \begin{pmatrix} R & S \\ S^T & R' \end{pmatrix}, \tag{2.4.72}$$

where

$$R = \| \langle z_j', z_k' \rangle \| ,$$
$$R' = \| \langle z_j'', z_k'' \rangle \| ,$$
$$S = \| \langle z_j', z_k'' \rangle \| . \tag{2.4.73}$$

Block representation is also possible for the other two matrices in (2.4.35); we will denote them by A_0, N_0.

Suppose that the following two conditions are satisfied:

$$A_0 = \begin{pmatrix} A & O \\ O & A \end{pmatrix}, \tag{2.4.74}$$

$$H_0 = \begin{pmatrix} R & S \\ -S & R \end{pmatrix} . \tag{2.4.75}$$

The latter of these implies that

$$R' = R ,$$
$$S^T = -S . \tag{2.4.76}$$

It is easily seen from (2.4.35), i.e. from the equation $A_0 H_0 + H_0 A_0^T = N_0$, that the matrix N_0 has the following block form

$$N_0 = \begin{pmatrix} N_1 & N_2 \\ -N_2 & N_1 \end{pmatrix}.$$ (2.4.77)

In other words, this matrix obeys relations similar to (2.4.76). The submatrices R, N_1, S, N_2 are then related by

$$AR + RA^T = N_1,$$ (2.4.78a)
$$AS + SA^T = N_2.$$ (2.4.78b)

We could also proceed in the opposite direction: to postulate equations (2.4.74, 77) and obtain (2.4.75) from them.

In the cases (2.4.74–77) it is expedient not to consider separately the real and imaginary parts of z_j and not to use $H_0, A_0,$ and N_0. It would be pertinent to introduce instead of them the matrix $H = \kappa^{-1} \| \langle z_j, z_k^* \rangle \|$, which can be readily shown to be equal to $2R - 2iS$. Besides, it is convenient to deal with the matrix

$$N = \kappa^{-1} \| \lim_{\tau \longrightarrow 0} \tau^{-1} \langle \Delta z_j \Delta z_k^* \rangle_{z^0} \|,$$ (2.4.79)

which, as it is easily seen, is equal to $2N_1 - 2iN_2$. By virtue of (2.4.78a), these matrices are related by

$$AH + HA^T = N,$$ (2.4.80)

which is essentially the same as (2.4.35).

Note that when dealing with complex quantities the formula $\kappa \| \langle B_\alpha, B_\beta \rangle \|^{-1} = \| \partial^2 \Psi(B)/\partial B_\alpha \partial B_\beta \|$, i.e. $- \| \varphi_{\alpha\beta} \|^{-1} = \| \psi_{\alpha\beta} \|$, is replaced by

$$\kappa \| \langle z_j, z_k^* \rangle \|^{-1} = \| \frac{\partial^2 \Psi(z, z^*)}{\partial z_j^* \partial z_k} \|.$$ (2.4.81)

We conclude by giving an example where the conditions (2.4.76) are fulfilled. Consider homogeneous fluctuations $y(\mathbf{r})$ in a space of arbitrary dimensions d. By carrying out a multidimensional Fourier transformation, we introduce the random spectrum

$$z(\mathbf{k}) = (2\pi)^{-d/2} \int \exp(i\mathbf{k} \cdot \mathbf{r}) y(\mathbf{r}) d\mathbf{r}.$$ (2.4.82)

The quantities $z(\mathbf{k})$ are an example of the complex parameters z_j, the subscript j here being the vector $\mathbf{k}$. Significantly, only quantities $z(\mathbf{k})$ for

$$\mathbf{k} \cdot \mathbf{s} > 0,$$ (2.4.83)

are independent variables and can play part of z_j. Here $\mathbf{s}$ is some arbitrary nonzero vector. The other quantities $z(\mathbf{k})$, $\mathbf{k} \cdot \mathbf{s} < 0$, are expressed via the first ones by the relation $z(-\mathbf{k}) = z^*(\mathbf{k})$, which is valid because the function $y(\mathbf{r})$ is real.

To prove (2.4.76) in this case, we will use the formula

$$\langle z(\mathbf{k}_1) z^T(\mathbf{k}_2) \rangle = G(\mathbf{k}_1) \delta(\mathbf{k}_1 + \mathbf{k}_2),$$ (2.4.84)

which is of the same type as [(A6.7) v.1] at $s = 2$. Substituting $z(\mathbf{k}) = z'(\mathbf{k}) + iz''(\mathbf{k})$ and considering (2.4.73), we can easily see that

$$\langle z(\mathbf{k}_1), z^T(\mathbf{k}_2)\rangle = [R - R' + i(S + S^T)]|_{\mathbf{k}_1\mathbf{k}_2}. \tag{2.4.85}$$

Let us now use (2.4.83). Owing to the condition $\mathbf{k}_1 \cdot \mathbf{s} > 0$ and $\mathbf{k}_2 \cdot \mathbf{s} > 0$, the function $\delta(\mathbf{k}_1 + \mathbf{k}_2)$ on the right of (2.4.84) is zero. Using also (2.4.85), we obtain $R - R' + i(S + S^T) = 0$, which gives (2.4.76), because the submatrices in it are real.

We can obtain from (2.4.84) for the example under consideration that $S = 0$. As for relation (2.4.74), it holds specifically in the case discussed in Sect. 2.5.

2.5 Example of Quasi-Frenergy Near a Nonequilibrium Phase Transition. The Bénard Instability

2.5.1 Basic Equations

The Bénard effect considered, for example, in [2.3] refers to the following phenomenon: When the gap between two horizontal parallel plates, whose temperatures T_1 and T_2 are different, is filled with a liquid at rest and when the temperature difference $T_1 - T_2$ increases very slowly, then convective motion of the liquid starts abruptly at a certain temperature difference, the eddy convective fluxes forming a regular periodic cell structure (the Bénard cells). A necessary condition for this convective instability of stationary liquid is a higher temperature of the lower plate. The liquid will then heat up at the lower plate, become lighter and rise. The upper colder layers of liquid will tend to sink.

Let the z-axis point upwards, and let the lower boundary of the liquid have the coordinate $z = 0$, and the upper one $z = l$. The constant temperature gradient in the liquid will then be

$$T_0(z) = T_1 - Mz. \tag{2.5.1}$$

Here $M > 0$ and $T_1 - T_2 = Ml$. In addition we introduce the small deviation $\vartheta(\mathbf{r})$ from T_0 by assuming that

$$T = T_0 + \vartheta. \tag{2.5.2}$$

For simplicity, we suppose that the density of the liquid is absolutely independent of pressure and only slightly dependent on temperature: $\rho = F(\mu T)$, where μ is a small parameter. Then the quantity β given by

$$\beta = -\rho^{-1}\partial\rho/\partial T \tag{2.5.3}$$

will be small, i.e. proportional to μ ($\beta = -\mu F'/F$), and so we can ignore the second derivative $\partial^2\rho/\partial T^2$. From (2.5.2,3) we have

$$\rho = \rho_0 + (\partial\rho/\partial T)\vartheta = \rho_0(1 - \beta\vartheta), \tag{2.5.4}$$

where $\rho_0 = [\rho]_{T=T_0}$.

We start off with the Navier-Stokes equation with the gravitational field taken into account

$$\dot{\mathbf{v}} + (\mathbf{v} \cdot \nabla)\mathbf{v} = -\rho^{-1}\nabla p + \mathbf{g} + \nu\nabla^2\mathbf{v} \tag{2.5.5}$$

and the heat-conduction equation with convection

$$\dot{T} + (\mathbf{v} \cdot \nabla)T = \chi \nabla^2 T, \tag{2.5.6}$$

where χ is the heat conduction coefficient; $\mathbf{g} = (0, 0, -g)$.

If in (2.5.5) we put $\mathbf{v} = 0$, and $\vartheta \equiv 0$, whereby ρ will coincide with $\rho_0 = [\rho]_{T_0} = \rho_1(1 + \beta M z)$ (by (2.5.1)), then the resultant equation will be

$$-\nabla p_0 + \mathbf{g}\rho_0 = 0. \tag{2.5.7}$$

It determines the unperturbed pressure

$$p_0(z) = p_1 - g\rho_1(z + \tfrac{1}{2}\beta M z^2). \tag{2.5.8}$$

Since β is small, we have $\beta\vartheta \ll 1$. Therefore, from (2.5.4) we have

$$\rho^{-1} = \rho_0^{-1}(1 - \beta\vartheta)^{-1} \approx \rho_0^{-1}(1 + \beta\vartheta). \tag{2.5.9}$$

If we then use (2.5.7,9) and substitute into (2.5.5) the equality $p = p_0 + p'$, which is a definition of the perturbed pressure p', we get

$$\dot{\mathbf{v}} + (\mathbf{v} \cdot \nabla)\mathbf{v} = -\rho_0^{-1}\nabla p' - \rho_0^{-1}\beta\vartheta\nabla(p_0 + p') + \nu\nabla^2\mathbf{v} \tag{2.5.10}$$

or, again using (2.5.7),

$$\dot{\mathbf{v}} + (\mathbf{v} \cdot \nabla)\mathbf{v} = -\rho_0^{-1}\nabla p' - \beta\mathbf{g}\vartheta - \rho_0^{-1}\beta\vartheta\nabla p' + \nu\nabla^2\mathbf{v}. \tag{2.5.11}$$

Substituting $T = T_1 - Mz + \vartheta$ into (2.5.6), we will find that ϑ obeys

$$\dot{\vartheta} - Mv_z + (\mathbf{v} \cdot \nabla)\vartheta = \chi\nabla^2\vartheta. \tag{2.5.12}$$

In addition to (2.5.11,12) we will use the incompressibility condition

$$\nabla \cdot \mathbf{v} = 0. \tag{2.5.13}$$

To be sure, the liquid under consideration cannot be treated as absolutely incompressible, otherwise the cause of the instability will vanish. It is exactly because the density changes, i.e. because there is a degree of compressibility, that terms with β appear in (2.5.11). But once we have obtained these terms, we can, to within reasonable accuracy, adopt the incompressibility condition.

2.5.2 The Linearized Langevin Equation

Let us return to the equation describing the process at hand. To solve the question of stability and to calculate the fluctuations near a stable state in a linear approximation we should linearize (2.5.11,12) in $\mathbf{v}, \vartheta, p'$. Discarding the nonlinear terms gives

$$\dot{\mathbf{v}} = -\rho_1^{-1}\nabla p' - \beta\mathbf{g}\vartheta + \nu\nabla^2\mathbf{v}, \tag{2.5.14}$$

$$\dot{\vartheta} = Mv_z + \chi\nabla^2\vartheta. \tag{2.5.15}$$

In the first of these equations we have substituted ρ_1 for ρ_0, using the inequality $\beta M l \ll 1$.

We will operate on both sides of (2.5.14) with the operator $-\nabla^{-2}$ curl curl. Since curl curl $\mathbf{a} = \nabla(\nabla \cdot \mathbf{a}) - \nabla^2\mathbf{a}$, by virtue of (2.5.13) we certainly will have $-\nabla^{-2}$ curl curl $\mathbf{v} = \mathbf{v}$. We thus arrive at the equation

$$\dot{v}_z = \beta g \nabla^{-2}(\nabla_x^2 + \nabla_y^2)\vartheta + \nu\nabla^2 v_z \tag{2.5.16}$$

for the z-component of $\dot{\mathbf{v}}$.

We turn now to the boundary conditions imposed on the solution of the above equations. At the boundaries $z = 0, z = l$, the temperature $T(\mathbf{r})$ may naturally be considered stable, and the z-component of the velocity may be taken to be zero. Then

$$\vartheta = 0, \quad v_z = 0 \tag{2.5.17}$$

for $z = 0$ and for $z = l$. We will further assume that at the boundaries the tangential stresses P_{xz}, P_{yz}, where $P_{ij} = -\eta(\nabla_i v_j + \nabla_j v_i)$, are zero. It follows that $\nabla_z v_x$ and $\nabla_z v_y$, and hence $\nabla_z(\nabla_x v_x + \nabla_y v_y)$ are zero. Using (2.5.13) gives

$$\nabla_z^2 v_z = 0 \quad \text{for} \quad z = 0 \quad \text{and} \quad z = l. \tag{2.5.18}$$

The boundary conditions (2.5.17,18) will be satisfied if v_z, ϑ are sought in the form

$$v_z = B_1(x, y, t)\sin(\pi n z / l),$$
$$\vartheta = B_2(x, y, t)\sin(\pi n z / l), \tag{2.5.19}$$

where n is an integer. Especially important is the value $n = 1$, which corresponds to the fundamental harmonic.

It is convenient to go over to the spectral representation, i.e. to carry out the Fourier transformation in x and y to obtain $-k_1^2 - k_2^2 = -k^2$ instead of $\nabla_x^2 + \nabla_y^2$. As a result, we obtain, from (2.5.15,16),

$$\dot{B}_1 = \beta g k^2[(\pi/l)^2 + k^2]^{-1}B_2 - \nu[(\pi/l)^2 + k^2]B_1 + \kappa^{1/2}\xi_1(\mathbf{k}, t),$$
$$\dot{B}_2 = MB_1 - \chi[(\pi/l)^2 + k^2]B_2 + \kappa^{1/2}\xi_2(\mathbf{k}, t), \tag{2.5.20}$$

where κ is a small parameter, and $\mathbf{k}$ is a two-dimensional vector. Here we have taken into account (2.5.19) at $n = 1$ and added the small fluctuations $\kappa^{1/2}\xi_1, \kappa^{1/2}\xi_2$, which are assumed to be Gaussian, to have zero mean and to be delta-correlated in time. In this case B_1, B_2 as functions of time represent a Markov process of the Fokker-Planck type.

2.5.3 Stationary Probability Density of Internal Parameters

Since the random functions $\xi_1(\mathbf{k}, t), \xi_2(\mathbf{k}, t)$ are stationary, homogeneous and delta-correlated in time, we have

$$\langle \xi_i(\mathbf{k}, t)\xi_j^*(\mathbf{k}', t') \rangle = S_{ij}(\mathbf{k})\delta(\mathbf{k} - \mathbf{k}')\delta(t - t'), \tag{2.5.21}$$

where $S_{ij}(\mathbf{k})$ are spectral densities.

We will now break-down the **k**-plane into elementary squares of the form

$$(q_{\alpha\beta})_1 \leq k_1 \leq (q_{\alpha\beta})_1 + \Delta k_1 \,,$$
$$(q_{\alpha\beta})_2 \leq k_2 \leq (q_{\alpha\beta})_2 + \Delta k_2 \,, \tag{2.5.22}$$

where $q_{\alpha\beta}$ are points in **k** specified by the breaking down. Averaging over an elementary square gives

$$\tilde{\xi}_i = \frac{1}{\Delta k_1 \Delta k_2} \int_{q_1}^{q_1 + \Delta k_1} \int_{q_2}^{q_2 + \Delta k_2} \xi_i(\mathbf{k}, t) dk_1 dk_2 \,,$$
$$\tilde{B}_i = \frac{1}{\Delta k_1 \Delta k_2} \int_{q_1}^{q_1 + \Delta k_1} \int_{q_2}^{q_2 + \Delta k_2} B_i(\mathbf{k}, t) dk_1 dk_2 \,, \tag{2.5.23}$$

where $\mathbf{q} = \mathbf{q}_{\alpha\beta}$. From (2.5.21), we get

$$\langle \tilde{\xi}_i(t), \tilde{\xi}_j^*(t') \rangle = S_{ij}(q_{\alpha\beta})(\Delta k_1 \Delta k_2)^{-1} \delta(t - t') \,. \tag{2.5.24}$$

The integral quantities obey

$$\dot{\tilde{B}}_1 = -\nu[(\pi/l)^2 + q^2]\tilde{B}_1 + \beta g q^2[(\pi/l)^2 + q^2]^{-1}\tilde{B}_2 + \kappa^{1/2}\tilde{\xi}_1 \,, \tag{2.5.25a}$$
$$\dot{\tilde{B}}_2 = M\tilde{B}_1 - \chi[(\pi/l)^2 + q^2]\tilde{B}_2 + \kappa^{1/2}\tilde{\xi}_2 \,, \tag{2.5.25b}$$

which are similar to (2.5.20). The variables $\text{Re}\{\tilde{B}_1(\mathbf{q})\}$, $\text{Im}\{\tilde{B}_1(\mathbf{q})\}$, $\text{Re}\{\tilde{B}_2(\mathbf{q})\}$, $\text{Im}\{\tilde{B}_2(\mathbf{q})\}$ ($\mathbf{q} = \mathbf{q}_{\alpha\beta}$) given by these equations form a Markov process and the Fokker-Planck equation can be written for their probability densities. The expression on the right-hand side of (2.5.25a) in front of the fluctuational term is nothing but the coefficient $K_1(\tilde{B})$ of this equation; and on the right-hand side of (2.5.25b) we have the coefficient $K_2(\tilde{B})$. Hence

$$\hat{A} = -\left\|\frac{\partial K_\alpha(\tilde{B})}{\partial \tilde{B}_\beta}\right\| \equiv \begin{pmatrix} a & c \\ d & b \end{pmatrix} \,, \tag{2.5.26}$$

where

$$a = \nu[(\pi/l)^2 + q^2] \,,$$
$$b = \chi[(\pi/l)^2 + q^2] \,,$$
$$c = -\beta g q^2[(\pi/l)^2 + q^2]^{-1} \,,$$
$$d = -M \,. \tag{2.5.27}$$

We have thus found one of the matrices that enter into (2.4.80). Another matrix, $\hat{N} = \| \text{Re}\{k_{ij}^0\} \|$, concerns the statistical behavior of the random functions $\tilde{\xi}_1, \tilde{\xi}_2$: $\langle \tilde{\xi}_i(t_1) \tilde{\xi}_j^*(t_2) \rangle = k_{ij}^0(\mathbf{q}) \delta(t_{12})$ (see (2.4.79)). By (2.5.24)

$$k_{ij}^0(\mathbf{q}) = S_{ij}(\mathbf{q})(\Delta k_1 \Delta k_2)^{-1} \,. \tag{2.5.28}$$

Like $\hat{A}$, it is dependent on $\mathbf{q} = \mathbf{q}_{\alpha\beta}$.

The third matrix in (2.4.80), $\hat{H}$ is related to correlator $\langle \tilde{B}_i \tilde{B}_j^* \rangle$ through $\| \langle \tilde{B}_i \tilde{B}_j^* \rangle \| = \kappa \hat{H}$ (see Sect. 2.4.6). Consequently, solving (2.4.80), we can find this correlator. The two-component form of this equation can be solved easily using the procedure described in Sect. 2.7.1. Using (2.7.5a), we will find the upper left element of $\hat{H}$, and hence $\langle |\, \tilde{B}_1\, |^2 \rangle$:

$$
\begin{aligned}
\langle |\, \tilde{B}_1\, |^2 \rangle \; &= \; \tfrac{1}{2}\kappa(a+b)^{-1}(ab-cd)^{-1} \\
&\times \{ k_{11}^0[b(a+b)-cd] - 2k_{12}^0 bc + k_{22}^0 c^2 \} \, .
\end{aligned}
\tag{2.5.29}
$$

Let us take a look at the factor $(a+b)^{-1}(ab-cd)^{-1}$ in the last expression. By (2.5.27) we have

$$
(a+b)^{-1}(ab-cd)^{-1} = l^6 \nu^{-1}\chi^{-1}(\nu+\chi)^{-1}[(\pi^2+l^2q^2)^3 - Rl^2q^2]^{-1}, \tag{2.5.30}
$$

where $R = \beta g M l^4/(\nu\chi)$ is the Rayleigh number. This factor, and hence $\langle |\, \tilde{B}_1\, |^2 \rangle$, becomes infinite at

$$
R = (\pi^2 + l^2q^2)^3/(lq)^2 \equiv f(l^2q^2) \, , \tag{2.5.31}
$$

thus suggesting that the stability of the nonconvectional state is lost. By minimizing the function $f(l^2q^2)$ in q^2 we find the value $q^2 \equiv k_c^2$, at which stability is first lost, and also the appropriate critical value R_c of the Rayleigh number at which a nonequilibrium phase transition occurs. We obtain

$$
\begin{aligned}
k_c^2 &= \tfrac{1}{2}\pi^2/l^2 \, , \\
R_c &= f(l^2k_c^2) = \tfrac{27}{4}\pi^4 \approx 657.5 \, .
\end{aligned}
\tag{2.5.32}
$$

Using the relationship $f(l^2k_c^2) - R_c = 0$, the expression $(\pi^2 + l^2q^2)^3 - Rl^2q^2 = l^2q^2[f(l^2q^2) - R]$, which enters (2.5.30), can be written as

$$
l^2q^2[f(l^2q^2) - f(l^2k_c^2) + R_c - R] \, . \tag{2.5.33}
$$

We will now turn to the critical region, i.e. the one where the deviations $R - R_c$, $q^2 - k_c^2$ are small. Using the smallness of $q^2 - k_c^2$, we can expand the function $f(l^2q^2)$ into a Taylor series at the point $l^2k_c^2$:

$$
\begin{aligned}
l^2q^2[f(l^2q^2) - f(l^2k_c^2)] &= l^2k_c^2[f'(l^2k_c^2)l^2(q^2 - k_c^2) \\
&+ \tfrac{1}{2}f''(l^2k_c^2)l^4(q^2 - k_c^2)^2] + f'(l^2k_c^2)l^4(q^2 - k_c^2)^2 \, .
\end{aligned}
\tag{2.5.34}
$$

We have discarded here the terms of order $(q^2 - k_c^2)^3$ and higher. Since the function $f(l^2q^2)$ has an extremum at $l^2k_c^2$, the terms with f' on the right-hand side vanish. Also, ignoring the error of order $(q^2 - k_c^2)(R - R_c)$, we will replace (2.5.33) by

$$
l^2k_c^2\left[\tfrac{1}{2}f''l^4(q^2 - k_c^2)^2 + R_c - R\right] \, . \tag{2.5.35}
$$

Likewise, with errors of the same order we can substitute k_c^2 for q^2 in the terms of the expression in braces in (2.5.29). As a result, with this error in the critical region from (2.5.29,30) we will have

$$
\langle |\, \tilde{B}_1\, |^2 \rangle = \kappa[g(R_c - R) + h(q^2 - k_c^2)^2]^{-1}(\Delta k_1 \Delta k_2)^{-1} \, , \tag{2.5.36}
$$

where

$$g = 2\nu\chi(\nu + \chi)l^{-4}k_c^2[S_{11}b^2 - 2S_{12}bc + S_{22}c^2]_{q^2=k_c^2}^{-1},$$
$$h = \tfrac{1}{2}f''l^4g \tag{2.5.37}$$

are constants. We have also used here (2.5.28), and written the expression for g bearing in mind that $ab - cd = 0$ at $q^2 = k_c^2$ and $R = R_c$. The constants g and h are positive since the matrix S_{ij} is positive definite (and assumed to be nondegenerate).

For Gaussian random forces ξ_1, ξ_2, the real and imaginary parts of the random functions $\tilde{B}_1$ and $\tilde{B}_2$ are given by the linear equations (2.5.25). Since the real and imaginary parts $\tilde{B}_{1r} = \mathrm{Re}\{\tilde{B}_1\}$ and $\tilde{B}_{1i} = \mathrm{Im}\{\tilde{B}_1\}$ have the same variance and are statistically independent, we have $\langle \tilde{B}_{1r}^2 \rangle = \langle \tilde{B}_{1i}^2 \rangle = \langle | \tilde{B}_1 |^2 \rangle / 2$. Accordingly, their joint probability density reads

$$
\begin{aligned}
w(\tilde{B}_{1r}, \tilde{B}_{1i}) &= \mathrm{const} \cdot \exp\left[-\frac{\tilde{B}_{1r}^2 + \tilde{B}_{1i}^2}{\langle | \tilde{B}_1 |^2 \rangle} \right] \\
&= \mathrm{const} \cdot \exp\left[-\frac{| \tilde{B}_1 |^2}{\langle | \tilde{B}_1 |^2 \rangle} \right]
\end{aligned}
\tag{2.5.38}
$$

or using (2.5.36),

$$
\begin{aligned}
&w(\tilde{B}_{1r}, \tilde{B}_{1i}) \\
&= \mathrm{const} \cdot \exp\{ -\kappa^{-1}[g(R_c - R) + h(q^2 - k_c^2)^2] \, | \tilde{B}_1 |^2 \, \Delta k_1 \Delta k_2 \}. \tag{2.5.39}
\end{aligned}
$$

This is the distribution of the complex quantity $\tilde{B}_1(\mathbf{q}_{\alpha\beta})$, which is the mean of $B_1(\mathbf{k})$, in one typical elementary square in the $\mathbf{k}$ plane. We will now proceed to compute the joint probability density of the means in the various squares.

Since $B_1(-\mathbf{k}) = B_1^*(\mathbf{k})$, the only independent quantities are $B_1(\mathbf{k})$ in the half-plane $\mathbf{k} \cdot \mathbf{s} > 0$, where $\mathbf{s}$ is an arbitrary vector. As a consequence, we must consider the joint probability density $w[\tilde{B}_{1r}(\mathbf{q}_{\alpha\beta}), \tilde{B}_{1i}(\mathbf{q}_{\alpha\beta}), \mathbf{q}_{\alpha\beta} \cdot \mathbf{s} > 0]$ of the means in the squares lying in one half-plane. It can be shown, by computing the correlators, that in the case of homogeneous fluctuations at any specific form of correlator $\langle B_1(\mathbf{r})B_1(\mathbf{r}') \rangle = f(\mathbf{r} - \mathbf{r}')$ the random quantities $B_{1r}(\mathbf{k}), B_{1i}(\mathbf{k})$ are independent of $B_{1r}(\mathbf{k}'), B_{1i}(\mathbf{k}')$ if $\mathbf{k} \neq \mathbf{k}'$; $\mathbf{k} \cdot \mathbf{k}' > 0$. Therefore, the means over squares in one half-plane are statistically independent and their joint probability density is found by multiplying together expressions of the type (2.5.39)

$$
\begin{aligned}
&w[\tilde{B}_{1r}(\mathbf{q}_{\alpha\beta}), \tilde{B}_{1i}(\mathbf{q}_{\alpha\beta}), \mathbf{q}_{\alpha\beta}\mathbf{s} > 0] \\
&= \mathrm{const} \cdot \exp\Bigg\{ -\frac{1}{\kappa} \sum_{\mathbf{q}_{\alpha\beta}\mathbf{s}>0} [g(R_c - R) + h(q_{\alpha\beta}^2 - k_c^2)^2] \\
&\quad \times | \tilde{B}_1(\mathbf{q}_{\alpha\beta}) |^2 \, \Delta k_1 \Delta k_2 \Bigg\}. \tag{2.5.40}
\end{aligned}
$$

If we pass to the limit $\Delta k_1 \longrightarrow 0$, $\Delta k_2 \longrightarrow 0$, we will obtain the functional probability density

$$w[B_{1r}(\mathbf{k}), B_{1i}(\mathbf{k}), \mathbf{k}\mathbf{s} > 0]$$

$$= \text{const} \cdot \exp\left\{-\frac{1}{\kappa}\int_{\mathbf{ks}>0}[g(R_c - R) + h(k^2 - k_c^2)^2] \mid B_1(\mathbf{k}) \mid^2 dk_1 dk_2\right\}$$

$$= \text{const} \cdot \exp\left\{-\frac{1}{2\kappa}\int[g(R - R_c) + h(k^2 - k_c^2)^2] \mid B_1(\mathbf{k}) \mid^2 d\mathbf{k}\right\}, \quad (2.5.41)$$

where on the right-hand side we now integrate over the entire $\mathbf{k}$ plane.

Comparing (2.5.41) and (2.3.3) when the vector $B = \{B_\alpha\}$ coincides with $B_1(\mathbf{k})$, i.e. at α having the sense of $\mathbf{k}$, we find the quasi-frenergy

$$\Psi[B_{1r}(\mathbf{k}), B_{1i}(\mathbf{k})] = \frac{1}{2}\int[g(R_c - R) + h(k^2 - k_c^2)^2] \mid B_1(\mathbf{k}) \mid^2 d\mathbf{k}. \quad (2.5.42)$$

If from the Fourier transform $B_1(\mathbf{k})$ we go over to the original $B_1(\mathbf{r})$, the quasi-frenergy becomes

$$\Psi[B_1(\mathbf{r})] = \frac{1}{2}\int\{g(R_c - R)B_1^2 + h[(\nabla^2 + k_c^2)B_1]^2\}d\mathbf{r}, \quad (2.5.43)$$

where $\mathbf{r}$ and ∇ are in two dimensions. It should be stressed that the expressions for quasi-frenergy are only valid in the critical region, at $R < R_c$, and in the Gaussian approximation.

2.5.4 A More General Expression for Quasi-Frenergy and Its Consequences

In order to obtain the quasi-frenergy determining the probability density $w[B_1(\mathbf{r})]$ at $R > R_c$, we should go beyond the framework of the Gaussian approximation. This amounts to adding to (2.5.43) a fourth-power term to yield

$$\Psi[B_1(\mathbf{r})] = \frac{1}{2}\int\{g(R_c - R)B_1^2 + h[(\nabla^2 + k_c^2)B_1]^2\}d\mathbf{r}$$
$$+ \int d(\mathbf{r}_1, \ldots, \mathbf{r}_4)B_1(\mathbf{r}_1)\ldots B_1(\mathbf{r}_4)d\mathbf{r}_1 \ldots d\mathbf{r}_4. \quad (2.5.44)$$

The determination of $d(\mathbf{r}_1, \mathbf{r}_2, \mathbf{r}_3, \mathbf{r}_4)$ is more cumbersome and so we will not provide it here. We will only indicate that d varies but little within the critical region, so that it can be determined at some value $R < R_c$, and then substituted into (2.5.44) at various R, including $R \geq R_c$.

At $R > R_c$ convective motion of the liquid sets in, and the relevant $B_1(\mathbf{r})$, by the H-theorem proved in Sect. 2.3.4, must correspond to the minimum of (2.5.44). Let us find this function. Expression (2.5.44) will be minimized in two stages. To begin with, we minimize the term

$$\frac{h}{2}\int[(\nabla^2 + k_c^2)B_1]^2 d\mathbf{r} \geq 0 \quad (2.5.45)$$

with the result that

$$(\nabla^2 + k_c^2)B_1 = 0. \quad (2.5.46)$$

Various functions may be a solution of this equation. Let us provide several examples. A simple solution of (2.5.46) is

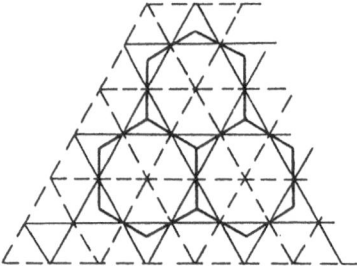

Fig. 2.7. Cells in which liquid circulates in isolation

$$B_1(\mathbf{r}) = A\cos(k_c\mathbf{n}\mathbf{r} + \varphi_0) \tag{2.5.47}$$

($|\mathbf{n}|^2 = 1$ and A, φ_0 are constants), to which a system of long eddies, parallel to one another, corresponds.

Another example of a solution is the function

$$B_1(\mathbf{r}) = A[\cos(k_c\mathbf{n}\mathbf{r} + \varphi_1) + \cos(k_c\mathbf{m}\mathbf{r} + \varphi_2)], \tag{2.5.48}$$

where $\mathbf{n}, \mathbf{m}$ are mutually perpendicular unit vectors. This function corresponds to a system of square cells in which the convention occurs.

For a third example we will take the function

$$B_1(\mathbf{r}) = A[\cos k_c\mathbf{n}_1 \cdot (\mathbf{r} - \mathbf{r}_0) + \cos k_c\mathbf{n}_2 \cdot (\mathbf{r} - \mathbf{r}_0) + \cos k_c\mathbf{n}_3 \cdot (\mathbf{r} - \mathbf{r}_0)], \tag{2.5.49}$$

where $\mathbf{n}_1, \mathbf{n}_2, \mathbf{n}_3$ are three unit vectors making an angle $2/3\pi : |\mathbf{n}_i \cdot \mathbf{n}_j| = 1/2, i \neq j$. Figure 2.7 shows (dotted line) the straight lines at which each of the cosine curves reaches a maximum. The solid lines mark their minimum (negative) values. At the intersections of the dotted lines the function $B_1(\mathbf{r})$ at $A > 0$ has maxima equal to $3A$. At these points and in their vicinity the convecting liquid rises. At a certain distance from the maxima on either side there are regions where the liquid sinks (negative B_1). Figure 2.7 also shows how the plane $\mathbf{r} = (x, y)$ is broken up into regular hexagons, in each of which the liquid contributes to the convection without leaving it. At the periphery of a hexagon the liquid sinks and in its center it rises. The motion will be reversed if in (2.5.49) $A < 0$. This form of convection has repeatedly been observed experimentally.

Other solutions of (2.5.46) are certainly possible. To answer the question of which of the solutions is more stable, we will have to go beyond the critical region and deal with the strongly nonlinear equations describing fully developed convection.

We will now go on to minimize the remaining expression

$$\Psi[B_1(r)] = \frac{g}{2}(R_c - R)\int B_1^2 d\mathbf{r} + \int d(\mathbf{r}_1, \dots, \mathbf{r}_4)B_1(\mathbf{r}_1)$$
$$\dots B_1(\mathbf{r}_4)d\mathbf{r}_1 \dots d\mathbf{r}_4. \tag{2.5.50}$$

To begin with, we will substitute $B_1(\mathbf{r}) = A\varphi(\mathbf{r})$ of the type (2.5.47–49) into it to obtain

$$\Psi[A] = -GS(R - R_c)A^2 + DSA^4 \,, \tag{2.5.51}$$

where

$$G = \frac{g}{2S} \int_S \varphi^2(\mathbf{r})d\mathbf{r} \,,$$

$$D = \frac{1}{S} \int_S \cdots \int_S d(\mathbf{r_1}, \ldots, \mathbf{r_4})\varphi(\mathbf{r_1}) \ldots \varphi(\mathbf{r_4})d\mathbf{r_1} \ldots d\mathbf{r_4} \,, \tag{2.5.52}$$

and S is the total surface area of the liquid layer. Minimizing (2.5.51) gives $A = (2GD)^{1/2}(R - R_c)^{1/2}/2$. In addition, using a formula of the type (2.3.3), we have the probability density for the amplitude:

$$w(A) = \text{const} \cdot \exp\{\kappa^{-1}[GS(R - R_c)A^2 - DSA^4]\} \,. \tag{2.5.53}$$

This resembles the amplitude distribution for the Van der Pol oscillator.

The amplitude A may also be assumed to be slowly varying as a function of $\mathbf{r} = (x, y)$ (more slowly than a cosine function). Then, substituting $B_1 = A\varphi$ and $(\nabla^2 + k_c^2)B_1 = \varphi\nabla^2 A + 2\nabla A \cdot \nabla\varphi$ into (2.5.44) gives

$$\Psi[A(\mathbf{r})] = \int \left\{ -G(R - R_c)A^2(\mathbf{r}) + \frac{h}{g}G(\nabla^2 A)^2 \right.$$

$$\left. + \sum_{i,j=1}^{2} H_{ij}(\nabla_i A)(\nabla_j A) + DA^4 \right\} d\mathbf{r} \,. \tag{2.5.54}$$

We have used here the fact that the quantities

$$G = \frac{g}{2S_0} \int_{S_0} \varphi^2 d\mathbf{r} \,,$$

$$D = \frac{1}{S_0} \int_{S_0} d(\mathbf{r_1}, \ldots, \mathbf{r_4})\varphi(\mathbf{r_1}) \ldots \varphi(\mathbf{r_4})d\mathbf{r_1} \ldots d\mathbf{r_4} \tag{2.5.55}$$

are weakly dependent on S_0 provided that the surface area S_0 is not too small. In (2.5.54) H_{ij} is given by

$$H_{ij} = \frac{2h}{S_0} \int_{S_0} \nabla_i\varphi\nabla_j\varphi d\mathbf{r} \,. \tag{2.5.56}$$

For the case (2.5.47) we have $H_{ij} = hk_c^2 n_i n_j$. For the cases (2.5.48,49) we can easily find $H_{ij} = hk_c^2\delta_{ij}$ and $H_{ij} = 3hk_c^2\delta_{ij}/2$ respectively. Expression (2.5.54) can be used to work out the correlators of amplitude fluctuations.

Instead of considering only the amplitude fluctuations, we can suppose that φ_0 in (2.5.47), or φ_1, φ_2 in (2.5.48), or $\mathbf{r}_0$ in (2.5.49), etc., are slowly fluctuating as well. Another approach is also possible: we can represent $B_1(\mathbf{r})$ in the form $A_0\varphi + \delta B_1$ and calculate the statistical behavior of the fluctuations of the deviation δB_1. In any event, it is convenient and effective to use quasi-frenergy to analyze the fluctuations.

It is to be noted that in the above derivation of the consequences of the stationary probability density $w[B_1(\mathbf{r})] = \text{const} \cdot \exp[-\Psi[B_1(\mathbf{r})]/\kappa]$, where $\Psi[B_1(\mathbf{r})]$ is given by (2.5.44), we have made some simplification that is justified at fairly small κ. The above method of deriving eqs. (2.5.51,52,54,56) from (2.5.44) is very simple but not well grounded since the formula $B_1(\mathbf{r}) = A\varphi(\mathbf{r})$, unlike the formula $B_1(\mathbf{r}) = A\varphi(\mathbf{r}) + \delta B(\mathbf{r})$, is invalid when fluctuations $\delta B(\mathbf{r})$ are taken into account. More rigorous method requiring more complex calculations is considered in Sect. A3.

Before we leave this subsection we note that the quasi-frenergy which characterizes the start of laser radiation has, in the critical region, a form similar to (2.5.44) [2.4].

2.6 Fluctuations Near One-Component Nonequilibrium Kinetic Phase Transitions

2.6.1 One-Component Second-Order Transitions

When the vector of the internal parameters B has only one component, the solution of (2.4.15,18,19) is trivial:

$$\Psi_2 \equiv \psi_{11} = -2k_{1,1}/k_{11}^0,\tag{2.6.1a}$$

$$\Psi_3 \equiv \psi_{111} = -(2/k_{11}^0)[k_{1,11} + \Psi_2 k_{11,1}^0 + \tfrac{1}{3}\Psi_2^2 k_{111}^0],\tag{2.6.1b}$$

$$\Psi_4 \equiv \psi_{1111} = -(2/k_{11}^0)[k_{1,111} + \tfrac{3}{2}\Psi_2(k_{11,11}^0 + k_{11,1}^0\Psi_3/\Psi_2)$$
$$+ \Psi_2\Psi_3 k_{111}^0 + \Psi_2^2 k_{111,1}^0 + \Psi_2^3 k_{1111}^0].\tag{2.6.1c}$$

Near a phase transition Ψ_2 is small. If we omit the terms containing Ψ_2, except in (2.6.1a) we will be able to use the simpler equations

$$\Psi_2 = -2k_{1,1}/k_{11}^0,$$
$$\Psi_3^c = -2k_{1,11}/k_{11}^0,$$
$$\Psi_4^c = -2k_{1,111}/k_{11}^0 - 3k_{11,1}^0\Psi_3/k_{11}^0.\tag{2.6.2}$$

The phase transition will be a second-order transition if $\Psi_3^c = 0$ (i.e. $k_{1,11}(\Theta_c) = 0$) and $\Psi_4^c > 0$. The probability density (2.4.64) will then be

$$w(B) = \text{const} \cdot \exp\left\{-\frac{1}{2\kappa}\left[\Psi_2(B - B_0)^2 + \frac{1}{12}\Psi_4(B - B_0)^4\right]\right\},\tag{2.6.3}$$

where $\Psi_4 = \Psi_4^c$. Changing the variable

$$y = (\Psi_4/12\kappa)^{1/4}(B - B_0),\tag{2.6.4}$$

we will get

$$w(y) = \text{const} \cdot \exp(-\alpha y^2 - y^4/2),\tag{2.6.5}$$

where

$$\alpha = 3^{1/2}\kappa^{-1/2}\Psi_4^{-1/2}\Psi_2.\tag{2.6.6}$$

The range of Θ where

$$2\alpha \sim 1,\ \text{i.e.}\,(12)^{1/2}\Psi_2 \sim \kappa^{1/2}\Psi_4^{1/2},\tag{2.6.7}$$

is said to be the critical region. The parameter fluctuations in it are anomalously large, i.e. they have the same order of magnitude as at the critical point $\Theta = \Theta_c$ itself.

At that point $\alpha = 0$ and the probability density (2.6.5) reads

$$w(y) = \text{const} \cdot \exp(-y^4/2) \, . \tag{2.6.8}$$

Using it gives

$$
\begin{aligned}
\langle y^2 \rangle &= \int_{-\infty}^{\infty} \exp(-y^4/2) y^2 dy \bigg/ \int_{-\infty}^{\infty} \exp(-y^4/2) dy \\
&= 2^{1/2} \Gamma(3/4) \Gamma(1/4) = 0,478 \ldots .
\end{aligned}
\tag{2.6.9}
$$

Hence, by (2.6.4),

$$
\begin{aligned}
\langle B, B \rangle &= (12\kappa/\Psi_4)^{1/2} \langle y^2 \rangle \\
&= 0,478 (12\kappa/\Psi_4)^{1/2} \sim \kappa^{1/2} \, .
\end{aligned}
\tag{2.6.10}
$$

Beyond the critical region we have

$$\langle B, B \rangle = \kappa (d^2\Psi/dB^2)^{-1} \sim \kappa \, . \tag{2.6.11}$$

This can also be found using the Gaussian approximation of the probability density (2.3.3). The variance of the fluctuations within the critical region, whose order equals that of (2.6.10), at small κ is much larger than (2.6.11).

Using the probability density (2.6.5) we can compute the behavior of the variance

$$
\begin{aligned}
\langle B, B \rangle &= (12\kappa/\Psi_4)^{1/2} \langle y^2 \rangle \\
&= (12\kappa/\Psi_4)^{1/2} \int_0^{\infty} \exp(-\alpha y^2 - y^4/2) y^2 dy \bigg/ \int_0^{\infty} \exp(-\alpha y^2 - y^4/2) dy
\end{aligned}
\tag{2.6.12}
$$

in the critical region. Applying eqs. (19.5.3) and (19.3.1) of Ref. 2.8, we easily find

$$
\begin{aligned}
\langle y^2 \rangle &= \int_0^{\infty} \exp(-\alpha x - x^2/2) x^{1/2} dx \bigg/ \int_0^{\infty} \exp(-\alpha x - x^2/2) x^{-1/2} dx \\
&= \frac{D_{-3/2}(\alpha)}{2 D_{-1/2}(\alpha)} \, ,
\end{aligned}
\tag{2.6.13}
$$

where $D_j(\alpha)$ is the parabolic cylindrical function.

The variation of $\langle y^2 \rangle$ with α calculated from (2.6.13) is plotted in Fig. 2.8.

2.6.2 Non-Gaussian Properties
of Fluctuations Near a Second-Order Transition

In the critical region (2.6.7) parameters fluctuate not only intensely but also in a non-Gaussian manner. In the one-component case the most important parameters characterizing the deviation of the distribution from Gaussian are the coefficients of asymmetry and excess

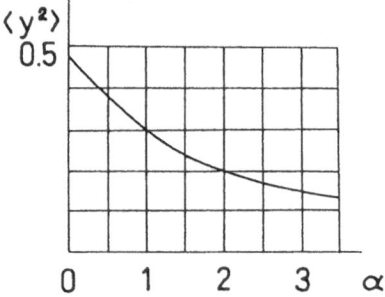

Fig. 2.8. Plot enabling one to find the dependence of $\langle (B - B_0)^2 \rangle = (12\kappa/\Psi_4)^{1/2}\langle y^2 \rangle$ on $\alpha = (3/\kappa)^{1/2}\Psi_4^{-1/2}\Psi_2$

$$\gamma_a = \frac{\langle B, B, B \rangle}{\langle B, B \rangle^{3/2}},$$

$$\gamma_e = \frac{\langle B, B, B, B \rangle}{\langle B, B \rangle^2}. \tag{2.6.14}$$

For the probability density (2.6.3) the coefficient of asymmetry is zero and the coefficient of excess coincides with the appropriate coefficient $\langle y, y, y, y \rangle / \langle y, y \rangle^2$ for (2.6.4). Using the formulas [(2.1.15) v.1], we can easily see that it can be written as

$$\gamma_e = \langle y^4 \rangle / \langle y^2 \rangle^2 - 3, \tag{2.6.15}$$

where

$$\langle y^4 \rangle = \int_0^\infty \exp(-\alpha x - x^2/2)x^{3/2}dx \bigg/ \int_0^\infty \exp(-\alpha x - x^2/2)x^{-1/2}dx. \tag{2.6.16}$$

Applying the same formula as in the derivation of (2.6.13) gives

$$\langle y^4 \rangle = 3/4 D_{-5/2}(\alpha)/D_{-1/2}(\alpha) \tag{2.6.17}$$

and hence

$$\gamma_e = 3\frac{D_{-5/2}(\alpha)D_{-1/2}(\alpha)}{D^2_{-3/2}(\alpha)} - 3 \tag{2.6.18}$$

For the critical point itself, by analogy with (2.6.9), we find

$$\langle y^4 \rangle = 2\Gamma(5/4)/\Gamma(1/4) = 0.5,$$

$$\gamma_e = -0.812. \tag{2.6.19}$$

The dependence γ_e vs. α (2.6.18) is represented in Fig. 2.9. It is seen that in the critical region the coefficient of excess is about minus unity, and so the fluctuations are strongly non-Gaussian.

Beyond the critical region, by the use of (2.3.17), we get

$$\gamma_e = -\kappa\frac{d^4\Psi}{dB^4}\left(\frac{d^2\Psi}{dB^2}\right)^{-2}. \tag{2.6.20}$$

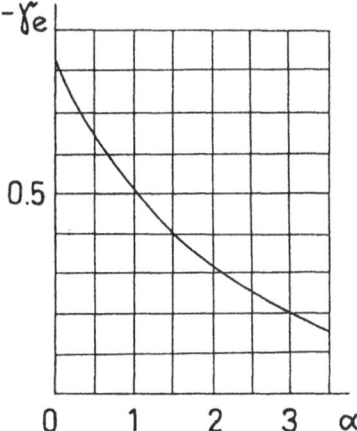

Fig. 2.9. The dependence of the coefficient of excess on $\alpha = (3/\kappa)^{1/2}\Psi_4^{-1/2}\Psi_2$

It follows that beyond the critical region $|\gamma_e|$ is of the order of κ, which is much less than the quantity presented above (assuming that $\kappa \ll 1$).

Before we move on to the first-order transitions, we will note that the approximation of the quasi-frenergy

$$\Psi(B) = \tfrac{1}{2}\Psi_2(B - B_0)^2 + \tfrac{1}{24}\Psi_4(B - B_0)^4 \qquad (2.6.21)$$

assumed in (2.6.3), at $\Psi_2 = a(\Theta - \Theta_c)$ is typical of the phenomenological Landau theory of equilibrium second-order transitions.

2.6.3 One-Component First-Order Transition

Suppose now that $\Psi_3^c \neq 0$. We will then have to turn to the expression

$$\Psi(B) \approx \Psi(B_0) + \tfrac{1}{2}\Psi_2 b^2 + \tfrac{1}{6}\Psi_3 b^3 + \tfrac{1}{24}\Psi_4 B^4 , \qquad (2.6.22)$$

where $b = B - B_0$, $\Psi_3 = \Psi_3^c$, $\Psi_4 = \Psi_4^c$. We would now like to find its extrema. Equating the derivative to zero yields

$$\Psi_2 b + \tfrac{1}{2}\Psi_3 b^2 + \tfrac{1}{6}\Psi_4 b^3 = 0 . \qquad (2.6.23)$$

At small b (We suppose that $\Psi_2 > 0$ and the system is in a state for which b is small) we can instead of that take a simpler equation

$$\Psi_2 b + \tfrac{1}{2}\Psi_3 b^2 = 0 , \qquad (2.6.24)$$

whose roots are

$$b_1 = 0 ,$$
$$b_2 = -2\Psi_2/\Psi_3 . \qquad (2.6.25)$$

The function

$$u(b) = \tfrac{1}{2}\Psi_2 b^2 + \tfrac{1}{6}\Psi_3 b^3 \tag{2.6.26}$$

is zero at $b = 0$ and it is equal to $2\Psi_2^3\Psi_3^{-2}/3$ at $b = -2\Psi_2/\Psi_3$. At small b, the function $u(b)$ differs only slightly from $\Psi(B) - \Psi(B_0)$, therefore the relative height of the "watershed" can be found using $u(b)$. At $b = 0$, $u(b)$ and $\Psi(B_0 + b)$ have a local minimum, which corresponds to a metastable state, and at $b = -2\Psi_2/\Psi_3 \equiv b_2$ lies the "watershed", its relative height being

$$\Delta u = \tfrac{2}{3}\Psi_2^3\Psi_3^{-2} > 0 \tag{2.6.27}$$

This quantity is positive since $\Psi_2 > 0$.

For the metastable state corresponding to $b = 0$ to exist for a more or less long period of time, the ratio

$$\lambda \equiv \Delta u/\kappa = \ln[w(0)/w(b_2)], \tag{2.6.28}$$

which determines the average lifetime of a system in that state, should not be too small. We will arbitrarily assume that a metastable state is "sufficiently stable" if

$$\Delta u/\kappa \geq 3, \tag{2.6.29}$$

hence

$$w(b_2) \leq \exp(-3)w(0) \approx w(0)/20. \tag{2.6.30}$$

Of course, instead of 3 we could have taken any other number, say 4 or 2.5. The following formulas would then have to be modified accordingly.

It follows from (2.6.27,28) that

$$\Psi_2 = (\tfrac{3}{2}\kappa\lambda)^{1/3}\Psi_3^{2/3}, \quad \lambda \geq 3. \tag{2.6.31}$$

Now let us look at the probability density

$$w(b) = \text{const} \cdot \exp\{-(2\kappa)^{-1}[\Psi_2 b^2 + \tfrac{1}{3}\Psi_3 b^3 + \tfrac{1}{12}\Psi_4 b^4]\} \tag{2.6.32}$$

of the type (2.4.64) in the critical region subject to the condition that $\lambda \sim 3$, but $\lambda > 3$. If now we introduce the variable

$$\begin{aligned} z &= (\Psi_2/\kappa)^{1/2}b \ \text{sign} \ \Psi_3 \\ &= (3\lambda/2)^{1/6}\kappa^{-1/3}\Psi_3^{1/3}b \ \text{sign} \ \Psi_3 \end{aligned} \tag{2.6.33}$$

and apply the equation $\lambda = 2\Psi_2^3\Psi_3^{-2}\kappa^{-1}/3$, instead of (2.6.31) we will have

$$w(z) = \text{const}\cdot\exp\{-\tfrac{1}{2}z^2 - \tfrac{1}{3}(6\lambda)^{-1/2}z^3 - \tfrac{1}{24}\kappa^{1/3}(2/(3\lambda))^{2/3}\Psi_3^{-4/3}\Psi_4 z^4\}, \tag{2.6.34}$$

where the small (because κ is small) term containing $\kappa^{1/3}$ can be discarded. True, (2.6.34) can only be used at "this side of the watershed", i.e., assuming

$$w(z) = \text{const} \cdot \exp[-\tfrac{1}{2}z^2 - \tfrac{1}{3}(6\lambda)^{-1/2}z^3] \quad \text{for} \quad z > -(6\lambda)^{1/2},$$

$$w(z) = 0 \quad \text{for} \quad z < -(6\lambda)^{1/2}. \tag{2.6.35}$$

This probability density must be used to work out the correlators $\langle z, z \rangle, \langle z, z, z \rangle$, $\langle z, z, z, z \rangle$, which are independent of κ and are functions of λ. This can conveniently be done using

$$J_k(\lambda) = \frac{1}{\sqrt{2\pi}} \int_{-\sqrt{6\lambda}}^{\infty} z^k \exp\left(-\frac{z^2}{2} - \frac{z^3}{3\sqrt{6\lambda}}\right) dz$$

$$= \frac{1}{\sqrt{2\pi}} \int_{-\sqrt{6\lambda}}^{\infty} \exp\left(-\frac{z^2}{2}\right) z^k \left(1 - \frac{z^3}{3\sqrt{6\lambda}} + \frac{z^6}{108\lambda} - \ldots\right) dz, \quad (2.6.36a)$$

$$(2\pi)^{-1/2} \int_{-\infty}^{\infty} \exp(-z^2/2) z^{2l} dz = (2l-1)!! . \quad (2.6.36b)$$

Knowing the correlators of z, from (2.6.33), we obtain the correlators

$$\begin{aligned}
\langle B, B \rangle &= \kappa^{2/3}(2/(3\lambda))^{1/3}\Psi_3^{-2/3}\langle z, z \rangle, \\
\langle B, B, B \rangle &= \kappa(2/(3\lambda))^{1/2}\Psi_3^{-1}\langle z, z, z \rangle, \\
\langle B, B, B, B \rangle &= \kappa^{4/3}(2/(3\lambda))^{2/3}\Psi_3^{-4/3}\langle z, z, z, z \rangle
\end{aligned} \quad (2.6.37)$$

of the initial parameter B. A rough estimate of the correlators of z that enter this expression can be obtained using the following asymptotic expressions, derived from (2.6.36):

$$\begin{aligned}
\langle z, z \rangle &= 1 + 2/(3\lambda) + O(\lambda^{-2}), \\
\langle z, z, z \rangle &= -2^{1/2}(3\lambda)^{-1/2}[1 + O(\lambda^{-1})], \\
\langle z, z, z, z \rangle &= 2/\lambda + O(\lambda^{-2}),
\end{aligned} \quad (2.6.38)$$

hence

$$\begin{aligned}
\gamma_a &= -(2/(3\lambda))^{1/2}[1 + O(\lambda^{-1})], \\
\gamma_e &= (2/\lambda)[1 + O(\lambda^{-1})].
\end{aligned} \quad (2.6.39)$$

Of course, for the extreme values of λ, which correspond to the boundaries of relative stability, we will need more accurate calculations.

It is seen from (2.6.37) that in the critical region the correlators of the internal parameter are of the order of $\kappa^{2/3}, \kappa, \kappa^{4/3}$, respectively. As a consequence, in the critical region the fluctuations of B as in the case of the second-order phase transition are anomalously large and substantially non-Gaussian, although the magnitude of latter is somewhat smaller than for the second-order phase transition. The estimate of $\langle B, B, B, B \rangle \sim \kappa^{4/3}$ for the first-order transition must be compared with the estimate $\langle B, B, B; B \rangle \sim \kappa$ for the second-order transition. The latter follows from (2.6.4) and from $\langle y, y, y, y \rangle \sim 1$ (see (2.6.17)). Beyond the critical region the correlators $\langle B, B \rangle, \langle B, B, B \rangle, \langle B, B, B, B \rangle$ have the order of κ, κ^2 and κ^3, respectively.

2.6.4 Intermediate Case of Kinetic Phase Transition

If Ψ_3 is small, or, more precisely, is of the order of $\kappa^{1/4}$, and also $\Psi_4 > 0$, then we have the intermediate case between first- and second-order transitions. The region of these values of Ψ_3 can be said to be intermediate. Changing the variable (2.6.4), we can then bring the probability density

$$
\begin{aligned}
w(B) \;=\; &\text{const} \cdot \exp\{-(2\kappa)^{-1}[\Psi_2(B - B_0)^2 \\
&+3^{-1}\Psi_3(B - B_0)^3 + (12)^{-1}\Psi_4(B - B_0)^4]\}
\end{aligned}
$$

into the form

$$
w(y) = \text{const} \cdot \exp(-\alpha y^2 - \sigma y^3 - \tfrac{1}{2}y^4), \tag{2.6.40}
$$

where

$$
\sigma = \tfrac{1}{6}(12/\Psi_4)^{3/4}\kappa^{-1/4}\Psi_3. \tag{2.6.41}
$$

When $\sigma \sim 1$, the mean square $\langle y^2 \rangle$ is given by

$$
\langle y^2 \rangle = \frac{\int_{-\infty}^{\infty} \exp(-\alpha y^2 - \sigma y^3 - \tfrac{1}{2}y^4)y^2 dy}{\int_{-\infty}^{\infty} \exp(-\alpha y^2 - \sigma y^3 - \tfrac{1}{2}y^4)dy}. \tag{2.6.42}
$$

This relation is more complicated than (2.6.13). The same is true of the higher stationary moments; the coefficients (2.6.14) are also found in a more complicated way. But since the probability density (2.6.40) is independent of κ explicitly, we have $\langle y^2 \rangle \sim 1$, $\langle y, y, y \rangle \sim 1$, $\langle y, y, y, y \rangle \sim 1$, and from (2.6.4) we immediately obtain

$$
\begin{aligned}
&\langle B, B \rangle \sim \kappa^{1/2}, \\
&\langle B, B, B \rangle \sim \kappa^{3/4}, \\
&\langle B, B, B, B \rangle \sim \kappa.
\end{aligned} \tag{2.6.43}
$$

The two- and four-fold correlators are of the same order as for the second-order transition, but the three-fold correlator for $\Psi_3 \neq 0$ is no longer zero.

2.6.5 Another Form of the First-Order Phase Transition

Suppose now that $\Psi_3 = 0$, but $\Psi_4 < 0$. Instead of (2.6.26), we will then use the function

$$
u(b) = \tfrac{1}{2}\Psi_2 b^2 + \tfrac{1}{24}\Psi_4 b^4 \quad (b = B - B_0). \tag{2.6.44}
$$

It has an extremum if

$$
\Psi_2 b + \tfrac{1}{6}\Psi_4 b^3 = 0, \tag{2.6.45}
$$

hence

$$
\begin{aligned}
&b_1 = 0, \\
&b_2 = (-6\Psi_2/\Psi_4)^{1/2}
\end{aligned} \tag{2.6.46}
$$

and

$$
\begin{aligned}
\Delta u &= u(b_2) \\
&= -\tfrac{3}{2}\Psi_2^2\Psi_4^{-1},
\end{aligned}
\tag{2.6.47}
$$

so that, according to the equation $\lambda = \Delta u/\kappa$, we have

$$
\begin{aligned}
\lambda &= \Delta u/\kappa \\
&= -\tfrac{3}{2}\kappa^{-1}\Psi_2^2/\Psi_4,
\end{aligned}
\tag{2.6.48}
$$

Introducing the variable $z = (\Psi_2/\kappa)^{1/2}b$, i.e.

$$
z = [\tfrac{2}{3}(\lambda/\kappa)\mid\Psi_4\mid]^{1/4}b,
\tag{2.6.49}
$$

and using (2.6.48), instead of the probability density $w(b) = \mathrm{const}\cdot\exp(-u(B)/\kappa)$, we will obtain the probability density

$$
w(z) = \mathrm{const}\cdot\exp[-z^2/2 + z^4/16\lambda]
\tag{2.6.50}
$$

to be utilized at "this side of the watershed", i.e. in the range $-2\lambda^{1/2} < z < 2\lambda^{1/2}$. Moments and correlators can be derived by the use of the integrals

$$
\begin{aligned}
J_k(\lambda) &= (2\pi)^{-1/2}\int_{-2\sqrt{\lambda}}^{2\sqrt{\lambda}} z^k\exp(-z^2/2 + z^4/16\lambda)dz \\
&= (2\pi)^{-1/2}\int_{-2\sqrt{\lambda}}^{2\sqrt{\lambda}}\exp(-z^2/2)z^k[1 + z^4/16\lambda + \tfrac{1}{2}(z^4/16\lambda)^2 + \ldots]dz.
\end{aligned}
\tag{2.6.51}
$$

Using these we find, in particular,

$$
\begin{aligned}
\langle z, z\rangle &= 1 + 3/(4\lambda) + O(\lambda^{-2}), \\
\langle z, z, z\rangle &= 0, \\
\langle z, z, z, z\rangle &= 3/(2\lambda) + O(\lambda^{-2})
\end{aligned}
\tag{2.6.52}
$$

and hence, by (2.6.49),

$$
\begin{aligned}
\langle B, B\rangle &= (3\kappa/2\lambda)^{1/2}\mid\Psi_4\mid^{-1/2}[1 + 3/4\lambda + O(\lambda^{-2})], \\
\langle B, B, B\rangle &= 0, \\
\langle B, B, B, B\rangle &= -(3\kappa/2\lambda)\mid\Psi_4\mid^{-1}(3/2\lambda + O(\lambda^{-2})), \\
\gamma_a &= 0, \\
\gamma_e &= (6/\lambda)(1 + O(\lambda^{-1})),
\end{aligned}
\tag{2.6.53}
$$

where $\lambda > 3$. At relatively small λ, close to the stability boundary, the correlators must be computed by (2.6.51) with a higher accuracy. The magnitude of fluctuations and their non-Gaussian character in the critical region $\lambda \sim 3$, $\lambda > 3$ will be the same here as for the second-order transition described in Sect. 2.6.1.

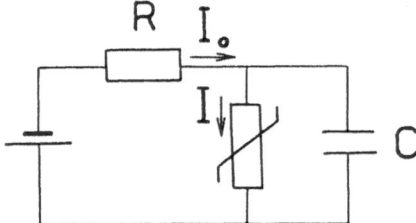

Fig. 2.10. Circuit with a tunnel diode, in which bistability is possible

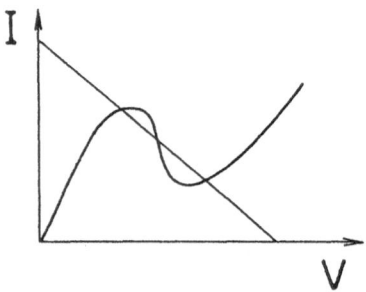

Fig. 2.11. Tunnel diode characteristic and the determination of bistable states

2.6.6 Example of a One-Component Transition: Tunnel Diode Circuit

Consider a circuit including a voltage source, a resistance R and a tunnel diode with a capacitance C connected in parallel with the diode (Fig. 2.10). The tunnel diode is known to have a falling $I - V$ characteristic (Fig. 2.11). We will approximate it by the polynomial

$$I = f(V) = d(V^3 - bV^2 + cV), \tag{2.6.54}$$

where $b, c, d > 0$, $b^2 - 3c > 0$. This function has its extrema at the points

$$V_1 = \tfrac{1}{3}[b - (b^2 - 3c)^{1/2}],$$
$$V_2 = \tfrac{1}{3}[b + (b^2 - 3c)^{1/2}], \tag{2.6.55}$$

the first of these corresponding to a local maximum, and the second to a local minimum.

If we denote by I_0 the current flowing through R, and by Q the charge on the capacitor, we will get the equations

$$\mathcal{E} = RI_0 + Q/C,$$
$$I = f(Q/C),$$
$$\dot{Q} = I_0 - I \tag{2.6.56}$$

describing the circuit. Hence

$$\dot{Q} = (\mathcal{E} - V)/R - f(V), \tag{2.6.57}$$

where $V = Q/C$. Account of the fluctuations ξ requires that we use, instead of (2.6.57), the equation

$$\dot{Q} = (\mathcal{E} - V)/R - f(V) + \xi(Q, t) \tag{2.6.58}$$

or the master equation that corresponds to it. We will then have

$$K_1(Q) = R^{-1}(\mathcal{E} - Q/C) - f(Q/C). \tag{2.6.59}$$

The stable point is found from

$$K_1(Q) = 0,$$
$$(\mathcal{E} - V)/R = f(V). \tag{2.6.60}$$

If $\mathcal{E}$ or R change, a nonequilibrium transition is possible. In that case, in accordance with (2.4.32), the equations

$$\partial \Psi / \partial B_1 = 0, \tag{2.6.61a}$$
$$\psi_{11} = 0 \tag{2.6.61b}$$

must hold simultaneously. The first of these, by virtue of (2.3.86a) is equivalent to equation $\kappa_1(x) = O(\kappa)$ or $K_1(B) = O(\kappa)$. Ignoring the small terms of the order of κ we will have $K_1(B) = 0$. Equation (2.6.61b) is, by (2.6.1), equivalent to relationship $k_{1,1} = 0$ or $dK_1(Q)/dQ = 0$ at $\Theta = \Theta_c$. So, at the critical point the equations

$$K_1(B) = 0,$$
$$dK_1(Q)/dQ = 0 \tag{2.6.62}$$

hold simultaneously. From (2.6.59,54) we get

$$\begin{aligned} K_1(Q) &= -d\{V^3 - bV^2 + [c + 1/(dR)]V - \mathcal{E}/(dR)\}, \\ V &= Q/C \end{aligned} \tag{2.6.63}$$

This expression can be rewritten as

$$K_1(Q) = -d(y^3 - 3py + 2q), \tag{2.6.64}$$

where

$$y = V - \tfrac{1}{3}b, \quad 3p = \tfrac{1}{3}b^3 - c - (dR)^{-1},$$
$$q = \tfrac{1}{6}b[c + (cR)^{-1}] - (b/3)^3 - \mathcal{E}(2dR)^{-1}. \tag{2.6.65}$$

Using (2.6.64), we write (2.6.62) as

$$y^3 - 3py + 2q = 0, \tag{2.6.66a}$$
$$y^2 - p = 0. \tag{2.6.66b}$$

These two equations only have a joint solution if

$$p^3 = q^2. \tag{2.6.67}$$

The solution is

$$y_c = q^{1/3}, \tag{2.6.68}$$

which can easily be verified by substitution. At $q < 0$ the root $q^{1/3}$ is understood to be $-(-q)^{1/3}$. Equation (2.6.66a), subject to condition (2.6.67), has the roots

$$y_1^c = y_2^c = -y_3^c/2 = q^{1/3}. \tag{2.6.69}$$

We will assume that $p = (b/3)^2 - [c + (dR)^{-1}]/3 > 0$ is fixed, and, i.e. q varies. Equation (2.6.67) yields two critical values $q_c = \pm p^{3/2}$.

In the range $-p^{3/2} < q < p^{3/2}$ two stable values of Q are possible, i.e. multistability takes place. Since, by (2.6.69,68), we have for any y

$$y^3 - 3py + 2q_c = (y - y_c)^2(y + 2y_c) \tag{2.6.70}$$

with $y_c = q_c^{1/3}$, eq. (2.6.66a) at arbitrary q can be written as

$$(y - y_c)^2(y + 2y_c) + 2(q - q_c) = 0. \tag{2.6.71}$$

Near the critical point the factor $y + 2y_c$ can be replaced by $3y_c$ (then the error is $O((y - y_c)^3)$, i.e. very small for $y = y_1, y_2$) and it follows from (2.6.71) that

$$3y_c(y - y_c)^2 = 2(q_c - q). \tag{2.6.72}$$

The equation $K_1(Q) = 0$, which takes the form (2.6.66a) and (in the neighborhood of the critical point) the form (2.6.72), determines the stable point y_1 or the stable value $Q_0 = CV_0 = C(y_1 + b/3)$. From (2.6.72), we obtain

$$y_1 = y_c + [2(q_c - q)/3y_c]^{1/2}. \tag{2.6.73}$$

Using (2.6.64), where $y = Q/C - b/3$, we can readily find the derivative $k_{1,1} = dK_1/dQ$ at the stable point y_1:

$$\begin{aligned} k_{1,1} &= C^{-1}dK_1/dy \\ &= -3dC^{-1}(y_1^2 - p) \\ &\approx -6dC^{-1}y_c(y_1 - y_c), \end{aligned} \tag{2.6.74}$$

Applying (2.6.73) gives

$$k_{1,1} = -2\sqrt{6}dC^{-1}y_c^{1/2} \,|\, q_c - q \,|^{1/2}. \tag{2.6.75}$$

We can also find easily the higher derivatives

$$\begin{aligned} k_{1,11} &= -6dC^{-2}y_c, \\ k_{1,111} &= -6dC^{-3}. \end{aligned} \tag{2.6.76}$$

Knowing the above derivatives (2.6.75,76) and applying the formula (2.6.1) where $k_{11}^0, k_{11,1}^0$ are predetermined by the properties of fluctuations, we can easily find Ψ_2, Ψ_3, Ψ_4. In this example $\Psi_3 \neq 0$, and thus there occurs the first-order transition described in Sect. 2.6.3, and we can apply the relationships provided there.

If $\mathcal{E} = 0$, then there is only one stable value, $Q = 0$, which corresponds to thermodynamic equilibrium. As $\mathcal{E}$ increases subject to the condition

$$\mathcal{E}/(2dR) < (b/6)[c + 1/(dR)] - (b/3)^3 - p^{3/2}, \tag{2.6.77}$$

i.e. $q > p^{3/2}$, the stable value shifts but remains the only one. This set of stable values can be referred to as a stable branch. At the critical value

$$\mathcal{E}_{c1} = 2dR\{(b/6)[c + 1/(dR)] - (b/3)^3 - p^{3/2}\}, \quad \text{i.e.} \quad q = p^{3/2}, \tag{2.6.78}$$

a second stable value, $Q = C(y_c + b/3)$, $y_c = q^{1/3}$, appears; it gives rise to a nonequilibrium branch. In the range

$$\frac{b}{6}\left(c + \frac{1}{dR}\right) - \left(\frac{b}{3}\right)^3 - p^{3/2} < \frac{\mathcal{E}}{2dR} < \frac{b}{6}\left(c + \frac{1}{dR}\right) - \left(\frac{b}{3}\right)^3 + p^{3/2},$$
$$\text{i.e.} - p^{3/2} < q < p^{3/2}, \tag{2.6.79}$$

there are both equilibrium and nonequilibrium branches. At the second critical value

$$\mathcal{E}_{c2} = 2dR\{(b/c)[c + 1/(dR)] - (b/3)^3 + p^{3/2}\}, \quad \text{i.e.} \quad q = -p^{3/2}, \tag{2.6.80}$$

the equilibrium branch disappears, and at $\mathcal{E} > \mathcal{E}_{c2}$ only the nonequilibrium branch remains. The presence of both equilibrium and nonequilibrium branches of stable states is characteristic of many nonequilibrium systems.

2.6.7 Vibrational Two-Component Case Reducible to One-Component Case

If the vector $\mathbf{B}$ of internal parameters has two components B_1, B_2, then, ignoring the fluctuations, the system is described by

$$\dot{x} = ax + cy + f_1(x, y),$$
$$\dot{y} = dx + by + f_2(x, y), \tag{2.6.81}$$

where $x = B_1 - B_1^0$, $y = B_2 - B_2^0$; f_1, f_2 contain the terms nonlinear in x, y.

In a two-component phase transition the following special cases are possible: (1) the matrix

$$-\hat{A} = \|\,k_{\alpha,\beta}\,\| = \begin{pmatrix} a & c \\ d & b \end{pmatrix} \tag{2.6.82}$$

can, at the critical point $\Theta = \Theta_c$, have one zero and one nonzero eigenvalue; (2) the matrix A can have two zero eigenvalues, and (3) the matrix A can have two purely imaginary eigenvalues

$$\alpha_1(\Theta_c) = i\omega \quad (\omega \neq 0),$$
$$\alpha_2(\Theta_c) = -i\omega. \tag{2.6.83}$$

In that case, $a + b = 0$, $\omega^2 = -a^2 - cd > 0$.

The last case, which we refer to as the oscillatory one, will be considered below.

At the critical point equations (2.6.81), if we discard the nonlinear terms, have the following solution:

$$x(t) = x_0 \cos \omega t + \omega^{-1}(ax_0 + cy_0) \sin \omega t \,,$$
$$y(t) = \omega^{-1}(dx_0 - ay_0) \sin \omega t + y_0 \cos \omega t \,, \qquad (2.6.84)$$

which can be verified by direct substitution. Here x_0, y_0 are the values of $x(t), y(t)$ at the initial time $t = 0$.

By virtue of the inequality $-cd > a^2$, the coefficients c and d must have different signs. To be more specific, we will assume that $c > 0$. Otherwise, we can interchange x and y; then c will be replaced by d and vice versa, and a will be replaced by b. We will introduce the function

$$E = cy^2 + (a - b)xy - dx^2 \,. \qquad (2.6.85)$$

We can easily find its time derivative by using (2.6.81)

$$\dot{E} = (a + b)E + [(a - b)y - 2dx]f_1 + [(a - b)x + 2cy]f_2 \,, \qquad (2.6.86)$$

According to this equation, near a critical point at small x, y the function E varies slowly as compared with parameters x and y, which vary relatively fast. In fact, near the critical point, $a + b$ is close to zero, and f_1, f_2 are far smaller than ax, cy, dx because x and y are small. Consequently, in that case we can apply the method of slowly varying amplitude. The "amplitude" A can be defined, say, by

$$E = cA^2 = (\gamma A)^2 \,, \qquad (2.6.87)$$

where $\gamma = c^{1/2}$. We will take as the initial time $t = 0$ the moment when the coordinate x is zero $(x = 0)$. Then, from (2.6.85,87) we find that the initial value of the amplitude coincides with y_0. In the approximation that corresponds to (2.6.84), the amplitude is constant. Since $x_0 = 0, y_0 = A$, these equations can be written as

$$x(t) = (c/\omega)A \sin \omega t \,,$$
$$y(t) = A[\cos \omega t - (a/\omega) \sin \omega t] \,. \qquad (2.6.88)$$

In a more rigorous treatment relying on the (2.6.81,86), instead of (2.6.88), we should use the equations

$$x(t) = c\omega^{-1}A(t) \sin(\omega t + \varphi) \,,$$
$$y(t) = A(t)[\cos(\omega t + \varphi) - a\omega^{-1} \sin(\omega t + \varphi)] \,, \qquad (2.6.89)$$

where $A(t)$ and $\varphi(t)$ are the slowly varying amplitude and phase.

Using (2.6.87), equation (2.6.86), where $a - b$ near the critical point can be replaced by $2a$, will be reduced to the form

$$\dot{A} = \tfrac{1}{2}(a + b)A + A^{-1}c^{-1}(ay - dx)f_1(x, y)$$
$$+ A^{-1}(ac^{-1}x + y)f_2(x, y) \,. \qquad (2.6.90)$$

Here x and y are given by (2.6.89).

From (2.6.89) we can readily find

$$\tan(\omega t + \varphi) = \omega x/(cy + ax) \,,$$
$$\omega t + \varphi = \arctan[\omega x/(cy + ax)] \,. \qquad (2.6.91)$$

From this and (2.6.81), we can also obtain the equation for the phase

$$
\begin{aligned}
\dot{\varphi} \;=\;& \omega[cA^2 + (a+b)xy]^{-1} \\
&\times[-(a+b)xy + yf_1(x,y) - xf_2(x,y)]\,.
\end{aligned}
\tag{2.6.92}
$$

Equations of the type (2.6.90) and (2.6.92), taken after the substitution of (2.6.89), are well known in the theory of nonlinear oscillations that are close to simple harmonic motions. This theory allows one to use them for deriving vibrationless equations for amplitude and phase in various approximations (see, e.g., [2.9]). In the first approximation, to derive the amplitude equation it is sufficient to substitute (2.6.89) into (2.6.90) and to average the right-hand side over the period $T_0 = 2\pi/\omega$. This yields

$$
\dot{A} = \tfrac{1}{2}(a+b)A + g(A)\,,
\tag{2.6.93}
$$

where

$$
g(A) = (cT_0A)^{-1}\int_0^T [(ay - dx)f_1(x,y) + (ax + cy)f_2(x,y)]dt\,.
\tag{2.6.94}
$$

In the integration here the amplitude A and phase φ are assumed to be constant. The function (2.6.94) contains terms nonlinear in A. Notice that $g(-A) = g(A)$.

If we take fluctuational forces into account, we must use, instead of (2.6.81),

$$
\begin{aligned}
\dot{x} &= ay + cy + f_1(x,y) + \xi(t)\,, \\
\dot{y} &= dx + by + f_2(x,y) + \eta(t)\,.
\end{aligned}
\tag{2.6.95}
$$

Now, instead of (2.6.93), we will have, by (2.6.89),

$$
\dot{A} = \tfrac{1}{2}(a+b)A + g(A) + \zeta(t)\,,
\tag{2.6.96}
$$

where, by (2.6.89),

$$
\begin{aligned}
\zeta(t) \;=\;& (cA)^{-1}[(ay - dx)\xi + (ax + cy)\eta] \\
\;=\;& c^{-1}(a\cos\Phi - \omega\sin\Phi)\xi(t) + \eta(t)\cos\Phi \quad (\Phi = \omega t + \varphi)\,.
\end{aligned}
\tag{2.6.97}
$$

The correlator of the random function $\zeta(t)$ can also be averaged over the period. In some cases, for deriving (2.6.93) or (2.6.96) we may need to apply higher approximations of the above-mentioned technique.

Changing from (2.6.95) to (2.6.96) reduces the two-component transition to a one-component transition. The kind of transition is determined by the sign of $g(A)$ at small A: if $g(A) < 0$ at very small A, the phase transition is second-order (this corresponds to soft excitation of oscillations); if $g(A) > 0$ the phase transition is first-order (hard excitation of oscillations). In the first of these cases, by using the equation (2.6.96) with $g(A)$ found for small amplitudes, we can examine not only the transition from the absence of oscillations to their presence, but also the reverse transition from the presence of self-excited oscillations to their absence.

In concluding this subsection we note that, strictly speaking, the mean $\langle\zeta\rangle$ of the random force in (2.6.96) is nonzero due to correlations between amplitude and phase in (2.6.97) (namely, we have $\langle\zeta\rangle = \text{const} \cdot A^{-1}$, const $\sim \kappa$). To avoid the influence of this term on Ψ, we should set $w(A) = \text{const} \cdot A\exp(-\Psi(A)/\kappa)$ instead of $w(A) = \text{const} \cdot \exp(-\Psi(A)/\kappa)$. Since allowance for $\langle\zeta\rangle$ introduces no significant changes at small κ, this mean value may be ignored.

2.6.8 Example of a Vibrational Phase Transition: Brusselator

As an example we will consider a three-molecular autocatalytic chemical reaction. Suppose that the following reactions occur:

$$A \xrightarrow[k_{-1}]{k_1} X \,,$$

$$B + X \xrightarrow[k_{-2}]{k_2} Y + D \,,$$

$$2X + Y \xrightarrow[k_{-3}]{k_3} 3X \,,$$

$$X \xrightarrow[k_{-4}]{k_4} G \,, \qquad\qquad (2.6.98)$$

where A, B, X, Y, D, G symbolize the reagents. Suppose also that the nonequilibrium is caused by the quick and continuous removal of the reaction products D and G, the system thus being open. Therefore, the reaction constants k_{-2}, k_{-4} may be taken to be zero. Furthermore, to simplify the reaction equations, it is expedient to put $k_{-1} = k_{-3} = 0$. The reactions (2.6.98) will then be described by

$$d[X]/dt = k_1[A] - (k_2[B] + k_4)[X] + k_3[X]^2[Y] \,,$$
$$d[Y]/dt = k_2[B][X] - k_3[X]^2[Y] \,. \qquad\qquad (2.6.99)$$

The symbols in square brackets stand for molar concentrations. Let $[A], [B]$ be maintained constant. This model has been given the name Brusselator, since it was put forward by the Brussels school.

Introducing the notation

$$B_1 = (k_3/k_4)^{1/2}[X] \,,$$
$$B_2 = (k_3/k_4)^{1/2}[Y] \,,$$
$$m = (k_2/k_4)[B] \,,$$
$$n = k_1(k_3/k_4^3)^{1/2}[A] \,,$$
$$\tau = k_4 t \,, \qquad\qquad (2.6.100)$$

we can reduce (2.6.99) to

$$dB_1/d\tau = n - (m+1)B_1 + B_1^2 B_2 \,,$$
$$dB_2/d\tau = mB_1 - B_1^2 B_2 \,. \qquad\qquad (2.6.101)$$

Next we equate to zero the right-hand sides to find the stationary point B^0:

$$B_1^0 = n \,,$$
$$B_2^0 = m/n \,. \qquad\qquad (2.6.102)$$

Denoting $x = B_1 - B_1^0, y = B_2 - B_2^0$ and substituting $B_1 = n + x$ and $B_2 = m/n + y$ into (2.6.101) gives

$$\dot{x} = (m-1)x + n^2y + [(m/n)x^2 + 2nxy + x^2y],$$
$$\dot{y} = -mx - n^2y - [(m/n)x^2 + 2nxy + x^2y], \qquad (2.6.103)$$

where the differentiation dot denotes the derivative with respect to τ. These equations are a special case of (2.6.81); the terms in brackets are nonlinear, they add up to f_1 and f_2. Thus, we have

$$a = m-1, \ b = -n^2, \ c = n^2 > 0, \ d = -m. \qquad (2.6.104)$$

The eigenvalues of (2.6.82) are

$$-\alpha_{1,2} = \tfrac{1}{2}(m-1-n^2) \pm i[n^2 - (\tfrac{1}{2}(m-1-n^2))^2]^{1/2}. \qquad (2.6.105)$$

This suggests that the critical point obeys the equation

$$m-1-n^2 = 0, \qquad (2.6.106)$$

and in it $\alpha_{1,2} = \pm i\omega$. At $m-1-n^2 < 0$ the stationary point (2.6.102) is stable, and at $m-1-n^2 > 0$ it is unstable. Loss of stability corresponds to the vibrational phase transition discussed in the previous subsection.

According to (2.6.85) and (2.6.87) the amplitude of vibrations can be defined by

$$
\begin{aligned}
A^2 &= y^2 + n^{-2}(n^2 + m - 1)xy + n^{-2}mx^2 \\
&\approx y^2 + 2xy + (1 + n^{-2})x^2.
\end{aligned} \qquad (2.6.107)
$$

Equation (2.6.90) then becomes

$$\dot{A} = \frac{m-1-n^2}{2}A + \frac{x}{n^2A}\left(\frac{m}{n}x^2 + 2nxy + x^2y\right). \qquad (2.6.108)$$

We have to substitute (2.6.88), i.e.

$$x = n^2\omega^{-1}A\sin\omega t,$$
$$y = A[\cos\omega t - ((m-1)/\omega)\sin\omega t] \approx A(\cos\omega t - n^2\omega^{-1}\sin\omega t), \qquad (2.6.109)$$

and average the right-hand side of (2.6.108) over the period. In so doing, the terms $(m/n)x^3 + 2nx^2y$ will have no effect on the result of the averaging. Thus we should average the expression

$$(n^2A)^{-1}x^3y = n^4\omega^{-3}A^3\sin^3\omega t(\cos\omega t - n^2\omega^{-1}\sin\omega t) \qquad (2.6.110)$$

to obtain $-3n^6\omega^{-4}A^3/8$. Therefore, (2.6.96) becomes

$$\dot{A} = \frac{m-1-n^2}{2}A - \tfrac{3}{8}n^6\omega^{-4}A^3 + \zeta. \qquad (2.6.111)$$

Suppose that the correlator $\langle\zeta(t_1), \zeta(t_2)\rangle$, after averaging, has the form

$$\langle\zeta(t_1), \zeta(t_2)\rangle = \kappa N\delta(t_1 - t_2), \qquad (2.6.112)$$

where N is a positive constant. Solving then the stationary Fokker-Planck equation, or (2.4.23), gives

$$w(A) = \text{const} \cdot \exp\left[-\frac{1}{\kappa N}\left(\frac{n^2 - m + 1}{2}A^2 + \frac{3}{16}n^6\omega^{-4}A^4\right)\right], \qquad (2.6.113)$$

i.e.

$$\Psi_2 = (n^2 - m + 1)(N)^{-1},$$
$$\Psi_3 = 0,$$
$$\Psi_4 = 9n_c^6/(2N\omega^4). \qquad (2.6.114)$$

The role of the parameter Θ may be played here by m or n [in the latter case $\Psi_2 = 2(n - n_c)/(n_c N)$]. This phase transition is the second-order transition discussed in Sect. 2.6.1; it is described by (2.6.10,12,18).

At $n^2 - m + 1 < 0$, but not too far from the critical point, the stable point A_0 is found from

$$N\partial\Psi(A)/\partial A = (n^2 - m + 1)A + \frac{3}{4}n^6\omega^{-4}A^3 = 0, \qquad (2.6.115)$$

i.e. it is

$$A_0 = 2 \cdot 3^{-1/2}n^{-3}\omega^2(m - 1 - n^2)^{1/2}. \qquad (2.6.116)$$

Expanding about this point gives

$$\begin{aligned}
w(A) &= \text{const} \cdot \exp\{-(\kappa N)^{-1}[(m - 1 - n^2)(A - A_0)^2 \\
&\quad + \frac{3}{4}n^6\omega^{-4}A_0(A - A_0)^3 + \frac{3}{16}n^6\omega^{-4}(A - A_0)^4]\}. \qquad (2.6.117)
\end{aligned}$$

Accordingly, the probability density of the type (2.4.64), which describes the stability of vibrations, has the form

$$\begin{aligned}
w(A) &= \text{const} \cdot \exp\{-(2\kappa)^{-1}[\Psi_2(n)(A - A_0)^2 \\
&\quad + \frac{1}{3}\Psi_3(n_c)(A - A_0)^3 + \frac{1}{12}\Psi_4(n_c)(A - A_0)^4]\}, \qquad (2.6.118)
\end{aligned}$$

where

$$\Psi_2(n) = 4(n_c - n)n_c/N,$$
$$\Psi_3(n_c) = 0,$$
$$\Psi_4(n_c) = 9n_c^6/(2N\omega^4). \qquad (2.6.119)$$

We see that the disappearance of oscillatory conditions is also a second-order transition as described in Sect. 2.6.1.

In this example the calculation of the constant N in (2.6.112) which describes the intensity of fluctuational inputs is a straightforward exercise. The fluctuational equations read

$$d[X]/dt = \ldots + \xi_0(t), \, d[Y]/dt = \ldots + \eta_0(t), \qquad (2.6.120)$$

where the dots denote the terms written out in (2.6.99). Since the random functions ξ_0, η_0 are shot noises, from (2.6.98) we have

$$\langle\xi_0(t_1)\xi_0(t_2)\rangle = (N_AV)^{-1}\{k_1[A] + (k_2[B] + k_4)[X] + k_3[X]^2[Y]\}\delta(t_{12}),$$
$$\langle\eta_0(t_1)\eta_0(t_2)\rangle = -\langle\xi_0(t_1)\eta_0(t_2)\rangle$$
$$= (N_AV)^{-1}\{k_2[B][X] + k_3[X]^2[Y]\}\delta(t_{12}). \qquad (2.6.121)$$

All the terms are taken with the plus sign because shot noises add up in the case of counterflows. The factor $(N_AV)^{-1}$ has emerged since $[X]$, $[Y]$ are molar concentrations and when one molecule appears or disappears they change by $\pm(N_AV)^{-1}$.

According to (2.6.100,102), we find from (2.6.121) the correlators of ξ, η in (2.6.95) as functions of the time $\tau = k_4t$:

$$\langle\xi(\tau_1),\xi(\tau_2)\rangle = 2\frac{k_1k_3k_4^2}{N_AV}\left(1 + \frac{k_2}{k_4}[B]\right)[A]\delta(\tau_{12}), \qquad (2.6.122a)$$

$$\langle\eta(\tau_1),\eta(\tau_2)\rangle = -\langle\xi(\tau_1),\eta(\tau_2)\rangle$$
$$= (N_AV)^{-1}k_1k_3k_4^2(1 + 2k_2k_4^{-1}[B])[A]\delta(\tau_{12}). \qquad (2.6.122b)$$

Making (2.6.97) explicit gives

$$\zeta = \xi(\omega^{-1}\sin\omega\tau + \cos\omega\tau) + \eta\cos\omega\tau, . \qquad (2.6.123)$$

Hence, using (2.6.122) we can easily find (2.6.112), where

$$\kappa N = \tfrac{1}{2}(N_AV)^{-1}k_1k_3k_4^2\{1 + 2\omega^{-2}(1 + k_2k_4^{-1}[B])\}[A]. \qquad (2.6.124)$$

The small parameter κ here is $(N_AV)^{-1}$ or some quantity proportional to it. If the system is exposed to external fluctuational inputs, the parameter κ can be larger.

2.7 Parameter Fluctuations Near a Two-Component Kinetic Phase Transition

2.7.1 Determining the Matrix $\psi_{\alpha\beta}$

In the two-component case parameter fluctuations in the nonequilibrium kinetic phase transition can also be studied using the methods considered in Sects. 2.5,6.

For the two-component vector $B = (B_1, B_2)$ the matrix $\psi_{\alpha\beta}$ has three independent elements; $\psi_{\alpha\beta\gamma}$ has four, and $\psi_{\alpha\beta\gamma\delta}$, five. We denote these as follows:

$$\Psi_{20} = \psi_{11}, \ \Psi_{11} = \psi_{12} = \psi_{21}, \ \Psi_{02} = \psi_{22};$$
$$\Psi_{30} = \psi_{111}, \ \Psi_{21} = \psi_{112}, \ \Psi_{12} = \psi_{122}, \ \Psi_{02} = \psi_{222};$$
$$\Psi_{40} = \psi_{1111}, \ \Psi_{31} = \psi_{1112}, \ \Psi_{22} = \psi_{1122}, \ \Psi_{13} = \psi_{1222}, \ \Psi_{04} = \psi_{2222}. \qquad (2.7.1)$$

The matrix

$$\| \psi_{\alpha\beta} \| = \begin{pmatrix} \Psi_{20} & \Psi_{11} \\ \Psi_{11} & \Psi_{02} \end{pmatrix} = \hat{H}^{-1} \qquad (2.7.2)$$

is found from the equations (2.4.15), which are equivalent to the set of linear equations (2.4.35). We now introduce the following notation for the matrix elements in (2.4.35):

$$\hat{A} = - \parallel k_{\alpha,\beta} \parallel \; = \begin{pmatrix} a & c \\ d & b \end{pmatrix}, \tag{2.7.3a}$$

$$\hat{H} = \begin{pmatrix} x & y \\ y & z \end{pmatrix}, \tag{2.7.3b}$$

$$\hat{N} = \parallel k^{0}_{\mu\nu} \parallel \; = \begin{pmatrix} 2\alpha & \beta \\ \beta & 2\gamma \end{pmatrix}. \tag{2.7.3c}$$

Substituting (2.7.3) into (2.4.35), we obtain the system of three equations

$$
\begin{array}{l}
ax + cy = \alpha, \\
dx + (a+b)y + cz = \beta, \\
dy + bz = \gamma,
\end{array}
\qquad
\begin{pmatrix} a & c & 0 \\ d & a+b & c \\ 0 & d & b \end{pmatrix}
\begin{pmatrix} x \\ y \\ z \end{pmatrix}
=
\begin{pmatrix} \alpha \\ \beta \\ \gamma \end{pmatrix}.
\tag{2.7.4}
$$

Solving it yields

$$x = \Delta^{-1}\{\alpha[b(a+b) - cd] - \beta bc + \gamma c^2\}, \tag{2.7.5a}$$

$$y = -\delta^{-1}(\alpha bd - \beta ab + \gamma ac), \tag{2.7.5b}$$

$$z = \Delta^{-1}\{ad^2 - \beta ad + \gamma[a(a+b) - cd]\}, \tag{2.7.5c}$$

where

$$\Delta = (a+b)(ab - cd). \tag{2.7.6}$$

We will now go over from $\hat{H}$ to the inverse matrix

$$
\begin{aligned}
\parallel \psi_{\alpha\beta} \parallel \; &= \begin{pmatrix} \Psi_{20} & \Psi_{11} \\ \Psi_{11} & \Psi_{02} \end{pmatrix} \\
&= \begin{pmatrix} x & y \\ y & z \end{pmatrix}^{-1} \\
&= (xz - y^2)^{-1} \begin{pmatrix} z & -y \\ -y & x \end{pmatrix}.
\end{aligned}
\tag{2.7.7}
$$

Using (2.7.5) and (2.7.7) gives

$$
\begin{aligned}
\Psi_{20} &= (a+b)G^{-1}\{\alpha d^2 - \beta ad + \gamma[a(a+b) - cd]\}, \\
\Psi_{11} &= (a+b)G^{-1}(\alpha bd - \beta ab + \gamma ac), \\
\Psi_{02} &= (a+b)G^{-1}\{\alpha[b(a+b) - cd] - \beta bc + \gamma c^2\},
\end{aligned}
\tag{2.7.8}
$$

where

$$
\begin{aligned}
G \; = \; &\alpha^2 d^2 - \beta^2 ab + \gamma^2 c^2 - \beta(\alpha d - \gamma c)(a - b) \\
&+ \alpha\gamma(2ab - 2cd + a^2 + b^2).
\end{aligned}
\tag{2.7.9}
$$

When at the critical point $\Theta = \Theta_c$ one of the eigenvalues α_1, α_2 of the matrix $\hat{A}$ vanishes, and the other does not, the expression $ab - cd$ tends to zero as $\Theta \longrightarrow \Theta_c$, and $a + b$ does not. In the oscillatory case, where the eigenvalues α_1, α_2 are purely imaginary, it is $a + b$ that tends to zero, not $ab - cd$. Lastly, when at the critical point both eigenvalues $\alpha_1(\Theta_c), \alpha_2(\Theta_c)$ are zero,

$$a + b \longrightarrow 0, \quad ab - cd \longrightarrow 0 \quad \text{for} \quad \Theta \longrightarrow \Theta_c. \tag{2.7.10}$$

The above properties of $a + b$ and $ab + cd$ enable us to somewhat simplify (2.7.8,9) near the critical point. So in the case (2.7.10), if

$$\alpha d - \beta a - \gamma c \neq 0 \quad \text{for} \quad \Theta = \Theta_c, \tag{2.7.11}$$

we will be able to use the formulas

$$\begin{aligned}
\Psi_{20} &= (a + b)d(\alpha d - \beta a - \gamma c)^{-1}, \\
\Psi_{11} &= (a + b)b(\alpha d - \beta a - \gamma c)^{-1}, \\
\Psi_{02} &= -(a + b)c(\alpha d - \beta a - \gamma c)^{-1}.
\end{aligned} \tag{2.7.12}$$

In the case (2.7.10), just as in the oscillatory case discussed in Sect. 2.6.7, the matrix (2.7.2) vanishes as the critical point is approached.

2.7.2 Calculation of Matrices of the Third Derivatives $\psi_{\alpha\beta\gamma}$

This matrix is given by (2.4.18), where instead of $\psi_{\alpha\mu}k_{\mu,\nu}\psi_{\nu\sigma}^{-1} = v_{\alpha\sigma}$ we can take the binomial on the right-hand side of (2.4.54). In that case, where all the matrix elements $\psi_{\alpha\beta}$ vanish at the critical point, in place of (2.4.18) we can apply the simpler equation

$$v_{\alpha\sigma}\psi_{\alpha\beta\gamma} + v_{\beta\sigma}\psi_{\sigma\alpha\gamma} + v_{\gamma\sigma}\psi_{\sigma\alpha\beta} = z_{\alpha\beta\gamma}, \tag{2.7.13}$$

where

$$z_{\alpha\beta\gamma} = \psi_{\alpha\mu}k_{\mu,\beta\gamma} + \psi_{\beta\mu}k_{\mu,\alpha\gamma} + \psi_{\gamma\mu}k_{\mu,\alpha\beta}. \tag{2.7.14}$$

Here the higher degrees of $\psi_{\alpha\beta}$ are discarded. In analogy with (2.6.3a), we will introduce the notation

$$\| v_{\alpha\sigma} \| = \begin{pmatrix} a_0 & c_0 \\ d_0 & b_0 \end{pmatrix}. \tag{2.7.15}$$

Substituting (2.7.15) into (2.7.13) and summing gives

$$2a_0 x_1 + c_0 x_2 = z_1, d_0 x_1 + (a_0 + \tfrac{1}{2}b_0)x_2 + c_0 x_3 = z_2,$$

$$d_0 x_2 + (b_0 + \tfrac{1}{2}a_0)x_3 + c_0 x_4 = z_3,$$

$$d_0 x_3 + 2b_0 x_4 = z_4 \tag{2.7.16}$$

by (2.7.1). Here

$$\begin{aligned}
x_1 &= \Psi_{30}, \ x_2 = 2\Psi_{21}, \ x_3 = 2\Psi_{12}, \ x_4 = \Psi_{03}, \\
z_1 &= \tfrac{2}{3}z_{111} = 2\Psi_{20}k_{1,11} + 2\Psi_{11}k_{2,11}, \\
z_2 &= z_{112} = 2\Psi_{20}k_{1,12} + 2\Psi_{11}k_{2,12} + \Psi_{02}k_{2,11} + \Psi_{11}k_{1,11}, \\
z_3 &= z_{122} = \Psi_{20}k_{1,22} + \Psi_{11}k_{2,22} + 2\Psi_{02}k_{2,12} + 2\Psi_{11}k_{1,12}, \\
z_4 &= \tfrac{2}{3}z_{222} = 2\Psi_{11}k_{1,22} + 2\Psi_{02}k_{2,22}.
\end{aligned} \tag{2.7.17}$$

The solution of the equations (2.7.16) is

$$x_i = \lim_{\Theta \to \Theta_c} D^{-1} M_{ji} z_j \,, \qquad (2.7.18)$$

where D, M_{ij} are respectively the determinant and the algebraical adjuncts of the matrix

$$\begin{pmatrix} 2a_0 & c_0 & 0 & 0 \\ d_0 & (a_0 + b_0/2) & c_0 & 0 \\ 0 & d_0 & (b_0 + a_0/2) & c_0 \\ 0 & 0 & d_0 & 2b_0 \end{pmatrix}. \qquad (2.7.19)$$

The limit in (2.7.18) comes from the fact that we are interested in the derivatives of quasi-frenergy at the critical point $\Theta \longrightarrow \Theta_c$. After a simple calculation, we arrive at

$$\begin{aligned}
D &= r[2(a_0 + b_0)^2 + r]\,, \\
M_{11} &= r(a_0 + 5b_0/2) + b_0^3, \ M_{44} = r(b_0 + 5a_0/2) + a_0^3\,, \\
M_{22} &= 2a_0(2b_0^2 + r), \ M_{33} = 2b_0(2a_0^2 + r)\,, \\
M_{12} &= -d_0(2b_0^2 + r), \ M_{34} = -d_0(2a_0^2 + r), \ M_{23} = -4a_0 b_0 d_0\,, \\
M_{13} &= 2b_0 d_0^2, \ M_{24} = 2a_0 d_0^2, \ M_{14} = -d_0^3\,,
\end{aligned} \qquad (2.7.20)$$

where

$$r = a_0 b_0 - c_0 d_0 \,. \qquad (2.7.21)$$

We obtain M_{ji} if we interchange c_0 and d_0 in the expression for M_{ij}.

Thus, equations (2.7.13) and (2.7.16) can be solved readily to yield $\psi_{\alpha\beta\gamma}$. The limiting process in (2.7.18) is carried out differently in different special cases.

2.7.3 Parameter Fluctuations in the Critical Region for Two-Component First-Order Transition

If the limiting process $\Theta \longrightarrow \Theta_c$ in (2.7.18) yields the zero derivatives $\Psi_{30}, \Psi_{21}, \Psi_{12}, \Psi_{03}$, then this kinetic phase transition will be a first-order transition.

At not-too-large deviations $b_\alpha = B_\alpha - B_\alpha^0$ we can apply the expression

$$w(b) = \text{const} \cdot \exp[-(2\kappa)^{-1}(\psi_{\alpha\beta} b_\alpha b_\beta + \tfrac{1}{3}\psi_{\alpha\beta\gamma} b_\alpha b_\beta b_\gamma)] \qquad (2.7.22)$$

which determines the stationary probability density of the parameters. Introducing the polar coordinates by

$$\begin{aligned}
b_1 &= \rho \sin\varphi \equiv \rho n_1\,, \\
b_2 &= \rho \cos\varphi \equiv \rho n_2\,,
\end{aligned} \qquad (2.7.23)$$

we can write the expression in (2.7.22) in square brackets as

$$-(2\kappa)^{-1}[g(\varphi)\rho^2 + \tfrac{1}{3}f(\varphi)\rho^3] \equiv -\kappa^{-1} U(\rho, \varphi)\,, \qquad (2.7.24)$$

where

$$g(\varphi) = \psi_{\alpha\beta} n_\alpha n_\beta$$
$$= \Psi_{20} \sin^2 \varphi + 2\Psi_{11} \sin \varphi \cos \varphi + \Psi_{02} \cos^2 \varphi \,, \qquad (2.7.25a)$$
$$f(\varphi) = \psi_{\alpha\beta\gamma} n_\alpha n_\beta n_\gamma$$
$$= \Psi_{30} \sin^3 \varphi + 3\Psi_{21} \sin^2 \varphi \cos \varphi + 3\Psi_{12} \sin \varphi \cos^2 \varphi + \Psi_{03} \cos^3 \varphi \,. \qquad (2.7.25b)$$

Let $f(\varphi) \neq 0$ for any φ and besides $g(\varphi) > 0$. We will now find out whether or not the function $U(\rho, \varphi)$ has "watershed" points, i.e., maxima with the radius ρ varying and the angle φ fixed. The maximum condition reads

$$\partial U(\rho, \varphi)/\partial\rho = g(\varphi)\rho + \tfrac{1}{2}f(\varphi)\rho^2 = 0 \,. \qquad (2.7.26)$$

Hence the coordinate of the "watershed" point is

$$\rho_w(\varphi) = -2g(\varphi)/f(\varphi) > 0 \,. \qquad (2.7.27)$$

Do such points exist, i.e., can $g(\varphi)/f(\varphi)$ be negative? It is seen from (2.7.25) that this ratio changes sign when the vector $\mathbf{n}$ does, and so watersheds do indeed exist. Using (2.7.27), we will find the height of the "watershed" to be

$$U(\rho_w(\varphi), \varphi) = \tfrac{2}{3}g^3(\varphi)/f^2(\varphi) \,. \qquad (2.7.28)$$

We would like to minimize this expression. Let a minimum occur at φ_m, then

$$\Delta U = \tfrac{2}{3}g^3(\varphi_m)/f^2(\varphi_m) = \min_\varphi \left[\tfrac{2}{3}g^3(\varphi)/f^2(\varphi)\right] \,. \qquad (2.7.29)$$

In analogy with (2.6.28) we will introduce the parameter of minimal relative height of the "watershed"

$$\lambda = \tfrac{2}{3}\kappa^{-1}g^3(\varphi_m)f^{-2}(\varphi_m) \,. \qquad (2.7.30)$$

We can arbitrarily refer to the state corresponding to the point B^0 as sufficiently stable if $\lambda \geq 3$ (or, say, $\lambda \geq 4$), and to the range of Θ, where $\lambda \sim 3$, but $\lambda > 3$ as a critical region.

In the critical region, instead of (2.7.22), we can take the probability density

$$w(b) = \text{const} \cdot \exp[-(2\kappa)^{-1}(\psi_{\alpha\beta}b_\alpha b_\beta + \tfrac{1}{3}\psi_{\alpha\beta\gamma}b_\alpha b_\beta b_\gamma)] \quad \text{for} \sum_\alpha b_\alpha^2 < \rho_w(\varphi_m) \,,$$

$$w(b) = 0 \quad \text{for} \sum_\alpha b_\alpha^2 > \rho_w(\varphi_m) \qquad (2.7.31)$$

and find the corresponding correlators. There is a less rigorous way of finding the correlators in the critical region. For simplicity, we can use the probability density (2.7.22), but represent it in the form of the expansion

$$w(b) = \text{const} \cdot \exp[-(2\kappa)^{-1}\psi_{\alpha\beta}b_\alpha b_\beta]$$
$$\times \{1 - (6\kappa)^{-1}\psi_{\alpha\beta\gamma}b_\alpha b_\beta b_\gamma + \tfrac{1}{2}[(6\kappa)^{-1}\psi_{\alpha\beta\gamma}b_\alpha b_\beta b_\gamma]^2 + \ldots\} \,. \qquad (2.7.32)$$

Using this expansion, or rather only the terms that are written out here we will have

$$\langle B_\alpha \rangle = B_\alpha^0 - \tfrac{1}{2}\kappa\psi_{\alpha\rho}^{-1}\psi_{\rho\sigma\tau}\psi_{\sigma\tau}^{-1},$$

$$\langle B_\alpha, B_\beta \rangle = \kappa\psi_{\alpha\beta}^{-1} + \tfrac{1}{2}\kappa^2\psi_{\alpha\rho}^{-1}\psi_{\rho\sigma\tau}\psi_{\mu\nu\lambda}(\psi_{\beta\mu}^{-1}\psi_{\sigma\nu}^{-1}\psi_{\tau\lambda}^{-1} + \psi_{\beta\sigma}^{-1}\psi_{\tau\mu}^{-1}\psi_{\nu\lambda}^{-1}),$$

$$\langle B_\alpha, B_\beta, B_\gamma \rangle = -\kappa^2\psi_{\alpha\mu}^{-1}\psi_{\beta\nu}^{-1}\psi_{\gamma\lambda}^{-1}\psi_{\mu\nu\lambda},$$

$$\langle B_\alpha, B_\beta, B_\gamma, B_\delta \rangle = 3\kappa^3\psi_{\alpha\mu}^{-1}\psi_{\beta\nu}^{-1}\psi_{\gamma\rho}^{-1}\psi_{\delta\sigma}^{-1}\{\psi_{\mu\nu\lambda}\psi_{\lambda\tau}^{-1}\psi_{\tau\rho\sigma}\}_{\text{sym}}. \qquad (2.7.33)$$

The factors κ^n here do not indicate the correct order of the correlators in the critical region. The fact is that the matrices $\psi_{\mu\nu}^{-1}$ in the critical region are large and they can be thought of as dependent on κ (at fixed λ). In order to bring out the true order of the terms in (2.7.33) in a graphic way, we will introduce the matrices

$$\tilde{\psi}_{\alpha\beta} = \psi_{\alpha\beta}/g(\varphi_m),$$

$$\tilde{\psi}_{\alpha\beta\gamma} = \psi_{\alpha\beta\gamma}/f(\varphi_m) \qquad (2.7.34)$$

"normalized" to unity, i.e. obeying

$$\tilde{\psi}_{\alpha\beta}n_\alpha(\varphi_m)n_\beta(\varphi_m) = 1,$$

$$\tilde{\psi}_{\alpha\beta\gamma}n_\alpha(\varphi_m)n_\beta(\varphi_m)n_\gamma(\varphi_m) = 1. \qquad (2.7.35)$$

The matrix $\tilde{\psi}_{\alpha\beta}$ is convenient because, unlike $\psi_{\alpha\beta}$, it does not tend to zero as $\Theta \longrightarrow \Theta_c$. In (2.7.33), we write $g(\varphi_m)\tilde{\psi}_{\alpha\beta}$, $f(\varphi_m)\tilde{\psi}_{\alpha\beta\gamma}$ instead of $\psi_{\alpha\beta}, \psi_{\alpha\beta\gamma}$. Using (2.7.30), we then have

$$\langle B_\alpha, B_\beta \rangle = (2\kappa^2)^{1/3}(3\lambda)^{-1/3} \mid f(\varphi_m) \mid^{-2/3} [\tilde{\psi}_{\alpha\beta}^{-1} - (3\lambda)^{-1}\tilde{\psi}_{\alpha\rho}^{-1}\tilde{\psi}_{\rho\sigma\tau}$$

$$\times \tilde{\psi}_{\mu\nu\lambda}(\tilde{\psi}_{\beta\mu}^{-1}\tilde{\psi}_{\sigma\nu}^{-1}\tilde{\psi}_{\tau\lambda}^{-1} + \tilde{\psi}_{\beta\sigma}^{-1}\tilde{\psi}_{\tau\mu}^{-1}\tilde{\psi}_{\nu\lambda}^{-1}) + O(\lambda^{-2})],$$

$$\langle B_\alpha, B_\beta, B_\gamma \rangle = -2\kappa(3\lambda)^{-1}f^{-1}(\varphi_m)\tilde{\psi}_{\alpha\mu}^{-1}\tilde{\psi}_{\beta\nu}^{-1}\tilde{\psi}_{\gamma\lambda}^{-1}\tilde{\psi}_{\mu\nu\lambda} + O(\lambda^{-2}),$$

$$\langle B_\alpha, B_\beta, B_\gamma B_\delta \rangle = 3\kappa^{4/3}2^{5/3}(3\lambda)^{-5/3} \mid f(\varphi_m) \mid^{-4/3} \tilde{\psi}_{\alpha\mu}^{-1}$$

$$\times \tilde{\psi}_{\beta\nu}^{-1}\tilde{\psi}_{\gamma\rho}^{-1}\tilde{\psi}_{\delta\sigma}^{-1}\{\tilde{\psi}_{\mu\nu\lambda}\tilde{\psi}_{\lambda\tau}^{-1}\tilde{\psi}_{\tau\rho\sigma}\}_{\text{sym}} + O(\lambda^{-8/3}). \qquad (2.7.36)$$

The quantity $f(\varphi_m)$ here is independent of κ, λ.

The formulas derived serve as a two-dimensional generalization of (2.6.37,38). Using the expansion (2.7.32), we can also obtain more exact relationships. It is seen from (2.7.36) that, just as in the one-component case, with two components in the critical region parameter fluctuations are abnormally large and strongly non-Gaussian, unlike the usual fluctuations, which occur far from the critical point. Formulas similar to (2.7.33,36) also hold when B has more than two components.

2.7.4 Two-Component Transition: A Special Case

When the matrix $\hat{A} = - \parallel k_{\alpha,\beta} \parallel$ has only one zero eigenvalue at the critical point, i.e. when

$$ab - cd = 0, \ a + b \neq 0 \quad \text{at} \quad \Theta = \Theta_c, \qquad (2.7.37)$$

the elements of $\psi_{\alpha\beta}$, as is seen from (2.7.8,9), do not tend to zero as the critical point is approached. This distinguishes this case from (2.7.10) and the oscillatory case. Although $\psi_{\alpha\beta}$ does not tend to zero, the inverse matrix $\psi_{\alpha\beta}^{-1}$ tends to infinity as $\Theta \longrightarrow \Theta_c$, which can be verified using (2.7.5,6). In the special case (2.7.37) we cannot use (2.7.14), instead we have to employ the complete expression

$$z_{\alpha\beta\gamma} = 3\{\psi_{\alpha\sigma}k_{\sigma,\beta\gamma}\}_{\text{sym}} + 3\{\psi_{\alpha\mu}\psi_{\beta\nu}k_{\mu\nu,\gamma}^0\}_{\text{sym}} + \psi_{\alpha\mu}\psi_{\beta\nu}\psi_{\gamma\lambda}k_{\mu\nu\lambda}^0 \qquad (2.7.38)$$

that enters into (2.4.18). Accordingly, $\psi_{\alpha\beta\gamma}$ are now influenced by the coefficients $K_{\alpha\beta\gamma}$ in the master equation, which is not the Fokker-Planck equation. This means that the Fokker-Planck approximation is now insufficient if the initial equation is not the Fokker-Planck equation.

By virtue of the expression $ab = cd$, at $\Theta = \Theta_c$ (2.7.8) take the form

$$\Psi_{20} = (a + b)G^{-1}(\alpha d^2 - \beta ad + \gamma a^2),$$
$$\Psi_{11} = (a + b)G^{-1}(\alpha bd - \beta ab + \gamma ac),$$
$$\Psi_{02} = (a + b)G^{-1}(\alpha b^2 - \beta bc + \gamma c^2). \qquad (2.7.39)$$

Since $b/d = c/a$ at the critical point (we will denote this ratio by s), from (2.7.39) we readily find

$$\Psi_{11} = s\Psi_{20},$$
$$\Psi_{02} = s^2\Psi_{20} \qquad (2.7.40)$$

for $\Theta = \Theta_c$. Similarly, from (2.7.5) we have

$$y/z \longrightarrow -s,$$
$$x/y \longrightarrow -s \qquad (2.7.41)$$

for $\Theta \longrightarrow \Theta_c$, where the rate at which these relations converge is the same. If we substitute (2.7.40) into the expression (2.7.25a), which defines $g(\varphi)$, we will have

$$g(\varphi) = \Psi_{20}(\sin\varphi + s\cos\varphi)^2. \qquad (2.7.42)$$

It follows that, although $\psi_{\alpha\beta}$ and $g(\varphi)$ generally do not tend to zero as $\Theta \longrightarrow \Theta_c$, the function $g(\varphi)$ does tend to zero at

$$\varphi_1 = -\arctan s,$$
$$\varphi_2 = \pm\pi - \arctan s. \qquad (2.7.43)$$

Of these two values we should choose the one for which $g(\varphi)/f(\varphi) < 0$ according to (2.7.27). This value coincides with φ_m in (2.7.30). For this value the height of the "watershed" will vanish as $\Theta \longrightarrow \Theta_c$, thus corresponding to a first-order transition.

A significant feature of this special case is that not at all c_1 and c_2, the correlators of the random variable $B_0 = c_1 B_1 + c_2 B_2$, have the same order of magnitude in κ. For the majority of values of c_1 and c_2 the correlators of B_0 have the same order as in (2.7.36)

$$\langle B_0, B_0 \rangle = c_\alpha c_\beta \langle B_\alpha, B_\beta \rangle \sim \kappa^{2/3},$$
$$\langle B_0, B_0, B_0 \rangle = c_\alpha c_\beta c_\gamma \langle B_\alpha, B_\beta, B_\gamma \rangle \sim \kappa,$$
$$\langle B_0, B_0, B_0, B_0 \rangle = c_\alpha c_\beta c_\gamma c_\delta \langle B_\alpha, B_\beta, B_\gamma, B_\delta \rangle \sim \kappa^{4/3}. \qquad (2.7.44)$$

A special situation arises at $c_2/c_1 = b/d$ or $c_2/c_1 = c/a$. The point is that the substitution of (2.7.33) into

$$c_\alpha c_\beta \langle B_\alpha, B_\beta \rangle \,,$$
$$c_\alpha c_\beta c_\gamma \langle B_\alpha B_\beta, B_\gamma \rangle, \ldots \qquad (2.7.45)$$

yields the combinations

$$
\begin{aligned}
c_\alpha \psi_{\alpha 1}^{-1} &= c_1[\psi_{11}^{-1} + (c_2/c_1)\psi_{12}^{-1}] \\
&= c_1[\psi_{11}^{-1} + (b/d)\psi_{12}^{-1}] \\
&= c_1[x + (b/d)y] \,, \qquad (2.7.46a) \\
c_\alpha \psi_{\alpha 2}^{-1} &= c_1[\psi_{12}^{-1} + (b/d)\psi_{22}^{-1}] \\
&= c_1[y + (b/d)z] \qquad (2.7.46b)
\end{aligned}
$$

or

$$
\begin{aligned}
c_\alpha \psi_{\alpha 1}^{-1} &= c_1[x + (c/a)y] \,, \\
c_\alpha \psi_{\alpha 2}^{-1} &= c_1[y + (c/a)z] \,. \qquad (2.7.47)
\end{aligned}
$$

Using (2.7.5,6), we can readily verify that these combinations do not tend to infinity as $ab - cd \longrightarrow 0$, i.e. as $\Theta \longrightarrow \Theta_c$. This suggests that the correlators $\langle B_0, B_0 \rangle$, $\langle B_0, B_0, B_0 \rangle$ at the above values of c_2/c_1 at the critical point have the order of κ and κ^2, respectively.

Thus, at $c_2/c_1 = b/d$ or $c_2/c_1 = c/a$ the quantity $B_0 = c_1 B_1 + c_2 B_2$ fluctuates in the critical region in the same manner as elsewhere. At other values of the ratio the critical fluctuations are much larger.

Note that in the case (2.7.37), it is convenient to reduce the number of variables, i.e., to go over to the one-component case and treat the quasi-frenergy as a function of one variable, say B_1. It is related to the one-dimensional distribution in the normal way: $w(B_1) = C \times \exp[-\Psi(B_1)/\kappa]$. Then the derivative $\psi_{11}' = [\partial^2 \Psi(B_1)/\partial B_1^2]_0$ will, by (2.7.5), be

$$\psi_{11}' = x^{-1} = (a+b)(ab-cd)[\alpha(b^2 + ab - cd) - \beta bc + \gamma c^2]^{-1} \,, \qquad (2.7.48)$$

i.e. it will tend to zero as $\Theta \longrightarrow \Theta_c$ and we will have an ordinary one-component kinetic phase transition, similar to those discussed in Sect. 2.6. The trick of reducing the number of variables has also been used in the example of Sect. 2.5. For the general case, it has been considered in [2.10].

2.7.5 Example of a Two-Component Kinetic Phase Transition

Consider a two-component multistable system – the open reactor of volume V with a continuous isobaric input and output of the reagent X. Inside the reactor occurs the first-order reaction $X \longrightarrow D$. The reaction product D goes out of the reactor together with X. The reaction is exothermic, i.e. it releases heat. Stirring maintains the concentration and temperature of the reagent at the same level throughout V. The reaction constant $k(T)$ is assumed to vary with temperature following the Arrhenius law

$$k(T) = L \exp(-E/RT), \tag{2.7.49}$$

where R is the gas constant, E the molar activation energy, and L a constant. So the reaction gives off heat with the result that the temperature increases, and so does the reaction rate. We thus see that the reaction can be self-enhanced, so that the system may be multistable.

Let the reagent with a concentration x_1 and temperature T_1 be fed to the reactor at a constant volume velocity q and let the mixture leave the reactor with the same velocity. The output concentration is x and the output temperature is T. The values of x and T are the same as within the reactor. For simplicity, we will assume that the reaction product D has the same heat capacity c_p as the reagent, therefore the output mixture has the same heat capacity as the input reagent. It can easily be seen that this process is described by

$$dx/dt = (q/V)(x_1 - x) - k(T)x + \xi, \tag{2.7.50a}$$
$$(Vc_p + C)dT/dt = qc_p(T_1 - T) + \Delta H V k(T)x + \eta. \tag{2.7.50b}$$

Equation (2.7.50b) describes the energy balance. Here ΔH is the heat of reaction, c_p is the heat capacity per unit volume, C is the total heat capacity of the reactor walls, and ξ, η are fluctuation terms having zero means.

With sufficiently large input and output the fluctuations can be thought of as having purely shot nature. It is a straightforward exercise to calculate their statistics using the ideal gas model. It is hardly necessary here to provide the results. We will confine ourselves to an examination of equations (2.7.50). Substituting (2.7.49) gives

$$\dot{x} = L[l(x_1 - x) - x \exp(-\mu/y)] + \xi,$$
$$\dot{y} = LV\Delta H[ml(y_1 - y) + x \exp(-\mu/y)] + \eta, \tag{2.7.51}$$

where

$$y = (Vc_p + C)T,$$
$$y_1 = (Vc_p + C)T_1,$$
$$l = q/(VL),$$
$$m = c_p[(Vc_p + C)\Delta H]^{-1},$$
$$\mu = E(Vc_p + C)/R. \tag{2.7.52}$$

We would now like to find the stationary points of the averaged equations, i.e. ones in which the fluctuations ξ and η are discarded. Equating the right-hand side to zero yields

$$l(x_1 - x) - x \exp(-\mu/y) = 0, \tag{2.7.53a}$$
$$ml(y_1 - y) + x \exp(-\mu/y) = 0. \tag{2.7.53b}$$

Adding these equations together gives

$$x - x_1 = m(y_1 - y) \tag{2.7.54}$$

at the stationary point. Substituting (2.7.54) into (2.7.53b), we get

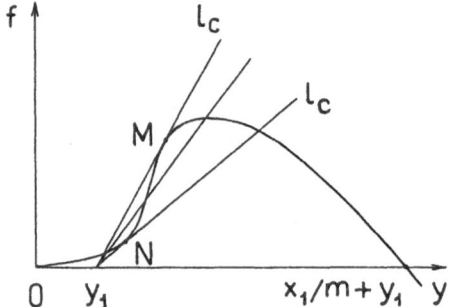

Fig. 2.12. Plot schematically illustrating the possibility of bistability in the open reactor

$$l(y - y_1) = (x_1/m + y_1 - y)\exp(-\mu/y). \qquad (2.7.55)$$

From this equation y can be found graphically. The functions of y on the left- and right-hand sides of this equation are schematically plotted in Fig. 2.12. The expression on the right-hand side is represented by the curve that at the origin of the coordinates is tangent to the abscissa axis. We will fix $y_1, \mu, x_1/m$ and change l, i.e. the slope of the straight line. If y_1 is taken to be sufficiently small, then depending on l, the number of roots may be different: one or three, i.e. multistability is possible. The critical value l_c, i.e. the value corresponding to a transition from mono- to multistability or vice versa, is one of the two values of l at which the straight line is tangent to the curve (the points of tangency being M and N). The value of y at which the tangency occurs will be denoted by y_c. Using (2.7.54) we can find the corresponding value $x_c = x_1 + m(y_1 - y_c)$.

It is easily seen that by (2.7.55) and because of the tangency the function

$$\chi(y) = l(y_1 - y) + (x_1/m + y_1 - y)\exp(-\mu/y) \qquad (2.7.56)$$

at the critical point has the expansion

$$\chi(y) = \tfrac{1}{2}\chi''(y_c)(y - y_c)^2 + \dots \quad \text{for} \quad l = l_c. \qquad (2.7.57)$$

It can be easily verified that, using (2.7.56), we can write (2.7.51) as

$$\dot{x} = L[(x_1 - x + my_1 - my)(l + \exp(-\mu/y)) - m\chi(y)] + \xi,$$
$$\dot{y} = LV\Delta H[m\chi(y) + (x - x_1 + my - my_1)\exp(-\mu/y)] + \eta. \qquad (2.7.58)$$

According to (2.7.54,57) the expressions in brackets vanish at $x = x_c, y = y_c$. Expanding these expressions in the deviations $b_1 = x - x_c, b_2 = y - y_c$, we will, by the use of (2.7.54,57), have

$$\dot{b}_1 = L\{-(b_1 + mb_2)[l_c + \exp(-\mu/y_c)$$
$$+ \mu y_c^{-2}\exp(-\mu/y_c)b_2] - \tfrac{1}{2}m\chi''b_2^2\} + \xi, \qquad (2.7.59a)$$

$$\dot{b}_2 = LV\Delta H[(b_1 + mb_2)\exp(-\mu/y_c)(1 + \mu y_c^{-2}b_2)$$
$$+ \tfrac{1}{2}m\chi''b_2^2] + \eta. \qquad (2.7.59b)$$

Keeping on the right-hand side the terms linear in b_i we can, from (2.7.3a), easily find the matrix $\hat{A}$ at the critical state

$$a = L[l_c + \exp(-\mu/y_c)], \quad c = ma,$$
$$d = -LV\Delta H \exp(-\mu/y_c), \quad b = md. \tag{2.7.60}$$

We see that in a given example, if the condition $a + b > 0$, i.e.

$$q_c(VL)^{-1}\exp(E/RT) + 1 > Vc_p/(Vc_p + C) \tag{2.7.61}$$

is met (and it is met automatically), at the critical point the matrix A has only one zero eigenvalue, or rather the condition (2.7.37) is satisfied. Consequently, we have here a first-order transition, as discussed in Sect. 2.7.4.

After having found $k^0_{\alpha\beta}, k^0_{\alpha\beta,\gamma}, k^0_{\alpha\beta\gamma}$ we can then use the equations (2.7.8,9,38) and (2.7.18,20) to find $\psi_{\alpha\beta}, \psi_{\alpha\beta\gamma}$ ($k_{\alpha,\beta\gamma}$ is easily found from (2.7.59)). It is expedient to carry out the simple, although somewhat unwieldy, calculations for given numerical values of $V, q, c_p, C, \Delta H$, etc. As a result, using (2.7.33), we can find the one-time stationary correlators for the processes $x(t)$ and $T(t)$ in the critical region.

2.7.6 Parameter Fluctuations Given by $\psi_{\alpha\beta\gamma\delta}$

If all the elements of $\psi_{\alpha\beta\gamma}(\Theta_c)$ are zero, then the kind of kinetic phase transition and the special features of the parameter fluctuations in a critical region are determined by the matrix $\psi_{\alpha\beta\gamma\delta}(\Theta_c)$, which, using (2.4.19), is found from

$$4\{v_{\alpha,\sigma}\psi_{\sigma\beta\gamma\delta}\}_{\text{sym}} = z_{\alpha\beta\gamma\delta}, \tag{2.7.62}$$

where

$$z_{\alpha\beta\gamma\delta} = 4\{\psi_{\alpha\mu}k_{\mu,\beta\gamma\delta}\}_{\text{sym}} \tag{2.7.63}$$

in the case (2.7.10) and the oscillatory case, and where

$$z_{\alpha\beta\gamma\delta} = 4\{\psi_{\alpha\mu}k_{\mu,\beta\gamma\delta}\}_{\text{sym}} + 6\{\psi_{\alpha\mu}\psi_{\beta\nu}k^0_{\mu\nu,\gamma\delta}\}_{\text{sym}}$$
$$+ 4\{\psi_{\alpha\mu}\psi_{\beta\nu}\psi_{\gamma\lambda}k^0_{\lambda\mu\nu,\delta}\}_{\text{sym}} + \psi_{\alpha\mu}\psi_{\beta\nu}\psi_{\gamma\lambda}\psi_{\delta\kappa}k^0_{\mu\nu\lambda\kappa} \tag{2.7.64}$$

in the case (2.7.37). Denoting

$$x_1 = \Psi_{40}, \quad x_2 = \Psi_{31}, \quad x_3 = 3\Psi_{22}, \quad x_4 = \Psi_{13}, \quad x_5 = \Psi_{04},$$
$$z_1 = z_{1111}/4, \quad z_2 = z_{1112}, \quad z_3 = z_{1122}/2, \quad z_4 = z_{1222}, \quad z_5 = z_{2222}/4 \tag{2.7.65}$$

and applying (2.7.15), we can write (2.7.62) as

$$a_0x_1 + c_0x_2 = z_1, \quad d_0x_1 + (3a_0 + b_0)x_2 + c_0x_3 = z_2,$$
$$d_0x_2 + \tfrac{1}{3}(a_0 + b_0)x_3 + c_0x_4 = z_3,$$
$$d_0x_3 + (a_0 + 3b_0)x_4 + c_0x_5 = z_4, \quad d_0x_4 + b_0x_5 = z_5. \tag{2.7.66}$$

These equations are solved easily. The roots can be represented in the form (2.7.18). The determinant and algebraic adjuncts that enter (2.7.18) are also deduced in a fairly simple way. In particular, the determinant has the form

$$D = (a_0 + b_0)[a_0 b_0 (a_0^2 + \tfrac{10}{3} a_0 b_0 + b_0^2)$$

$$+ c_0 d_0 (a_0^2 - \tfrac{10}{3} a_0 b_0 + b_0^2) + \tfrac{4}{3} c_0^2 d_0^2] \,. \tag{2.7.67}$$

On the right-hand side of (2.7.18) we can immediately cancel $a + b$, since this factor appears both in $\psi_{\alpha\beta}$ and z_i, by (2.7.8), and in $a_0 + b_0$, because (2.4.54) yields

$$a_0 + b_0 = a + b - \psi_{11} k_{11}^0 - 2\psi_{12} k_{12}^0 - \psi_{22} k_{22}^0 \,. \tag{2.7.68}$$

In the oscillatory case and in the case (2.7.10) this cancellation removes the 0/0 type uncertainty corresponding to the critical point.

Next we can write the probability density of parameter fluctuations. It has the form

$$w(B) = \mathrm{const} \cdot \exp(-U(b)/\kappa) \,, \tag{2.7.69}$$

where

$$b_\alpha = B_\alpha - B_\alpha^0 \,,$$
$$U(b) = \tfrac{1}{2} \psi_{\alpha\beta} b_\alpha b_\beta + \tfrac{1}{24} \psi_{\alpha\beta\gamma\delta} b_\alpha b_\beta b_\gamma b_\delta \,. \tag{2.7.70}$$

Using the polar coordinates (2.7.23), we reduce (2.7.70) to the form

$$U(\rho, \varphi) = \tfrac{1}{2} g(\varphi) \rho^2 + \tfrac{1}{24} h(\varphi) \rho^4 \,, \tag{2.7.71}$$

where $g(\varphi)$ is again given by (2.7.25a), and

$$\begin{aligned} h(\varphi) &= \psi_{\alpha\beta\gamma\delta} n_\alpha n_\beta n_\gamma n_\delta \\ &= \Psi_{40} \sin^4 \varphi + 4\Psi_{31} \sin^3 \varphi \cos \varphi \\ &\quad + 6\Psi_{22} \sin^2 \varphi \cos^2 \varphi + 4\Psi_{13} \sin \varphi \cos^3 \varphi + \Psi_{04} \cos^4 \varphi \,. \end{aligned} \tag{2.7.72}$$

The phase transition in question is a second-order transition if

$$h(\varphi) > 0 \quad \text{for all} \quad \varphi \,, \tag{2.7.73}$$

and a first-order transition, if

$$\min_{\varphi} h(\varphi) < 0 \,. \tag{2.7.74}$$

In the case (2.7.73) for the critical value of Θ at $B = B^0$ we have a stable stationary state, the fluctuations being described by the probability density

$$w(\rho, \varphi) d\rho d\varphi = \mathrm{const} \cdot \exp[-(24\kappa)^{-1} h(\varphi) \rho^4] \rho d\rho d\varphi \tag{2.7.75}$$

derived from (2.7.69,71). (We do not consider here the special case (2.7.37) when this does not hold.) Using (2.7.75) we can find the correlator

$$\begin{aligned} \langle B_\alpha, B_\beta \rangle &= \langle b_\alpha b_\beta \rangle \\ &= \langle \rho^2 n_\alpha n_\beta \rangle \\ &= N^{-1} \int \rho^2 n_\alpha n_\beta \exp(-h(\varphi) \rho^4 / (24\kappa)) \rho d\rho d\varphi \,, \end{aligned} \tag{2.7.76}$$

where

$$N = \int \exp(-h(\varphi)\rho^4/(24\kappa))\rho d\rho d\varphi \,. \tag{2.7.77}$$

Integrating with respect to ρ yields

$$\langle B_\alpha, B_\beta \rangle = 2(6\kappa/\pi)^{1/2} \int_{-\pi}^{\pi} n_\alpha(\varphi)n_\beta(\varphi)h^{-1}(\varphi)d\varphi \Big/ \int_{-\pi}^{\pi} h^{-1/2}(\varphi)d\varphi \,. \tag{2.7.78}$$

We see that this correlator has the order of $\kappa^{1/2}$. In an analogous way we find

$$\langle b_\alpha b_\beta b_\gamma b_\delta \rangle = 12\kappa \int_{-\pi}^{\pi} n_\alpha(\varphi)n_\beta n_\gamma n_\delta h^{-3/2}(\varphi)d\varphi \Big/ \int_{-\pi}^{\pi} h^{-1/2}(\varphi)d\varphi \,. \tag{2.7.79}$$

Hence at the critical point $\langle B_\alpha, B_\beta; B_\gamma, B_\delta \rangle$ has the order of κ. The above correlators are of this order at other points of the critical region (of Θ), which is defined by the condition

$$\kappa \max_\varphi [|\, h(\varphi)\,|\, /g^2(\varphi)] \sim 1\,. \tag{2.7.80}$$

At $\Theta \neq \Theta_c$, as a result of integration with respect to ρ, the expressions for the correlators will include parabolic cylindrical functions, just as in (2.6.13,17). Significantly, the excess coefficients in the critical region (2.7.80) are about unity. Accordingly, as in the one-dimensional case, the parameter fluctuations in the critical region are abnormally large and strongly non-Gaussian.

In a part of the critical region at a certain distance from $\Theta = \Theta_c$ the correlators can be worked out by expanding into a Taylor series in (2.7.69) using

$$\begin{aligned}
w(b) \;=\; & \text{const} \cdot \exp(-(2\kappa)^{-1}\psi_{\alpha\beta}b_\alpha b_\beta)\{1 - (24\kappa)^{-1}\psi_{\alpha\beta\gamma\delta}b_\alpha b_\beta b_\gamma b_\delta \\
& + \tfrac{1}{2}((24\kappa)^{-1}\psi bbbb)^2 + \dots\}\,.
\end{aligned} \tag{2.7.81}$$

If in the braces we keep only $1 - (24\kappa)^{-1}\psi bbbb$, then we will obtain

$$\begin{aligned}
\langle B_\alpha, B_\beta \rangle &= \langle b_\alpha b_\beta \rangle \\
&= \kappa\psi_{\alpha\beta}^{-1} - \tfrac{1}{2}\kappa^2\psi_{\alpha\mu}^{-1}\psi_{\beta\nu}^{-1}\psi_{\mu\nu\rho\sigma}\psi_{\rho\sigma}^{-1}\,,
\end{aligned} \tag{2.7.82a}$$

$$\langle B_\alpha, B_\beta, B_\gamma, B_\delta \rangle = -\kappa^3\psi_{\alpha\mu}^{-1}\psi_{\beta\nu}^{-1}\psi_{\gamma\rho}^{-1}\psi_{\delta\sigma}^{-1}\psi_{\mu\nu\rho\sigma}\,. \tag{2.7.82b}$$

In the critical region the matrix $\psi_{\alpha\beta}^{-1}$ is large and its value can be taken to be associated with κ. Therefore, the factors κ^n in (2.7.82) do not indicate the true order of terms in κ. To find this order requires the use of (2.7.80) or, which is about the same, the relationships

$$\psi_{\alpha\beta}\psi_{\gamma\delta} \sim \kappa\psi_{\alpha\beta\gamma\delta} \tag{2.7.83a}$$

or

$$\kappa\psi_{\alpha\mu}^{-1}\psi_{\beta\gamma}^{-1}\psi_{\mu\nu\gamma\delta} \sim 1 \tag{2.7.83b}$$

($\psi_{\alpha\beta}$ is of the order of $(\kappa\psi_{\alpha\beta\gamma\delta})^{1/2}$). Correspondingly, both terms in (2.7.82a) have the same order, $\kappa^{1/2}$, and the term on the right-hand side of (2.7.82b) has the order κ, which is in agreement with the factor κ in (2.7.79) and $\kappa^{1/2}$ in (2.7.78).

Let us now take a look at the case (2.7.74), i.e., the case of a kinetic phase transition of the first kind, where the point $B = B^0$ becomes unstable as Θ_c is approached. We will fix the angle φ and find the value of ρ at which the function (2.7.71) has a maximum corresponding to the "watershed". Equating the derivative with respect to ρ to zero gives

$$g(\varphi)\rho + \tfrac{1}{6}h(\varphi)\rho^3 = 0 \,; \tag{2.7.84}$$

hence

$$\begin{aligned} \rho_{\max}^2 &= -6g(\varphi)/h(\varphi) > 0 \,, \\ \rho_{\max} &= [-6g(\varphi)/h(\varphi)]^{1/2} \end{aligned} \tag{2.7.85}$$

(we only consider those values of φ at which the above inequality holds).

Substituting (2.7.85) into (2.7.71), we find the height of the "watershed"

$$\Delta U(\varphi) = \tfrac{3}{2}g^2(\varphi) \mid h(\varphi) \mid^{-1} \,. \tag{2.7.86}$$

Let this quantity be minimal at φ_m:

$$\Delta U(\varphi_m) = \tfrac{3}{2}\min_{\varphi}[g^2(\varphi) \mid h(\varphi) \mid^{-1}] \,. \tag{2.7.87}$$

Hence the minimum relative height of the "watershed" is

$$\lambda = 3(2\kappa)^{-1}g^2(\varphi_m)/ \mid h(\varphi_m) \mid \,. \tag{2.7.88}$$

As before, we will arbitrarily believe that the point $B = B^0$ is still stable at $\lambda \geq 3$, the critical region being given by $\lambda \sim 3$, but $\lambda > 3$ (the relationship $\lambda \sim 3$ only differs from (2.7.80) in a numerical factor).

If we introduce the matrices normalized to unity

$$\begin{aligned} \tilde{\psi}_{\alpha\beta} &= \psi_{\alpha\beta}/g(\varphi_m) \,, \\ \tilde{\psi}_{\alpha\beta\gamma\delta} &= \psi_{\alpha\beta\gamma\delta}/h(\varphi_m) \,, \end{aligned} \tag{2.7.89}$$

i.e. matrices with the properties

$$\begin{aligned} \tilde{\psi}_{\alpha\beta}n_\alpha n_\beta &= 1 \,, \\ \tilde{\psi}_{\alpha\beta\gamma\delta}n_\alpha n_\beta n_\gamma n_\delta &= 1 \,, \end{aligned} \tag{2.7.90}$$

then the relationships (2.7.82) can, from (2.7.88), be written as

$$\begin{aligned} \langle B_\alpha, B_\beta \rangle &= (3\kappa/2\lambda)^{1/2} \mid h(\varphi_m) \mid^{-1/2} \{\tilde{\psi}_{\alpha\beta}^{-1} + (3/4\lambda)\tilde{\psi}_{\alpha\mu}^{-1}\tilde{\psi}_{\beta\nu}^{-1}\tilde{\psi}_{\mu\nu\rho\sigma}\tilde{\psi}_{\rho\sigma}^{-1} \\ &\quad + O(\lambda^{-2})\} \,, \\ \langle B_\alpha, B_\beta, B_\gamma, B_\delta \rangle &= -\kappa\{(9/4\lambda^2)h^{-1}(\varphi_m)\tilde{\psi}_{\alpha\mu}^{-1}\tilde{\psi}_{\beta\nu}^{-1}\tilde{\psi}_{\gamma\rho}^{-1}\tilde{\psi}_{\delta\sigma}^{-1}\tilde{\psi}_{\mu\nu\rho\sigma} \\ &\quad + O(\lambda^{-3})\} \,. \end{aligned} \tag{2.7.91}$$

These equations are a many-dimensional generalization of (2.6.53). By keeping more terms in (2.7.81) we can achieve a higher accuracy in the parameter of "proximity to the critical point" λ^{-1}. Many of the formulas in this subsection, say (2.7.82,91), are applicable to any number of components of the internal parameter B.

Concluding remark: As the number of components of the vector B increases, so does the number of the various kinds of kinetic phase transition. Generally speaking, to each type of bifurcation there corresponds a kinetic phase transition. Of course, more complex types of kinetic phase transition are analyzed in a more complicated way that just shown.

2.8 Notes on References to Chapter 2

The various examples of open systems have been considered in many papers, e.g. in the work of *Prigogine* and his co-authors [2.2,3,11], and also in the investigations of other authors [2.4,10,12,13]. A number of examples and results concerning open systems are given in [2.14].

The generating equation (2.2.5) for Markov open systems was derived in [2.4]. The generating equation for completely open non-Markov systems was obtained in [2.15].

In proving the H-theorem of Sect. 2.3.5 we have followed the method suggested in [2.16,17]. In [2.18] this H-theorem was used for deriving the inequality of the type (2.3.38). Another proof of (2.3.38) is given in Sect. 2.3.4.

The concept of a nonequilibrium kinetic phase transition, which is introduced in this book, is very close to the concept of "catastrophe" in the catastrophe theory suggested by *Thom* [2.19]. But the application of the theory given in the papers of *Haken, Ebeling* and others [2.20–23] is closer to our treatment.

Nonequilibrium kinetic phase transitions have been studied in many papers, in particular, in [2.3,10–12,21–23].

3. The Kirchhoff's Form
of Fluctuation-Dissipation Relations

The theory given below can be regarded as the nonlinear nonequilibrium thermo-dynamics of radiation. We will derive universal relations suitable for wave motion described by linear equations in the presence of bodies that interact, linearly or nonlinearly, with the waves. These relations are a generalization of the well-known Kirchhoff's law to the case of any degree of coherence of waves, and also to the case of a possible nonlinear interaction of bodies with waves.

It is significant that the linear or nonlinear interaction of waves with bodies (reflection, refraction, absorption) is described phenomenologically, as is charac-teristic of nonlinear thermodynamics in general. Specifically, it is described using certain functions $U_{1,2}, U_{1,23}, U_{1,234}$. Furthermore, functions are introduced that de-scribe the correlators of fluctuations of waves. Various functions are connected by universal relations. The set of those relations, which are referred to as FDRs of the Kirchhoff type, in many respects resembles the sets of FDRs of the first, second or third kinds, although there are some differences. The derivation of these relations needs the FDRs of the third kind derived earlier.

3.1 Functions Describing Linear and Nonlinear Scattering, Reflection and Absorption of Waves

3.1.1 Incident and Transmitted Waves in One Particular Case

The Kirchhoff form of the FDRs describes the mechanisms of processes, including fluctuational ones, that occur in the case of linear and nonlinear reflection and refraction (generally, scattering) of waves and also in the case of their absorption. We suppose that the waves are travelling in a medium described by linear equations until the moment of scattering and after it, with the nonlinear effects being caused only by the scattering body. The waves may be of quite varied nature, i.e. they may be electromagnetic, acoustic or of yet some other kind. The types of scattering process may also be various.

Figure 3.1 shows the simplest case of scattering. Here strictly one-dimensional wave carriers, straight lines for example, are connected to some body ("black box"). The body scatters the approaching waves, and the scattered waves are emitted along the same lines. Alternatively, waveguides, along which the waves of various

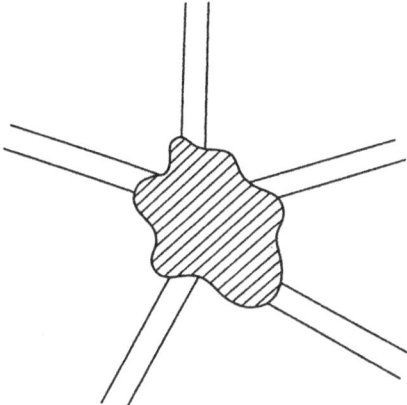

Fig. 3.1. Schematic picture of the body to which the one-dimensional wave carriers are connected. The body transforms the radiation delivered to it through these carriers

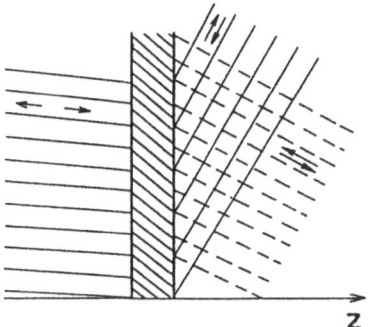

Fig. 3.2. Layer of substance which can transform (reflect, refract and absorb) the incident radiation

types, i.e. the various modes, can travel, may be connected to the scattering body instead of the one-dimensional wave carriers.

Another case of scattering is shown in Fig. 3.2. Here an infinite homogeneous plane-parallel layer of some substance linearly and nonlinearly scatters the approaching waves. This case is also possible when waves travelling in three-dimensional space are scattered by a finite body.

Let us begin with the case shown in Fig. 3.1 and consider a one-dimensional wave system. Let z be its longitudinal coordinate. As a consequence of the linearity and one-dimensionality of the wave system, the corresponding Hamiltonian is of the form

$$\mathcal{H}[u(z), \pi(z)] = \frac{1}{2} \int [m_0^{-1} \pi^2(z) + \kappa_0 (\partial u(z)/\partial z)^2] dz \,. \tag{3.1.1}$$

Here $u(z)$ is the wave field (or displacement), and $\pi(z)$ are the momenta conjugate to $u(z)$. In the quantum case we have the following commutation relations

$$[u(z), \pi(z')] = i\hbar\delta(z - z').\tag{3.1.2}$$

Knowing the Hamiltonian and applying the Hamilton equations, quantum or classical, we readily find

$$\dot{u} = \pi/m_0,\tag{3.1.3a}$$
$$\dot{\pi} = \partial f(z)/\partial z.\tag{3.1.3b}$$

Here $f(z) = \kappa_0 \partial u(z)/\partial z$. As is easily seen, instead of (3.1.3) one can take

$$\dot{f} = \kappa_0 \partial \dot{u}/\partial z,$$
$$\ddot{u} = m_0^{-1}\partial f/\partial z.\tag{3.1.4}$$

Using (3.1.3a) and the definition of the function f, from (3.1.2) we have

$$[f(z), \dot{u}(z')] = i\hbar v^2\delta'(z - z'),\tag{3.1.5}$$

where $v = (\kappa_0/m_0)^{1/2}$ is the wave velocity.

The quantity $f(z) = \kappa_0 \partial u(z)/\partial z$ has the meaning of the force conjugate to u, i.e. the product fu represents energy. This can be verified if we take into consideration the dimensions of (3.1.5). Since z has the dimensions of length, $\delta'(z - z')$ has the dimensions of $(\text{length})^{-2}$ and $v^2\delta'(z - z')$ has the dimensions of $(\text{time})^{-2}$. The Planck constant has the dimensions of energy multiplied by time. Therefore, the right-hand side of (3.1.5) has the dimensions of energy divided by time. From this it follows that $f\dot{u}$ has the dimensions of energy divided by time and fu has the meaning of energy.

In the case of a long electrical line, i.e. of a system of two parallel wires, the potential difference $V(z)$ plays the role of the force $f(z)$, and the electrical current $I(z)$ plays the role of $\dot{u}(z)$. Instead of (3.1.4,5) we will have

$$\dot{V} = \frac{1}{C_0}\frac{\partial I}{\partial z},$$
$$\dot{I} = \frac{1}{L_0}\frac{\partial V}{\partial z},\tag{3.1.6}$$

$$[V(z), I(z')] = i\hbar v^2\delta'(z - z').\tag{3.1.7}$$

Here $v^2 = (L_0 C_0)^{-1}$; C_0 and L_0 are capacity and inductance both calculated per unit length. Equations (3.1.6) are called the telegrapher's equations. Parallel with wave velocity one can introduce the wave resistance

$$R = (m_0\kappa_0)^{1/2}$$

or

$$R = (L_0/C_0)^{1/2}.\tag{3.1.8}$$

The solution of (3.1.6) can be written as

$$V(z,t) = 2^{-1/2}R^{1/2}[g_-(t + z/v) + g_+(t - z/v)],$$
$$I(z,t) = (2R)^{-1/2}[g_-(t + z/v) - g_+(t - z/v)].\tag{3.1.9}$$

This may be verified by direct substitution. Here g_- is the wave travelling in the negative z-direction, and g_+ is the wave travelling along $+z$. If the contact point of the conducting line with the scattering body corresponds to $z = 0$ and the line is located at $z > 0$, the wave g_- approaches the body and may be called the incident or approaching wave and denoted by g^a. The wave g_+ is the transmitted or receding one and may be denoted by g^r.

Solving (3.1.9) for $g^a = g_-$ and $g^r = g_+$ and using (3.1.7), we easily find the commutation relations

$$[g^a(t), g^r(t')] = 0, \tag{3.1.10a}$$

$$[g^a(t), g^a(t')] = [g^r(t), g^r(t')]$$
$$= i\hbar\delta'(t - t'). \tag{3.1.10b}$$

Here $\delta'(z/v) = v^2\delta'(z)$.

If we substitute (3.1.9) into

$$W = \tfrac{1}{2} \int [L_0 J^2(z) + C_0 V^2(z)]dz, \tag{3.1.11}$$

which is analogous to (3.1.1), we arrive at

$$\begin{aligned} W &= \tfrac{1}{2}v^{-1} \int \{[g^a(t + z/v)]^2 + [g^r(t - z/v)]^2\}dz \\ &= \tfrac{1}{2} \int \{[g^a(t)]^2 + [g^r(t)]^2\}dt. \end{aligned} \tag{3.1.12}$$

Introducing the Fourier transform

$$g^a(\omega) = (2\pi)^{-1/2} \int_{-\infty}^{\infty} \exp(-i\omega t)g^a(t)dt \tag{3.1.13}$$

and the analogous relation for $g^r(\omega)$, we get

$$W = \tfrac{1}{2} \int_{-\infty}^{\infty} [g^a(\omega)g^a(-\omega) + g^r(\omega)g^r(-\omega)]d\omega. \tag{3.1.14}$$

Let us now remember that the system under study consists of several one-dimensional wave carriers. The above consideration may be applied to each of them. Therefore, we will have several incident and transmitted waves and several functions $g_s^a(t), g_s^r(t)$, $s = 1, 2, \ldots$. Then the total energy will be equal to the sum of the energies of the type (3.1.12,14):

$$\begin{aligned} W &= \tfrac{1}{2} \sum_s \int_{-\infty}^{\infty} \{[g_s^a(t)]^2 + [g_s^r(t)]^2\}dt \\ &= \tfrac{1}{2} \sum_s \int_{-\infty}^{\infty} [g_s^a(\omega)g_s^a(-\omega) + g_s^r(\omega)g_s^r(-\omega)]d\omega. \end{aligned} \tag{3.1.15}$$

When the waveguides are in contact with the scattering body, the subscript s should be replaced by the pair (i, l). Here i is the waveguide number and l is the mode number for each waveguide. We suppose that (3.1.15) holds as before. This formula gives the exact definition of the numerical functions (or operational ones in the quantum case) $g_s^a(t)$ and $g_s^r(t)$ or $g_s^a(\omega)$ and $g_s^r(\omega)$. By virtue of (3.1.10) the commutation relations

$$
\begin{aligned}
[g_s^a(\omega), g_{s'}^a(\omega')] &= [g_s^r(\omega), g_{s'}^r(\omega')] \\
&= -\hbar\omega\delta(\omega + \omega')\delta_{ss'},
\end{aligned}
\tag{3.1.16}
$$

are valid for the Fourier transform (3.1.13) taken at the different values of ω and s.

3.1.2 Waves in Three-Dimensional Space

Let us consider now the scattering layer shown in Fig. 3.2 in the case of an electromagnetic field. We place the origin of the coordinates on the boundary of the layer and direct the z-axis, i.e. r_3, out of the layer and normal to it. For $z > 0$ the field $\mathbf{E}(\mathbf{r})$ is represented by

$$
E_\alpha(\mathbf{r}) = (\mu/\varepsilon)^{1/4}(2\pi)^{-3/2} \int \exp(i\mathbf{k}\mathbf{r})g_\alpha(\mathbf{k})d\mathbf{k}, \quad \alpha = 1, 2, 3.
\tag{3.1.17}
$$

Here

$$
\begin{aligned}
&g_\alpha(\mathbf{k}) \\
&= (2\pi)^{-3/2}(\varepsilon/\mu)^{1/4} \int_0^\infty dz \exp(-ik_3 z) \int dx dy \exp[-i(k_1 x + k_2 y)]E_\alpha(\mathbf{r}).
\end{aligned}
\tag{3.1.18}
$$

If we take into account the variation of $\mathbf{E}(\mathbf{r})$ in time, we should replace (3.1.17) by

$$
\begin{aligned}
E_\alpha(\mathbf{r}, t) &= 2^{-1/2}(\mu/\varepsilon)^{1/4}(2\pi)^{-3/2} \int \{\exp[i(\mathbf{k}\mathbf{r} + kvt)]g_\alpha^a(\mathbf{k}) \\
&\quad + \exp[i(\mathbf{k}\mathbf{r} - kvt)]g_\alpha^r(\mathbf{k})\}d\mathbf{k}.
\end{aligned}
\tag{3.1.19}
$$

Here

$$
k = |k_1^2 + k_2^2 + k_3^2|^{1/2} \operatorname{sign}k_3,
\tag{3.1.20}
$$

so that the factor $\exp[i(\mathbf{k}\mathbf{r} + kvt)]$ describes the incident wave, and the factor $\exp[i(\mathbf{k}\mathbf{r} - kvt)]$ describes the transmitted wave; $v = (\varepsilon\mu)^{-1/2} > 0$ is the wave velocity, and $(\mu/\varepsilon)^{1/4} = R^{1/2}$, where R is the wave resistance. Comparing (3.1.17) with (3.1.19) at $t = 0$ yields

$$
g_\alpha(\mathbf{k}) = 2^{-1/2}[g_\alpha^a(\mathbf{k}) + g_\alpha^r(\mathbf{k})].
\tag{3.1.21}
$$

In order to separate the quantity $g_\alpha(\mathbf{k})$ into these two terms, one must take into account the expansion of the type (3.1.17) for the field $\mathbf{H}(\mathbf{r})$ and use the Maxwell equations. However, there is no need to make corresponding calculations here.

Let us introduce the unit vector $\mathbf{n} = \mathbf{k}/k$. By virtue of (3.1.20) its z-component, n_3, is nonnegative. Using this vector, we can rewrite $\mathbf{k}\mathbf{r} \pm kvt$ as $k(\mathbf{n}\mathbf{r} \pm vt)$.

Transferring to spherical coordinates k, ϑ, φ, we have

$$
d\mathbf{k} = k^2 d|k| d\Omega = k^2 d|k| \sin\vartheta d\vartheta d\varphi.
\tag{3.1.22}
$$

The angle ϑ between $\mathbf{n}$ and the z-axis conventionally extends from 0 to π and we have $k > 0$. Here, however, we restrict the values of ϑ to $0 < \vartheta < \pi/2$, but assume that negative values of k are possible (they are needed in order that the argument $\omega = kv$ in (3.1.13,15) can take negative values). Then

$$\int \ldots d\mathbf{k} = \int_{n_3 > 0} d\Omega \int_{-\infty}^{\infty} k^2 dk \ldots \tag{3.1.23}$$

and (3.1.19) can be written as

$$
\begin{aligned}
E_\alpha(\mathbf{r}, t) \\
= (R/2)^{1/2} (2\pi)^{-3/2} \int_{n_3 > 0} d\Omega \int_{-\infty}^{\infty} dk\, k^2 \{ \exp[ik(\mathbf{nr} + vt)] g_\alpha^a(k, \mathbf{n}) \\
+ \exp[ik(\mathbf{nr} - vt)] g_\alpha^r(k, \mathbf{n}) \},
\end{aligned}
\tag{3.1.24}
$$

where $z > 0$, $g_\alpha^a(k, \mathbf{n}) = g_\alpha^a(k\mathbf{n}) = g_\alpha^a(\mathbf{k})$ and $g_\alpha^r(k, \mathbf{n}) = g_\alpha^r(k\mathbf{n}) = g_\alpha^r(\mathbf{k})$.

The field $\mathbf{E}(\mathbf{r}, t)$ that exists on the other side of the scattering layer, i.e. for $z < -d$, may be represented in an analogous way (d is the layer thickness). By analogy with (3.1.24) we will have

$$
\begin{aligned}
E_\alpha(\mathbf{r}, t) \\
= (R/2)^{1/2} (2\pi)^{-3/2} \int_{n_3 < 0} d\Omega \int_{-\infty}^{\infty} dk\, k^2 \{ \exp[ik(\mathbf{nr} + vt)] g_\alpha^a(k, \mathbf{n}) \\
+ \exp[ik(\mathbf{nr} - vt)] g_\alpha^r(k, \mathbf{n}) \},
\end{aligned}
\tag{3.1.25}
$$

for $z < -d$. The formulas (3.1.24,25) can be combined if the integration is made over the whole solid angle, which is equal to 4π, without limiting the values of n_3.

Let us consider the field energy on one side of the layer

$$W_1^{(+)} = \frac{1}{2} \int_{z>0} [\varepsilon E^2 + \mu H^2] d\mathbf{r} . \tag{3.1.26}$$

Avoiding a consideration of the magnetic field, we will use the simpler formula

$$W_1^{(+)} = \int_{z>0} \varepsilon E^2 d\mathbf{r} \tag{3.1.27}$$

(which is valid when the longitudinal field is absent) and neglect the cross products of the type $g^a g^r$.

Owing to (3.1.17) we have from (3.1.27)

$$W_1^{(+)} = (\mu\varepsilon)^{1/2} \int g_\alpha(\mathbf{k}) g_\alpha(-\mathbf{k}) d\mathbf{k} \tag{3.1.28}$$

or, if we substitute (3.1.21),

$$W_1^{(+)} = \frac{1}{2} v^{-1} \int [g_\alpha^a(\mathbf{k}) g_\alpha^a(-\mathbf{k}) + g_\alpha^r(\mathbf{k}) g_\alpha^r(-\mathbf{k})] d\mathbf{k} . \tag{3.1.29}$$

Transforming this integral according to (3.1.23) gives

$$
\begin{aligned}
W_1^{(+)} &= \frac{1}{2} v^{-1} \int_{n_3 > 0} d\Omega \int_{-\infty}^{\infty} [g_\alpha^a(k, \mathbf{n}) g_\alpha^a(-k, \mathbf{n}) \\
&\quad + g_\alpha^r(k, \mathbf{n}) g_\alpha^r(-k, \mathbf{n})] k^2 dk .
\end{aligned}
\tag{3.1.30}
$$

The field energy on the other side of the layer is determined by the analogous integral for values of $n_3 < 0$. The total energy is given by the integral

$$W_1 = \tfrac{1}{2}v^{-1} \int d\Omega \int_{-\infty}^{\infty} [g_\alpha^a(k,\mathbf{n})g_\alpha^a(-k,\mathbf{n}) + g_\alpha^r(k,\mathbf{n})g_\alpha^r(-k,\mathbf{n})]k^2 dk \quad (3.1.31)$$

over all possible values of n_3. The expression just obtained has the form (3.1.15) if we regard s as a pair $(\mathbf{n},\alpha)$ and determine the temporal Fourier transforms $g_{\mathbf{n}\alpha}^a(\omega)$ and $g_{\mathbf{n}\alpha}^r(\omega)$ as follows

$$g_{\mathbf{n}\alpha}^a(\omega) = \omega v^{-2}g_\alpha^a(\omega/v,\mathbf{n}),$$
$$g_{\mathbf{n}\alpha}^r(\omega) = \omega v^{-2}g_\alpha^r(\omega/v,\mathbf{n}). \quad (3.1.32)$$

Parallel with (3.1.32) we can consider the temporal functions

$$\begin{aligned} g_{\mathbf{n}\alpha}^a(t) &= (2\pi)^{-1/2} \int \exp(i\omega t)g_{\mathbf{n}\alpha}^a(\omega)d\omega \\ &= (2\pi)^{-1/2} \int \exp(ikvt)g_\alpha^a(k,\mathbf{n})k\,dk. \end{aligned} \quad (3.1.33)$$

Analogous expressions can be written for $g_{\mathbf{n}\alpha}^r(t)$.

Since the electromagnetic waves are transverse, the vectors $g_\alpha^{a,r}(k,\mathbf{n})$, $g_{\mathbf{n}\alpha}^{a,r}(\omega)$ have to be normal to $\mathbf{n}$, i.e.

$$g_\alpha^{a,r}(k,\mathbf{n})n_\alpha = 0,$$
$$g_{\mathbf{n}\alpha}^{a,r}(\omega)n_\alpha = 0. \quad (3.1.34)$$

Analysis shows that in the quantum case we have

$$\begin{aligned}[g_\alpha^a(k,\mathbf{n}),g_\beta^a(k',\mathbf{n}')] &= -[g_\alpha^r(k,\mathbf{n}),g_\beta^r(k',\mathbf{n}')] \\ &= -\hbar k v^2 \delta(k\mathbf{n}+k'\mathbf{n}')\tilde{\delta}_{\alpha\beta} \end{aligned} \quad (3.1.35)$$

with $\tilde{\delta}_{\alpha\beta} = \delta_{\alpha\beta} - n_\alpha n_\beta$. Using

$$\delta(k\mathbf{n}+k'\mathbf{n}') = k^{-2}\delta(k+k')\delta(\mathbf{n}-\mathbf{n}'), \quad (3.1.36)$$

where $\delta(\mathbf{n}-\mathbf{n}')$ is determined by

$$\int \delta(\mathbf{n}-\mathbf{n}')\varphi(\mathbf{n}')d\Omega' = \varphi(\mathbf{n}) \quad (3.1.37)$$

$(d\Omega' = d\mathbf{n}')$, from (3.1.35,32) we easily obtain

$$\begin{aligned}[g_{\mathbf{n}\alpha}^a(\omega),g_{\mathbf{n}'\beta}^a(\omega')] &= [g_{\mathbf{n}\alpha}^r(\omega),g_{\mathbf{n}'\beta}^r(\omega')] \\ &= -\hbar\omega\delta(\omega+\omega')\delta(\mathbf{n}-\mathbf{n}')\tilde{\delta}_{\alpha\beta}. \end{aligned} \quad (3.1.38)$$

Therefore, in the present case (3.1.16) holds for $s = (\mathbf{n},\alpha)$.

3.1.3 Introduction of Scattering Functions $U_{1,2,\dots n}$

The process of scattering is characterized by functions $U_{1,2\dots n}$, which are defined by

$$g_1^r = U_{1,2}g_2^a + \tfrac{1}{2}U_{1,23}g_2^a g_3^a + \tfrac{1}{6}U_{1,234}g_2^a g_3^a g_4^a + \dots. \quad (3.1.39)$$

Therefore, if the scattering functions and incident waves are known, we can find the transmitted waves. The subscripts $1, 2, \ldots$ in (3.1.39) mean the pairs $(s_1, t_1), (s_2, t_2), \ldots$ in the time representation and the pairs $(s_1, \omega_1), (s_2\omega_2), \ldots$ in the spectral representation. Summation and integration are carried out over repeated subscripts. It is possible to use the equivalent forms when $1, 2, \ldots$ mean $\alpha_1, \mathbf{n}_1, k_1, \ldots$ or $\alpha_1, k_1, \ldots$ or $\alpha_1, \mathbf{r}_1, \ldots$. However, to be definite, we will use the temporal form.

Equation (3.1.39) has a phenomenological macroscopic meaning when fluctuations are not considered. If the fluctuations of functions or operators g_1^a, g_1^r are considered, then (3.1.39) must be written as

$$\langle g_1^a \rangle = U_{1,2}\langle g_2^a \rangle + \tfrac{1}{2}U_{1,23}\langle g_2^a, g_3^a \rangle + \ldots . \tag{3.1.40}$$

It follows from the definition of the scattering functions that they satisfy the causality conditions, which are written as

$$U_{s_1, s_2 \ldots s_m}(t_1; t_2, \ldots, t_m) = 0 \quad \text{for} \quad t_1 < \max(t_2, \ldots, t_m) \tag{3.1.41}$$

in the time representation. They also satisfy a symmetry condition of the type $U_{1,23} = U_{1,32}$.

3.1.4 Scattering Functions and Absorption of Energy

The scattering functions describe not only the scattering of waves, but also the absorption of energy. According to (3.1.12) the total wave energy can be represented as

$$W = W^a + W^r , \tag{3.1.42}$$

where

$$W^a = \tfrac{1}{2} \sum_s \int [g_s^a(t)]^2 dt ,$$
$$W^r = \tfrac{1}{2} \sum_s \int [g_s^r(t)]^2 dt . \tag{3.1.43}$$

The latter equations are related to the time representation. In the arbitrary representation they may be written as

$$W^a = \tfrac{1}{2} J_{12} g_1^a g_2^a ,$$
$$W^r = \tfrac{1}{2} J_{12} g_1^r g_2^r . \tag{3.1.44}$$

The matrix J_{12} has the following form in the time and spectral representations respectively:

$$J_{12} = \delta_{s_1 s_2}\delta(t_1 - t_2) , \tag{3.1.45a}$$
$$J_{12} = \delta_{s_1 s_2}\delta(\omega_1 + \omega_2) , \tag{3.1.45b}$$

i.e. it coincides with the metric tensor appearing in [(5.2.18) v.1] (also see [(5.2.19) v.1]).

According to (3.1.39) the energy of the transmitted waves has the form

$$W^r = \tfrac{1}{2}[J_{12}U_{1,3}U_{2,4}g_3^a g_4^a + J_{12}U_{1,3}U_{2,45}g_3^a g_4^a g_5^a + \ldots]. \qquad (3.1.46)$$

The difference $W^a - W^r$ is absorbed energy. It must hold that

$$W^r \leq W^a. \qquad (3.1.47)$$

Here the equality sign refers to the case of no absorption, and the sign $<$ refers to the case of absorption. If the scattering is purely linear, from (3.1.46,47) we have

$$W^a - W^r = \tfrac{1}{2}(J_{12}g_1^a g_2^a - J_{34}U_{3,1}U_{4,2}g_1^a g_2^a) \geq 0. \qquad (3.1.48)$$

This means that the matrix $J_{12} - J_{34}U_{3,1}U_{4,2}$ is nonnegative definite:

$$J_{12} - J_{12}U_1 U_2 = \text{nonnegative definite} \qquad (3.1.49a)$$

or

$$J - U^T J U = \text{nonnegative definite}. \qquad (3.1.49b)$$

When absorption is absent, we have $U^T J U = J$. In conclusion of this section we note that using (3.1.45) we can write (3.1.16) in the form

$$
\begin{aligned}
[g_1^a, g_2^a] &= [g_1^r, g_2^r] \\
&= i\hbar p_1 J_{12}.
\end{aligned}
\qquad (3.1.50)
$$

Here, as before, p_1 is $\partial/\partial t_1$ in the time representation and $i\omega_1$ in the spectral representation.

3.1.5 Simplified Description of Wave Scattering

Let us introduce the functions $u_s^a(\omega)$ and $u_s^r(\omega)$ which have the meaning of the average power of incident and transmitted waves for a fixed frequency ω and subscript s.

The precise definition of $u_s^{a,r}(\omega)$ will be given later (Sect. 3.2.3). For the present, with some inaccuracy let

$$u_s^a(\omega) = C \langle [g_s^a(\omega), g_s^a(-\omega)]_+ \rangle, \qquad (3.1.51a)$$

$$u_s^r(\omega) = C \langle [g_s^r(\omega), g_s^r(-\omega)]_+ \rangle, \qquad (3.1.51b)$$

where the proportionality constant C is the same in both formulas. In the case of the simplified description of the scattering and absorption processes only the redistribution of energy is taken into account, and instead of (3.1.39)

$$u_1^r = R_{1,2}u_2^a + \tfrac{1}{2}R_{1,23}u_2^a u_3^a + \ldots \qquad (3.1.52)$$

should be used. Here $1, 2 \ldots$ have the meaning of pairs $(s_1, \omega_1), (s_2, \omega_2), \ldots$. The functions $R_{1,2\ldots}$ defined by (3.1.52) may be called the energetic scattering functions. The description of wave scattering given by (3.1.52) is certainly incomplete and approximate since the coherence properties of the waves are not considered. However, exactly this conception of the redistribution of energy is implied (in the case of linear scattering) when the ordinary Kirchhoff law is formulated. The content of this law will be considered in Sect. 3.2.3.

Integrating $R_{s',s}(\omega', \omega)$ with respect to ω' and summing over s', we obtain the reflection coefficient

$$R_s(\omega) = \sum_{s'} \int_0^\infty R_{s',s}(\omega',\omega) d\omega', \tag{3.1.53}$$

which corresponds to fixed values of ω and s. The difference

$$A_s(\omega) = 1 - R_s(\omega) \tag{3.1.54}$$

is the spectral absorptivity. It indicates what fraction of the incident wave energy is absorbed in the case of linear scattering at given values of ω and s.

For linear scattering, in the spectral representation (3.1.39) assumes the form

$$g_{s_1}^r(\omega_1) = \sum_{s_2} \int U_{s_1,s_2}(\omega_1,\omega_2) g_{s_2}^a(\omega_2) d\omega_2. \tag{3.1.55}$$

Substituting (3.1.55) into (3.1.51b) gives

$$\begin{aligned} u_{s_1}^r(\omega_1) &= \sum_{s_2 s_3} \int d\omega_2 d\omega_3 U_{s_1,s_2}(\omega_1,\omega_2) U_{s_1,s_3}(-\omega_1,-\omega_3) \\ &\times C\langle [g_{s_2}^a(\omega_2), g_{s_3}^a(-\omega_3)]\rangle. \end{aligned} \tag{3.1.56}$$

This precise (for linear scattering) equation must be compared with the approximate equation (3.1.52), which for linear scattering takes the form $u_1^r = R_{1,2} u_2^a$, i.e.,

$$u_{s_1}^r(\omega_1) = \sum_{s_2} \int d\omega_2 R_{s_1,s_2}(\omega_1,\omega_2) C\langle [g_{s_2}^a(\omega_2), g_{s_2}^a(-\omega_2)]_+\rangle. \tag{3.1.57}$$

In the framework of the applicability of (3.1.52) the expressions on the right-hand sides of (3.1.56,57) must be approximately equal

$$\begin{aligned} C\sum_{s_2 s_3} \int d\omega_2 d\omega_3 U_{s,s_2}(\omega,\omega_2) U_{s,s_3}(-\omega,-\omega_3)\langle [g_{s_2}^a(\omega_2), g_{s_3}^a(-\omega_3)]_+\rangle \\ = C\sum_{s_2} \int_0^\infty d\omega_2 R_{s,s_2}(\omega,\omega_2)\langle [g_{s_2}^a(\omega_2), g_{s_2}^a(-\omega_2)]_+\rangle. \end{aligned} \tag{3.1.58}$$

This equation is by no means always valid, so that (3.1.52) cannot always be used.

3.2 Linear and Quadratic FDRs (Generalized Kirchhoff's Laws)

3.2.1 Reciprocal Relation

We will derive the Kirchhoff form of FDR for the case where (3.1.9) holds for each fixed s. The results will then be of a general nature since the relationships of the type (3.1.9), or rather

$$\begin{aligned} h_s(z_s,t) &= (R_s/2)^{1/2} [g_s^a(t+z_s/v_s) + g_s^r(t-z_s/v_s)], \\ J_s(z_s,t) &= (2R_s)^{-1/2} [g_s^a(t+z_s/v_s) - g_s^r(t-z_s/v_s)], \end{aligned} \tag{3.2.1}$$

are valid in the general case. The subscript s may here be composite in nature (e.g. it may be a pair $\alpha, \mathbf{n}$). In (3.2.1) h_s are the thermodynamic forces conjugate to the internal thermodynamic parameters $B_s = \int J_s(t) dt$.

Suppose that the wave carriers are connected to a body at the points $z_s = 0$. Putting $z_s = 0$ in (3.2.1) gives

$$h_s(t) = (R_s/2)^{1/2}[g_s^a(t) + g_s^r(t)],$$
$$J_s(t) = (2R_s)^{-/2}[g_s^a(t) - g_s^r(t)]. \qquad (3.2.2)$$

The short form of these relations is

$$h_1 = 2^{-1/2}S_{12}^{-1}(g_2^a + g_2^r), \qquad (3.2.3a)$$
$$J_1 = 2^{-1/2}S_{12}(g_2^a - g_2^r), \qquad (3.2.3b)$$

where the subscript 1 denotes (s_1, t_1), etc., and

$$S_{12} = R_{s_1}^{-1/2}\delta_{s_1 s_2}\delta(t_{12}). \qquad (3.2.4)$$

It should be borne in mind, however, that the equations (3.2.3) are only valid in the time representation and are not invariant under the transformations [(5.2.9) v.1]. This is because, as pointed out in Sect. 5.2.2 v.1, the force h_1 is contravariant if B_1 and J_1 are assumed to be covariant. To make the formulas (3.2.3) invariant, we should substitute into them the metric tensor g_{12} that enters into [(5.2.18) v.1]. In this chapter we will denote the metric tensor by J_{12} in accordance with (3.1.45). In addition to J_{12} we also have the contravariant metric tensor $J^{12} = J_{12}^{-1}$; in the spectral representation $J^{12} = J_{12}$. Assuming g_1^a, g_1^r to be covariant and using the contravariant superscripts, we have

$$h^1 = 2^{-1/2}J^{12}(S^{-1})_2^3(g_3^a + g_3^r), \qquad (3.2.5a)$$
$$J_1 = 2^{-1/2}S_1^2(g_2^a - g_2^r) \qquad (3.2.5b)$$

instead of (3.2.3a,b). If no other representation besides temporal and spectral is considered, then the contravariant superscripts can be replaced by subscripts, so that (3.2.5a) will become

$$h_1 = 2^{-1/2}J_{12}S_{23}^{-1}(g_3^a + g_3^r) \equiv 2^{-1/2}J_1 S_1^{-1}(g_1^a + g_1^r). \qquad (3.2.6)$$

Next we can follow two procedures: either to use (3.2.5,6), or apply the simpler equations (3.2.3) of the time representation and go over to the general representation after obtaining the main results. We will take the second route.

The forces in (3.2.2,3) are applied to a scatterer, and the fluxes $J_s(t) = \dot{B}_s(t)$ are real fluxes on the scatterer's surface that correspond to changes in the internal parameters B_s of the scattering body. The functions $h_s(t)$ and $J_s(t)$ are related by the universal relationships [(5.2.8) and (5.6.8) v.1]. In the latter $Z_{1,2...m}$ are impedances characterizing the scatterer. Substituting (3.2.2,3) into [(5.6.8) v.1] yields

$$S_1^{-1}(g_1^a + g_1^r) = Z_1 S_1(g_1^a - g_1^r)$$
$$+ 2^{-3/2}Z_{1,23}S_2 S_3(g_2^a - g_2^r)(g_3^a - g_3^r) + \dots . \qquad (3.2.7)$$

If we confine ourselves to a linear approximation, we will have to omit the terms with $Z_{1,23}$, etc. Solving the linear relationship for g_1^r gives

$$g_1^{\Gamma} = (S_1 Z_1 S_1 + I_1)^{-1}(S_1 Z_1 S_1 - I_1)g_1^a\,, \tag{3.2.8}$$

where I_1 is the unit (identity) matrix.

Comparing (3.2.8) with (3.1.39), we find $U_{1,2}$, or rather express it in terms of $Z_{1,2}$. The corresponding equation in matrix form reads

$$U = (SZS + I)^{-1}(SZS - I) \tag{3.2.9a}$$

or

$$U = I - 2(SZS + I)^{-1}\,. \tag{3.2.9b}$$

Let us subject the resultant expression to time conjugation. Since S has the form $R_{s_1}^{-1/2}\delta_{s_1 s_2}\delta(t_{12})$, where R_s are positive numbers, the time conjugation operation leaves it unchanged, i.e. $S^{\text{t.c.}} = S$ (it is also symmetrical: $S^T = S$). It follows from (3.2.9) that

$$U^{\text{t.c.}} = I - 2(SZ^{\text{t.c.}}S + I)^{-1}\,. \tag{3.2.10}$$

The impedance $Z_{1,2}$, however, has the property $Z^{\text{t.c.}} = Z^T$, which is equivalent to the property [(5.6.24) v.1]. Accordingly, from (3.2.10)

$$U^{\text{t.c.}} = I - 2(SZ^T S + I)^{-1} = U^T\,, \tag{3.2.11}$$

i.e.

$$U_{1,2}^{\text{t.c.}} = U_{2,1}\,. \tag{3.2.12}$$

In arbitrary representation this relation should, of course, be somewhat changed. In fact, it is seen from the covariant representation

$$g_1^{\Gamma} = U_{1,}^{\cdot 2}g_2^a + \tfrac{1}{2}U_{1,}^{\cdot 23}g_2^a g_3^a + \cdots \tag{3.2.13}$$

of (3.1.39) that in the linear scattering matrix $U_{1,}^{\cdot 2}$ the first index is covariant and the second is contravariant. Since time conjugation does not change the covariant subscript into the contravariant one, nor vice versa, the relation $(U_{1,}^{\cdot 2})^{\text{t.c}} = U_{2,}^{\cdot 1}$ is impossible. Equation (3.2.12) must be replaced by the reciprocal relation

$$J_{24}(U_{1,}^{\cdot 4})^{\text{t.c.}} = J_{13}U_{2,}^{\cdot 3} \tag{3.2.14}$$

or

$$U_{2,1}^{\text{t.c}} = U_{1,2} \tag{3.2.15}$$

at

$$U_{1,2} = U_{1,}^{\cdot 3}J_{32}\,. \tag{3.2.16}$$

If the contravariant indices are subscripts, as in (3.2.6), then we should correct (3.2.12) by replacing it by the relation $U_{1,2}^{\text{t.c}}J_2 = U_{2,1}J_1$. We will arrive at this equation if we go through the same procedure starting with (3.2.5,6).

3.2.2 The Linear FDR

We will introduce the random forces $\mathcal{E}_s(t)$. The random fluxes $J_1 = \dot{B}_1$ are given by

$$J_1 = Y_{1,2}(h_2 + \mathcal{E}_2) + \tfrac{1}{2}Y_{1,23}(h_2 + \mathcal{E}_2)(h_3 + \mathcal{E}_3) + \dots , \qquad (3.2.17)$$

which is equivalent to [(5.6.26) v.1] in the approximation [(5.6.29,49) v.1]. Solving (3.2.17) for $h + \mathcal{E}$ yields

$$h_1 + \mathcal{E}_1 = Z_1 J_1 + \tfrac{1}{2} Z_{1,23} J_2 J_3 + \dots . \qquad (3.2.18)$$

In the linear approximation we have

$$h_1 + \mathcal{E}_1 = Z_1 J_1 . \qquad (3.2.19)$$

The variables h, J, g^a, g^r in (3.2.3) may be treated both as random and (in the quantum case) as operator variables. Substitution of (3.2.3) into (3.2.19) gives

$$g_1^a + g_1^r + 2^{1/2} S_1 \mathcal{E}_1 = S_1 Z_1 S_1 (g_1^a - g_1^r) . \qquad (3.2.20)$$

Solving this for g_1^r, we find

$$g_1^r = (S_1 Z_1 S_1 + I_1)^{-1} [(S_1 Z_1 S_1 - I_1) g_1^a - 2^{1/2} S_1 \mathcal{E}_1] . \qquad (3.2.21)$$

Hence, using (3.2.9), we obtain

$$g_1^r = U_1 g_1^a - 2^{-1/2}(I_1 - U_1) S_1 \mathcal{E}_1 . \qquad (3.2.22)$$

We will now use this to find the correlator of the transmitted wave. If the incident wave g_1^a is assumed to be fixed, the conditional correlator will be

$$U_{12} \equiv \langle g_1^r, g_2^r \rangle_{g^a} = \tfrac{1}{2}(I_1 - U_1)(J_2 - U_2) S_1 S_2 \langle \mathcal{E}_1, \mathcal{E}_2 \rangle . \qquad (3.2.23)$$

The correlator $L_{12} = \langle \mathcal{E}_1, \mathcal{E}_2 \rangle$ is given by [(5.6.35) v.1], so that

$$U_{12} = \tfrac{1}{2} kT\Theta_2^-(I_1 - U_1) S_1 (Z_{1,2} + Z_{2,1}) S_2 (J_2 - U_2^T) . \qquad (3.2.24)$$

If we take into account the expression $SZS = (I-U)^{-1}(I+U)$, which follows from (3.2.9), then

$$\begin{aligned}(I - U)S(Z + Z^T)S(I - U^T) &= (I+U)(I - U^T) + (I-U)(I + U^T) \\ &= 2(I - UU^T), \qquad (3.2.25)\end{aligned}$$

and so, from (3.2.24), we have

$$U_{12} = kT\Theta_2^-(I_{12} - U_{13}U_{23}) \equiv kT\Theta_2^-(1 - U_1 U_2) I_{12} . \qquad (3.2.26)$$

It is easily seen that the covariant form, i.e., one suitable for any representation, of the last equation is written as

$$U_{12} = kT\Theta_2^-(1 - U_1 U_2) J_{12} , \qquad (3.2.27)$$

or in more detail,

$$U_{12} = kT\Theta_2^- (J_{12} - U_{1,}^3 U_{2,}^4 J_{34}).$$ (3.2.28)

From (3.2.27) we can also find the symmetrized correlator

$$
\begin{aligned}
U_{12}^{\text{sym}} &= \tfrac{1}{2}(U_{12} + U_{21}) \\
&= kT\Theta_2(1 - U_1 U_2)J_{12}.
\end{aligned}
$$ (3.2.29)

Using (3.2.22), we can also obtain the unconditional correlator

$$
\begin{aligned}
\langle g_1^\tau, g_2^\tau \rangle &= U_1 U_2 \langle g_1^a, g_2^a \rangle + 2^{-1}(I_1 - U_1)(I_2 - U_2)S_1 S_2 \langle \mathcal{E}_1, \mathcal{E}_2 \rangle \\
&\quad - 2^{-1/2}U_1(I_2 - U_2)S_2 \langle g_1^a, \mathcal{E}_2 \rangle \\
&\quad - 2^{-1/2}(I_1 - U_1)S_1 U_2 \langle \mathcal{E}_1, g_2^a \rangle.
\end{aligned}
$$ (3.2.30)

The correlators $\langle g^a, \mathcal{E} \rangle$ vanish here owing to g^a and $\mathcal{E}$ being statistically independent. The fact is that in the stochastic representation [(5.6.64) v.1] the functions $\xi^{(\sigma)}$ are innovations, i.e., they describe new fluctuations, which are statistically independent of whatever has occurred before. In the case at hand they emerge from the scattering body and are independent of the incident waves g^a. In the linear approximation, [(5.6.64) v.1] reduce to $\mathcal{E}_1 = \sum_\sigma S_{12}^{(\sigma)} \xi_2^{(\sigma)}$. Consequently, in this approximation g^a and $\mathcal{E}$ are statistically independent. Discarding the cross-correlators, (3.2.30) yields

$$
\begin{aligned}
\langle g_1^\tau, g_2^\tau \rangle &= U_1 U_2 \langle g_1^a, g_2^a \rangle + \langle g_1^\tau, g_2^\tau \rangle \\
&= U_1 U_2 \langle g_1^a, g_2^a \rangle + kT\Theta_2^- (1 - U_1 U_2)J_{12}.
\end{aligned}
$$ (3.2.31)

Here we have used (3.2.23,27).

From this expression we can find the equilibrium correlator $\langle g_1^a, g_1^a \rangle_0$ corresponding to the fixed temperature T. We will assume that all the incident waves correspond to equilibrium thermal fluctuations at T. The scatterer has the same temperature. All the wave carriers will then be in thermal equilibrium, and the transmitted waves, too, will represent equilibrium fluctuations corresponding to the same temperature. Since both directions are equivalent at thermal equilibrium, the fluctuating waves travelling in each direction must have the same statistical properties. Therefore,

$$\langle g_1^\tau, g_2^\tau \rangle_0 = \langle g_1^a, g_2^a \rangle_0.$$ (3.2.32)

Using this, we find from (3.2.31)

$$(1 - U_1 U_2)\langle g_1^a, g_2^a \rangle_0 = kT\Theta_2^- (1 - U_1 U_2)J_{12}.$$ (3.2.33)

It follows that

$$
\begin{aligned}
\langle g_1^a, g_2^a \rangle_0 &= kT\Theta_2^- J_{12} \\
&= kT\Theta_1^+ J_{12}.
\end{aligned}
$$ (3.2.34)

Consequently, the symmetrized correlator has the form

$$\frac{1}{2}\langle[g_1^a, g_2^a]_+\rangle_0 = kT\Theta_1 J_{12}. \tag{3.2.35}$$

We now assume that the incident waves are equilibrium fluctuations for the temperature T_0, which does not coincide with the temperature T of the scatterer. Using (3.2.31), we will then have, by (3.2.34)

$$\langle g_1^r, g_2^r \rangle = kT_0\Theta_2^-(T_0)U_1U_2J_{12} + kT\Theta_2^-(T)(1 - U_1U_2)J_{12}. \tag{3.2.36}$$

Likewise, for the symmetrized correlator

$$\frac{1}{2}\langle[g_1^r, g_2^r]_+\rangle = U_1U_2 kT_0\Theta_1(T_0)J_{12} + (1 - U_1U_2)kT\Theta_1(T)J_{12}. \tag{3.2.37}$$

In terms of the conventional notation

$$\begin{aligned} \Theta_1(T) &= (i\hbar p_1/kT)\Gamma_1(T) \\ &= [i\hbar p_1/(2kT)]\coth[i\hbar p_1/(2kT)]. \end{aligned} \tag{3.2.38}$$

It is also useful to provide the spectral form of $\Theta_1(T)J_{12}$:

$$\Theta_1 J_{12} = [\hbar\omega_1/(2kT)]\coth[\hbar\omega_1/(2kT)]\delta(\omega_1 + \omega_2)\delta_{s_1 s_2}. \tag{3.2.39}$$

Before we leave this subsection we note that we can derive from (3.2.36) the mean commutator

$$\begin{aligned} \langle[g_1^r, g_2^r]\rangle &= kT(\Theta_1^+ - \Theta_1^-)(U_1U_2 + 1 - U_1U_2)J_{12} \\ &= i\hbar p_1 J_{12}, \end{aligned} \tag{3.2.40}$$

which is in accordance with (3.1.50).

3.2.3 Spectral Density of Emitted Waves. Conventional Form of Kirchhoff's Law

Equation (3.2.22) can be written in a more compact form

$$g_1^r = U_1 g_1^a + g_1^{\text{em}}, \tag{3.2.41}$$

where $U_1 g_1^a$ are scattered waves and g_1^{em} are emitted waves. In that case, (3.2.29) can be treated as the symmetrized correlator of emitted waves

$$\frac{1}{2}\langle[g_1^{\text{em}}, g_2^{\text{em}}]_+\rangle = kT(1 - U_1U_2)\Theta_1 J_{12}. \tag{3.2.42}$$

If we introduce the spectral density of emitted waves

$$S_{ss'}^{\text{em}}(\omega) = \int \exp(-i\omega t_{12}) \cdot \frac{1}{2}\langle[g_s^{\text{em}}(t_1), g_{s'}^{\text{em}}(t_2)]_+\rangle dt_{12}, \tag{3.2.43}$$

we will have

$$\frac{1}{2}\langle[g_s^{\text{em}}(\omega_1), g_{s'}^{\text{em}}(\omega_2)]_+\rangle = S_{ss'}^{\text{em}}(\omega_1)\delta(\omega_1 + \omega_2) \tag{3.2.44}$$

in analogy with [(A6.2) v.1]. Substituting (3.2.44) into the left-hand side of the spectral form of (3.2.42) gives

$$S_{ss'}^{em}(\omega_1)\delta(\omega_1 + \omega_2) = kT(1 - U_1 U_2)\Theta_1 J_{12}. \qquad (3.2.45)$$

According to (3.2.39), we can cancel out $\delta(\omega_1 + \omega_2)$ here.

Now let us consider the energy of the emitted waves. It is seen from (3.1.15) that $\sum_s [g_s^r(t)]^2/2$ has the meaning of energy, or rather of the power of receding, i.e. of emitted, scattered and reflected waves. Likewise,

$$N^{em} = \frac{1}{2}\sum_s \langle [g_s^{em}(t)]^2 \rangle \qquad (3.2.46)$$

is the mean power of emitted waves. If we now write the inverse transformation of (3.2.43) and put in it $t_{12} = 0$, we readily find that

$$\langle [g_s^{em}(t)]^2 \rangle = \frac{1}{2\pi}\int_{-\infty}^{\infty} S_{ss}^{em}(\omega)d\omega = \frac{1}{\pi}\int_0^{\infty} S_{ss}^{em}(\omega)d\omega. \qquad (3.2.47)$$

Hence,

$$N^{em} = \frac{1}{2\pi}\sum_s \int_0^{\infty} S_{ss}^{em}(\omega)d\omega. \qquad (3.2.48)$$

It is only natural to interpret the quantity

$$u_s^{em}(\omega) = \frac{1}{2\pi}S_{ss}^{em}(\omega) \qquad (3.2.49)$$

as the density of emitted power at fixed s and $\omega > 0$. Due to (3.2.44), (3.2.49) can be written as

$$u_s^{em}(\omega) = [4\pi\delta(0)]^{-1}\langle [g_s^{em}(\omega), g_s^{em}(-\omega)]_+ \rangle \qquad (3.2.50)$$

to establish the constant in (3.1.51).

Letting $s = s'$, $\omega_1 = -\omega_2$, in (3.2.45) dividing by $\delta(0)$ and using a more detailed notation, we will get

$$\begin{aligned} S_{ss}^{em}(\omega) &= f(\omega) - [\delta(0)]^{-1}\sum_{s_3}\int d\omega_3 d\omega_4 U_{ss_3}(\omega, \omega_3) \\ &\quad \times U_{ss_3}(-\omega, -\omega_4)f(\omega_3)\delta(\omega_3 - \omega_4), \end{aligned} \qquad (3.2.51)$$

where

$$f(\omega) = \frac{1}{2}\hbar\omega\coth(\frac{1}{2}\beta\hbar\omega) \qquad (3.2.52)$$

by (3.2.39). According to (3.2.35), the function (3.2.39) multiplied by kT is the correlator of equilibrium fluctuations. Using equation (3.1.58) for incident waves undergoing equilibrium fluctuations, we can transform the second term in (3.2.51) by expressing it through $R_{1,2}$:

$$S_{ss}^{em}(\omega) = f(\omega) - \sum_{s'}\int_0^{\infty} d\omega' R_{s,s'}(\omega, \omega')f(\omega'). \qquad (3.2.53)$$

For $\delta(0)$ in the previous equations to be meaningful, we should use the approximation of the delta function which is associated with the selection of a finite time interval of integration in (3.1.15). Substituting (3.2.53) into (3.2.49) and taking into account the form of $f(\omega)$, we will have

$$u_s^{\mathrm{em}}(\omega) = \frac{\hbar\omega}{4\pi}\coth\frac{\hbar\omega}{2kT} - \sum_{s'}\int_0^\infty d\omega' R_{s,s'}(\omega,\omega')\frac{\hbar\omega'}{4\pi}\coth\frac{\hbar\omega'}{2kT}\,.\tag{3.2.54}$$

Integrating this with respect to frequency and summing up over s, we find, from (3.2.48),

$$\begin{aligned}N^{\mathrm{em}} \equiv \sum_s\int_0^\infty u_s^{\mathrm{em}}(\omega)d\omega &= \sum_s\int_0^\infty \frac{\hbar\omega}{4\pi}\coth\left(\frac{\hbar\omega}{2kT}\right)[1-R_s(\omega)]d\omega\\ &= \sum_s\int_0^\infty \frac{\hbar\omega}{4\pi}\coth\left(\frac{\hbar\omega}{2kT}\right)A_s(\omega)d\omega\,,\end{aligned}\tag{3.2.55}$$

where we have considered (3.1.53,54). The resultant equations (3.2.54,55) are possible formulations of Kirchhoff's law. Equation (3.2.55) is the integral form of the law. Its differential form follows from (3.2.54) if we assume that $R_{ss'}(\omega,\omega')$ has the form

$$R_{ss'}(\omega,\omega') = R_{s'}(\omega')\delta_{ss'}\delta(\omega-\omega')\,.\tag{3.2.56}$$

Then, by (3.2.54),

$$u_s^{\mathrm{em}}(\omega)/A_s(\omega) = (\hbar\omega/4\pi)\cot(\hbar\omega/2kT)\,.\tag{3.2.57}$$

It follows that $u_s^{\mathrm{em}}(\omega)/A_s(\omega)$ is independent of s, not sensitive to the properties of the body, and equal to a universal function of T and ω.

3.2.4 Three-Dimensional Form of Kirchhoff's Law

When deriving (3.2.51) from (3.2.45), we assumed that $\delta_{ss} = 1$. This condition does not hold for three-dimensional scattering discussed in Sect. 3.1.2, where instead of s we have $(\alpha, \mathbf{n})$ and $J_{12} = \delta_{\alpha_1\alpha_2}\delta(\mathbf{n}_1-\mathbf{n}_2)\delta(\omega_1+\omega_2)$, so that $\delta_{ss} = \delta(\mathbf{n}-\mathbf{n}) = \infty$. In that case, we should slightly modify the derivation of Kirchhoff's law.

In the three-dimensional case, instead of (3.1.51), it is advisable to define the powers of incident and transmitted waves as follows:

$$\begin{aligned}u_{\alpha\mathbf{n}}^a(\omega) &= Cn_z\langle[g_{\alpha\mathbf{n}}^a(\omega),g_{\alpha\mathbf{n}}^a(-\omega)]_+\rangle\,,\\ u_{\alpha\mathbf{n}}^r(\omega) &= Cn_z\langle[g_{\alpha\mathbf{n}}^r(\omega),g_{\alpha\mathbf{n}}^r(-\omega)]_+\rangle\,,\end{aligned}\tag{3.2.58}$$

where $\omega > 0$. Considering (3.1.32) these equations can also be written in the form

$$\begin{aligned}u_\alpha^a(\mathbf{k}) &= C\frac{k_z}{k}\langle[g_\alpha^a(\mathbf{k}),g_\alpha^a(-\mathbf{k})]_+\rangle\,,\\ u_\alpha^r(\mathbf{k}) &= C\frac{k_z}{k}\langle[g_\alpha^r(\mathbf{k}),g_\alpha^r(-\mathbf{k})]_+\rangle\,.\end{aligned}\tag{3.2.59}$$

At $k_z > 0$, $u_\alpha^r(\mathbf{k})$ correspond to the waves transmitted to the right of the scattering layer; and at $k_z < 0$, to the waves reflected to the left (Fig. 3.2).

Using (3.2.59) and the expression $g_1^r = U_{1,2}g_2^a$, we can easily find, as in Sect. 3.1.5, that for (3.1.52) to hold in the case of linear scattering the following equation must be valid:

$$C \frac{k_z}{k} \sum_{\beta,\gamma} \int dk_2 dk_3 U_{\alpha,\beta}(\mathbf{k}, \mathbf{k}_2) U_{\alpha\gamma}(-\mathbf{k}, -\mathbf{k}_3) \langle [g_\beta^a(\mathbf{k}_2), g_\gamma^a(-\mathbf{k}_3)]_+ \rangle$$

$$= C \sum_{\beta} \int dk_2 R_{\alpha,\beta}(\mathbf{k}, \mathbf{k}_2)(k_z/k)_2 \langle [g_\beta^a(\mathbf{k}_2), g_\beta^a(-\mathbf{k}_2)]_+ \rangle . \qquad (3.2.60)$$

This is used instead of (3.1.58).

Let us introduce the spatial spectral density

$$G_{\alpha\beta}^{(+)}(\mathbf{k}) = \int \exp(i\mathbf{k}\mathbf{r}_{12}) \cdot \tfrac{1}{2} \langle [g_\alpha^{\mathrm{em}}(\mathbf{r}_1), g_\beta^{\mathrm{em}}(\mathbf{r}_2)]_+ \rangle^{(+)} d\mathbf{r}_{12} \qquad (3.2.61)$$

for emitted waves in the region to the right of the scattering layer. The superscript (+) on the correlator $\langle [g_\alpha^{\mathrm{em}}(\mathbf{r}_1), g_\beta^{\mathrm{em}}(\mathbf{r}_2)]_+ \rangle^{(+)}$ signifies that $z_1 + z_2 > 0$. In a similar way, we can introduce the spectral density

$$G_{\alpha\beta}^{(-)}(\mathbf{k}) = \int \exp(i\mathbf{k}\mathbf{r}_{12}) \cdot \tfrac{1}{2} \langle [g_\alpha^{\mathrm{em}}(\mathbf{r}_1), g_\beta^{\mathrm{em}}(\mathbf{r}_2)]_+ \rangle^{(-)} d\mathbf{r}_{12} , \qquad (3.2.62)$$

which corresponds to the waves emitted on the opposite side of the scattering layer.

We will define the power emitted from a unit surface as the z-component of the averaged Poynting vector $\mathbf{S} = [\mathbf{E} \times \mathbf{H}]$ of the emitted waves. We can readily find that at $z > 0$, i.e. to the right of the scattering layer,

$$N^{(+)} \equiv \langle S_z \rangle_{z>0} = \tfrac{1}{2}(2\pi)^{-3} \sum_\alpha \int G_{\alpha\alpha}^{(+)}(\mathbf{k}) \frac{k_z}{k} d\mathbf{k}$$

$$= (2\pi)^{-3} \sum_\alpha \int_{k_z>0} G_{\alpha\alpha}^{(+)}(\mathbf{k}) \frac{k_z}{k} d\mathbf{k} . \qquad (3.2.63)$$

Likewise, to the left of the layer,

$$N^{(-)} \equiv \langle S_z \rangle_{z<-d} = \tfrac{1}{2}(2\pi)^{-3} \sum_\alpha \int G_{\alpha\alpha}^{(-)}(\mathbf{k}) \frac{k_z}{k} d\mathbf{k}$$

$$= (2\pi)^{-3} \sum_\alpha \int_{k_z<0} G_{\alpha\alpha}^{(-)}(\mathbf{k}) \frac{k_z}{k} d\mathbf{k} . \qquad (3.2.64)$$

Adding (3.2.63) and (3.2.64) together, we find the total emitted power per unit surface

$$N^{\mathrm{em}} = (2\pi)^{-3} \sum_\alpha \left\{ \int_{k_z>0} G_{\alpha\alpha}^{(+)}(\mathbf{k}) \frac{k_z}{k} d\mathbf{k} + \int_{k_z<0} G_{\alpha\alpha}^{(-)}(\mathbf{k}) \frac{k_z}{k} d\mathbf{k} \right\} . \qquad (3.2.65)$$

This equation replaces (3.2.48). It suggests that the radiated power density for fixed α and $\mathbf{k}$ should be defined as

$$u_\alpha^{\mathrm{em}}(\mathbf{k}) = (2\pi)^{-3} \begin{cases} G_{\alpha\alpha}^{(+)}(\mathbf{k}) \frac{k_z}{k} & \text{for } k_z > 0, \\ G_{\alpha\alpha}^{(-)}(\mathbf{k}) \frac{k_z}{k} & \text{for } k_z < 0. \end{cases} \qquad (3.2.66)$$

Formulas (3.2.61,62) imply the equation

$$G_{\alpha\beta}^{(\pm)}(\mathbf{k})\delta(\mathbf{k} + \mathbf{k}') = \tfrac{1}{2} \langle [g_\alpha^{\mathrm{em}}(\mathbf{k}), g_\beta^{\mathrm{em}}(\mathbf{k}')] \rangle^{(\pm)} , \qquad (3.2.67)$$

similar to (3.2.44). Considering (3.2.67) we see that (3.2.66) is consistent with (3.2.59), i.e. is written as

$$
u_\alpha^{\rm em}(\mathbf{k}) = (16\pi^3\delta(0))^{-1}\frac{k_z}{k}\left\{\begin{array}{ll}\langle[g_\alpha^{\rm em}(\mathbf{k}),g_\alpha^{\rm em}(-\mathbf{k})]_+\rangle^{(+)} & \text{for}\quad k_z>0,\\ \langle[g_\alpha^{\rm em}(\mathbf{k}),g_\alpha^{\rm em}(-\mathbf{k})]_+\rangle^{(-)} & \text{for}\quad k_z<0.\end{array}\right. \quad (3.2.68)
$$

For $\delta(0)$ to be finite we should use the delta function approximation corresponding to finite volumes.

In the case under consideration, (3.2.42) reads

$$
\tfrac{1}{2}\langle[g_{\alpha_1\mathbf{n}_1}^{\rm em}(\omega_1),g_{\alpha_2\mathbf{n}_2}^{\rm em}(\omega_2)]_+\rangle
$$
$$
= k_BT(1-U_1U_2)\Theta_1\tilde\delta_{\alpha_1\alpha_2}\delta(\mathbf{n}_1-\mathbf{n}_2)\delta(\omega_1+\omega_2)\,. \quad (3.2.69)
$$

Using (3.1.32,36), we can readily see that this is equivalent to

$$
\tfrac{1}{2}\langle[g_\alpha^{\rm em}(\mathbf{k}_1),g_\beta^{\rm em}(\mathbf{k}_2)]_+\rangle^{(\pm)} = k_BT(1-U_1U_2)v\Theta_1\tilde\delta_{\alpha\beta}\delta(\mathbf{k}_1+\mathbf{k}_2)\,. \quad (3.2.70)
$$

Putting $\beta=\alpha$ and $k_2=-k_1$ and substituting the resultant expression into (3.2.68) yields

$$
u_\alpha^{\rm em}(\mathbf{k}) = (8\pi^3)^{-1}k_BTv(k_z/k)\left\{\Theta(ikv)-[\delta(0)]^{-1}\sum_{\beta\gamma}\int d\mathbf{k}_2 d\mathbf{k}_3 U_{\alpha,\beta}(\mathbf{k},\mathbf{k}_2)\right.
$$
$$
\left.\times\, U_{\alpha,\gamma}(-\mathbf{k},-\mathbf{k}_3)\Theta(ik_2v)\tilde\delta_{\beta\gamma}\delta(\mathbf{k}_2-\mathbf{k}_3)\right\}\,. \quad (3.2.71)
$$

Hence using equation (3.2.60) taken for the equilibrium incident waves, we obtain

$$
u_\alpha^{\rm em}(\mathbf{k}) = (8\pi^3)^{-1}v\left\{(k_z/k)(\tfrac{1}{2}\hbar kv)\coth(\tfrac{1}{2}\hbar\beta kv)\right.
$$
$$
\left.-\sum_\gamma\int dk' R_{\alpha,\gamma}(\mathbf{k},\mathbf{k}')(k_z'/k')\,(\tfrac{1}{2}\hbar k'v)\coth(\tfrac{1}{2}\hbar\beta k'v)\right\}\,. \quad (3.2.72)
$$

If this expression is integrated with respect to $\mathbf{k}$ and summed over α, i.e. over the various wave polarizations, we will obtain the total power (3.2.65). Integration can conveniently be carried out using spherical coordinates and then the variable $\omega=kv$. Using also (3.1.53,54) and assuming that $A_\gamma(\omega\mathbf{n})=A(\omega,\mathbf{n})$ is independent of γ, we obtain

$$
N^{\rm em} = \frac{s\hbar}{16\pi^3v^2}\int\omega^3\coth\left(\frac{\hbar\omega}{2k_BT}\right)n_zA(\omega,\mathbf{n})d\omega d\Omega\,, \quad (3.2.73)
$$

where s is the number of different possible polarizations, which for transverse electromagnetic waves is two, by (3.1.34). The last equation is the integral form of Kirchhoff's law. Other forms can also be derived that are similar to (3.2.54,57). The differential form is

$$
\frac{u^{\rm em}(\omega,\mathbf{n})}{A(\omega,\mathbf{n})} = \frac{s\hbar\omega^3}{16\pi^3v^2}\coth\frac{\hbar\omega}{2k_BT}n_z\,, \quad (3.2.74)
$$

where the right-hand side includes the universal function, and

$$
u^{\rm em}(\omega,\mathbf{n}) = \sum_\alpha u_\alpha^{\rm em}((\omega/v)\mathbf{n})\,. \quad (3.2.75)
$$

3.2.5 First Quadratic FDR

Solving (3.2.7) for g^r in the linear–quadratic approximation gives

$$g_1^r = U_1 g_1^a + 2^{-3/2}(S_1 Z_1 S_1 + I_1)^{-1} S_1 Z_{1,23}(I_2 - U_2)(I_3 - U_3)g_2^a g_3^a. \quad (3.2.76)$$

Comparison of this with (3.1.39) yields

$$U_{1,23} = 2^{-1/2}(S_1 Z_1 S_1 + I_1)^{-1} S_1 Z_{1,23}(I_2 - U_2)(I_3 - U_3). \quad (3.2.77)$$

If we then take (3.2.9) into account and treat S as a unit matrix (for short), we will have

$$U_{1,23} = 2^{-3/2}(I_1 - U_1)Z_{1,23}(I_2 - U_2)(I_3 - U_3). \quad (3.2.78)$$

Let us now take into account the fluctuations caused by scattering. Substituting (3.2.3) into (3.2.18) and again solving the resultant equation for g_1^r, we will have, by (3.2.9,78),

$$
\begin{aligned}
g_1^r &= U_1 g_1^a - 2^{-1/2}(I_1 - U_1)\mathcal{E}_1 + 2^{-1}U_{1,23}(g_2^a + 2^{-1/2}\mathcal{E}_2) \\
&\quad \times (g_3^a + 2^{-1/2}\mathcal{E}_3) + \dots.
\end{aligned} \quad (3.2.79)
$$

The random forces $\mathcal{E}$ that enter this expression are given by the stochastic representation [(5.7.58) v.1]. Substituting (3.2.3b) into [(5.7.58) v.1] and using (3.2.79), we can readily find

$$\mathcal{E}_1 = M_1 + \sum_\sigma [T_{12}^{(\sigma)}\xi_2^{(\sigma)} + 2^{-1/2}T_{123}^{(\sigma)}\xi_2^{(\sigma)}(I_3 - U_3)(g_3^a + 2^{-1/2}\mathcal{E}_3)] + \dots. \quad (3.2.80)$$

Substituting this into (3.2.79) gives the following useful formula:

$$
\begin{aligned}
g_1^r &= U_1 g_1^a - 2^{-1/2}(I_1 - U_1)\sum_\sigma [T_1^{(\sigma)}\xi_1^{(\sigma)} + 2^{-1/2}T_{123}^{(\sigma)}\xi_2^{(\sigma)}(I_3 - U_3) \\
&\quad \times (g_3^a + 2^{-1/2}\sum_\tau T_3^{(\tau)}\xi_3^{(\tau)})] + 2^{-1}U_{1,23}(g_2^a + 2^{-1/2}\sum_\sigma T_2^{(\sigma)}\xi_2^{(\sigma)}) \\
&\quad \times (g_3^a + 2^{-1/2}\sum_\tau T_3^{(\tau)}\xi_3^{(\tau)}).
\end{aligned} \quad (3.2.81)
$$

We have discarded the small constant $M_1 \sim kT$ here, since it does not influence the results below.

We will use the equation obtained to compute correlators of receding waves. Recall that $\xi^{(\sigma)}$ in (3.2.81) are independent random functions with zero means and correlators [(5.6.65) v.1]. In the quantum case they are operator functions. In the quantum case g_1^a on the right-hand side of (3.2.81) is also an operator function. But instead of an operator function we will substitute the numerical function, which we will denote by the same symbol and whose values we will fix. After g^a has been fixed, we can proceed to find the conditional correlators

$$\langle g_1^r, g_2^r \rangle_{g^a} = U_{12} + U_{12,3}g_3^a + \dots, \quad (3.2.82a)$$
$$\langle g_1^r, g_2^r, g_3^r \rangle_{g^a} = U_{123} + \dots. \quad (3.2.82b)$$

For $\langle g_1^{\text{r}}, g_2^{\text{r}}\rangle_{g^{\text{a}}}$ to be found in the first approximation, to which we are going to confine ourselves, we will only have to keep the terms of the order of kT that contain correlators $\langle \xi_1^{(\sigma)}, \xi_2^{(\sigma)}\rangle$. In that approximation we will only have to keep in (3.2.81) the terms linear in ξ. We readily arrive at

$$U_{12,3} = 2^{-3/2}(I_1 - U_1)(I_2 - U_2)\sum_\sigma (T_{143}^{(\sigma)}T_2^{(\sigma)}R_{42}^{(\sigma)}$$

$$+T_{243}^{(\sigma)}T_1^{(\sigma)}R_{14}^{(\sigma)})(I_3 - U_3) - 2^{-1}(I_2 - U_2)U_{1,43}\sum_\sigma T_4^{(\sigma)}T_2^{(\sigma)}R_{42}^{(\sigma)}$$

$$-2^{-1}(I_1 - U_1)U_{2,43}\sum T_1^{(\sigma)}T_4^{(\sigma)}R_{14}^{(\sigma)}. \tag{3.2.83}$$

Besides that

$$U_{12} = 2^{-1}(I_1 - U_1)(I_2 - U_2)\sum_\sigma T_1^{(\sigma)}T_2^{(\sigma)}R_{12}^{(\sigma)}. \tag{3.2.84}$$

We now use [(5.6.68b) v.1]. According to [(5.7.59) and (5.6.100) v.1] it takes the form

$$\sum_\sigma (T_{143}^{(\sigma)}T_2^{(\sigma)}R_{42}^{(\sigma)} + T_{243}^{(\sigma)}T_1^{(\sigma)}R_{14}^{(\sigma)}) = Z_{12,3}. \tag{3.2.85}$$

Using (3.2.84,85), we can reduce (3.2.83) to the form

$$\begin{aligned}U_{12,3} &= 2^{-3/2}(I_1 - U_1)(I_2 - U_2)Z_{12,3}(I_3 - U_3)\\ &\quad -U_{1,43}(I_4 - U_4)^{-1}U_{42} - U_{2,43}(I_4 - U_4)^{-1}U_{14}.\end{aligned} \tag{3.2.86}$$

Substituting [(5.6.101b) v.1] and (3.2.26) yields

$$\begin{aligned}\beta U_{12,3} &= 2^{-3/2}(1 - U_1)(1 - U_2)[\Theta_2^- Z_{1,23} + \Theta_1^+ Z_{2,13}\\ &\quad -(p_1\Theta_2^- + p_2\Theta_1^+)p_3^{-1}Z_{3,12}^{\text{t.c.}}](1 - U_3) - \Theta_2^- U_{1,43}(1 - U_4)^{-1}\\ &\quad \times(I_{42} - U_{45}U_{25}) - \Theta_1^+ U_{2,43}(1 - U_4)^{-1}(I_{14} - U_{15}U_{45})\end{aligned} \tag{3.2.87}$$

or, by (3.2.78),

$$\begin{aligned}\beta U_{12,3} &= (1 - U_2)(1 - U_2^T)^{-1}\Theta_2^- U_{1,23} + (1 - U_1)(1 - U_1^T)^{-1}\Theta_1^+ U_{2,31}\\ &\quad -(p_1\Theta_2^- + p_2\Theta_1^+)p_3^{-1}U_{3,12}^{\text{t.c.}} - (1 - U_2 U_2^T)(1 - U_2^T)^{-1}\Theta_2^- U_{1,23}\\ &\quad -(1 - U_1 U_1^T)(1 - U_1^T)^{-1}\Theta_1^+ U_{2,13}.\end{aligned} \tag{3.2.88}$$

Hence after some algebra we obtain

$$\beta U_{12,3} = -\Theta_2^- U_2 U_{1,23} - \Theta_1^+ U_1 U_{2,13} - (p_1\Theta_2^- + p_2\Theta_1^+)p_3^{-1}U_{3,12}^{\text{t.c}}. \tag{3.2.89}$$

This is exactly the first quadratic FDR of the Kirchhoff type, which corresponds to the time representation. From this we can readily find the covariant form

$$\begin{aligned}\beta U_{12,}^3 &= -\Theta_2^- U_{2,}^4 J_{45} U_1^{53} - \Theta_1^+ U_{1,}^4 J_{45} U_2^{53}\\ &\quad -(p_1\Theta_2 + p_2\Theta_1)p_3^{-1}J_{14}J_{25}J^{36}(U_{6,}^{45})^{\text{t.c.}},\end{aligned} \tag{3.2.90}$$

This relation, which is valid in any representation, has the short form

$$\begin{aligned}\beta U_{12,3} &= -\Theta_2^- U_2 J_2 U_{1,23} - \Theta_1^+ U_1 J_1 U_{2,13}\\ &\quad -(p_1\Theta_2^- + p_2\Theta_1^+)p_3^{-1}J_1 J_2 J_3 U_{3,12}^{\text{t.c.}}\end{aligned} \tag{3.2.91}$$

in which no superscripts are used. In the nonquantum case, (3.2.91) reads

$$U_{12,3} = -kT(U_2 J_2 U_{1,23} + U_1 J_1 U_{2,13} - J_1 J_2 J_3 U_{3,12}^{\text{t.c.}}). \tag{3.2.92}$$

3.2.6 Second Quadratic FDR

To obtain U_{123} in (3.2.82b), we should put $g^a = 0$ in (3.2.81) and calculate $\langle g_1^r, g_2^r, g_3^r \rangle$ keeping only the terms of the order of $(kT)^2$ that contain $R_{klm}^{(\sigma)}$ or products of the type $R_{kl}^{(\sigma)} R_{mn}^{(\tau)}$. In the quantum case, we obtain, using (3.2.81) and [(5.6,65) v.1]

$$
\begin{aligned}
U_{123} = \; & -2^{-3/2}(1-U_1)(1-U_2)(1-U_3)\sum_\sigma T_1^{(\sigma)}T_2^{(\sigma)}T_3^{(\sigma)}R_{123}^{(\sigma)} \\
& -2^{-5/2}(1-U_1)(1-U_2)(1-U_3)\sum_{\sigma,\tau}[T_{145}^{(\sigma)}(1-U_5)(T_2^{(\sigma)}R_{42}^{(\sigma)}T_3^{(\tau)}R_{53}^{(\tau)} \\
& +T_3^{(\sigma)}R_{43}^{(\sigma)}T_2^{(\tau)}R_{52}^{(\tau)}) + T_{245}^{(\sigma)}(1-U_5)(T_1^{(\sigma)}R_{14}^{(\sigma)}T_3^{(\tau)}R_{53}^{(\tau)} \\
& +T_3^{(\sigma)}R_{43}^{(\sigma)}T_1^{(\tau)}R_{15}^{(\tau)}) + T_{345}^{(\sigma)}(1-U_5)P_{12}T_1^{(\sigma)}R_{14}^{(\sigma)}T_2^{(\tau)}R_{25}^{(\tau)}] \\
& +2^{-2}\sum_{\sigma,\tau}[(1-U_2)(1-U_3)U_{1,45}T_4^{(\sigma)}T_2^{(\sigma)}T_5^{(\tau)}T_3^{(\tau)}R_{42}^{(\sigma)}R_{53}^{(\tau)} \\
& +(1-U_1)(1-U_3)U_{2,45}T_4^{(\sigma)}T_1^{(\sigma)}T_5^{(\tau)}T_3^{(\tau)}R_{14}^{(\sigma)}R_{53}^{(\tau)} \\
& +(1-U_1)(1-U_2)U_{3,45}T_4^{(\sigma)}T_1^{(\sigma)}T_5^{(\tau)}T_2^{(\tau)}R_{14}^{(\sigma)}R_{25}^{(\tau)}].
\end{aligned}
\tag{3.2.93}
$$

Recalling that $\sum_\sigma T_1^{(\sigma)}T_2^{(\sigma)}T_3^{(\sigma)}R_{123}^{(\sigma)} = Z_{123}$ (see [(5.6.68c) v.1]) and using (3.2.84, 85) and the relationships

$$
\sum_\sigma T_{143}^{(\sigma)}T_2^{(\sigma)}R_{42}^{(\sigma)} = Z_{\bar{1}2,3}^- \equiv Q_{12,3}^- p_3^{-1},
$$

$$
\sum_\sigma T_{243}^{(\sigma)}T_1^{(\sigma)}R_{14}^{(\sigma)} = Z_{21,3}^+ \equiv Q_{21,3}^+ p_3^{-1}
\tag{3.2.94}
$$

which by [(5.7.59) v.1] are equivalent to [(5.6.77) v.1], we arrive at

$$
\begin{aligned}
U_{123} = \; & -2^{-3/2}(1-U_1)(1-U_2)(1-U_3)Z_{123} \\
& -2^{-3/2}[(1-U_1)(1-U_2)Z_{12,5}U_{53} \\
& +(1-U_1)(1-U_3)(Z_{13,5}^- U_{52} + Z_{31,5}^+ U_{25}) \\
& +(1-U_2)(1-U_3)Z_{23,5}U_{15}] \\
& +U_{1,45}(1-U_4)^{-1}(1-U_5)^{-1}U_{42}U_{53} \\
& +U_{2,45}(1-U_4)^{-1}(1-U_5)^{-1}U_{14}U_{53} \\
& +U_{3,45}(1-U_4)^{-1}(1-U_5)^{-1}U_{14}U_{25}.
\end{aligned}
\tag{3.2.95}
$$

We will now substitute [(5.6.101) v.1] and the equations

$$
Z_{12,3}^\pm = kT\Theta_2^\pm (Z_{1,23} - p_1 p_3^{-1} Z_{3,12}^{t.c.})
\tag{3.2.96}
$$

which follow from [(5.6.85,10) v.1] and from the connection between $Z_{12,3}^\pm$ and $Q_{12,3}^\pm$ given by (3.2.94). If now we use, in addition, (3.2.26,78), we will get

$$
\begin{aligned}
\beta^2 U_{123} = \; & X_2 X_3 \Theta_2^- \Theta_3^- U_{1,23} + X_1 X_3 \Theta_1^+ \Theta_3^- U_{2,13} + X_1 X_2 \Theta_1^+ \Theta_2^+ U_{3,12} \\
& +X_1 \Theta_2^- \Theta_3^- U_{1,23}^{t.c.} + X_2 \Theta_1^+ \Theta_3^- U_{2,13}^{t.c.} + X_3 \Theta_1^+ \Theta_2^+ U_{3,12}^{t.c.} \\
& -N_3 \Theta_3^- [X_2 \Theta_2^- U_{1,23} + X_1 \Theta_1^+ U_{2,13} - (p_1 \Theta_2^- + p_2 \Theta_1^+)p_3^{-1} U_{3,12}^{t.c.}]
\end{aligned}
$$

$$
\begin{aligned}
-N_2[&X_3\Theta_2^-\Theta_3^-U_{1,23} + X_1\Theta_1^+\Theta_2^+U_{3,12} \\
&-(p_1\Theta_2^-\Theta_3^- + p_3\Theta_1^+\Theta_2^+)p_2^{-1}U_{2,13}^{\text{t.c.}}] - N_1\Theta_1^+[X_3\Theta_3^-U_{2,31} \\
&+X_2\Theta_2^+U_{3,12} - (p_2\Theta_3^- + p_3\Theta_2^+)p_1^{-1}U_{1,23}^{\text{t.c.}}] + N_2N_3\Theta_2^-\Theta_3^-U_{1,23} \\
&+N_1N_3\Theta_1^+\Theta_3^-U_{2,13} + N_1N_2\Theta_1^+\Theta_2^+U_{3,12} \,,
\end{aligned}
\tag{3.2.97}
$$

where

$$
\begin{aligned}
X_l &= (1 - U_l)(1 - U_l^T)^{-1}\,, \\
N_l &= (1 - U_lU_l^T)(1 - U_l^T)^{-1}\,.
\end{aligned}
\tag{3.2.98}
$$

In (3.2.97), the terms containing expressions of the type $(p_k\Theta_l^\pm\Theta_m^\pm + p_l\Theta_k^\pm\Theta_m^\pm)p_m^{-1}$ can be simplified by using the identity

$$
p_1\Theta_2^-\Theta_3^- + p_2\Theta_1^+\Theta_3^- + p_3\Theta_1^+\Theta_2^+ = 0 \quad \text{for} \quad p_1 + p_2 + p_3 = 0\,,
\tag{3.2.99}
$$

which is equivalent to [(5.2.104) v.1]. Collecting like terms in (3.2.98) with the help of the expression $X_l - N_l = -U_l$, we get

$$
\beta^2 U_{123} = \Theta_2^-\Theta_3^-(U_2U_3U_{1,23} - U_1U_{1,23}^{\text{t.c.}}) + \Theta_1^+\Theta_3^-(U_1U_3U_{2,13} - U_2U_{2,13}^{\text{t.c.}})
$$
$$
+\Theta_1^+\Theta_2^+(U_1U_2U_{3,12} - U_3U_{3,12}^{\text{t.c.}})\,.
\tag{3.2.100}
$$

This is the second quadratic FDR. To obtain the covariant form, the expression $U_2U_3U_{1,23} - U_1U_{1,23}^{\text{t.c.}}$ will have to be replaced by

$$
U_{2,}^4 J_{45}U_{3,}^6 J_{67}U_{1,}^{57} - U_{1,}^4 J_{25}J_{36}(U_{4,}^{56})^{\text{t.c.}}\,.
\tag{3.2.101}
$$

Using the form similar to (3.2.91), we will have

$$
\begin{aligned}
\beta^2 U_{123} = \ &\Theta_2^-\Theta_3^-(U_2J_2U_3J_3U_{1,23} - U_2J_2U_3U_{1,23}^{\text{t.c.}}) \\
&+\Theta_1^+\Theta_3^-(U_1J_1U_3J_3U_{2,13} - U_2J_1J_3U_{2,13}^{\text{t.c.}}) \\
&+\Theta_1^+\Theta_2^+(U_1J_1U_2J_2U_{3,12} - U_3J_1J_2U_{3,12}^{\text{t.c.}})\,.
\end{aligned}
\tag{3.2.102}
$$

In the nonquantum limit the operators $\Theta^\pm$ tend to unity and the FDR obtained reads

$$
\beta^2 U_{123} = P_{(123)}(U_2J_2U_3J_3U_{1,23} - U_1J_2J_3U_{1,23}^{\text{t.c.}})\,.
\tag{3.2.103}
$$

3.2.7 Modified Stochastic Representation

Instead of (3.2.81) we can use the following stochastic representation

$$
g_1^\tau = U_1g_1^a + \tfrac{1}{2}U_{1,23}g_2^ag_3^a + \sum_\sigma(\zeta_1^{(\sigma)} + s_{123}^{(\sigma)}\zeta_2^{(\sigma)}g_3^a)\,,
\tag{3.2.104}
$$

which is similar to [(5.6.64) v.1]. Comparison of (3.2.104) with (3.2.81) gives

$$
\zeta_1^{(\sigma)} = -2^{-1/2}(1 - U_1)[T_1^{(\sigma)}\xi_1^{(\sigma)} + 2^{-1}T_{123}^{(\sigma)}\xi_2^{(\sigma)}(1 - U_3)\sum_\tau T_3^{(\tau)}\xi_3^{(\tau)}]
$$
$$
+2^{-2}U_{1,23}\sum_{\sigma,\tau}T_2^{(\sigma)}T_3^{(\tau)}\xi_2^{(\sigma)}\xi_3^{(\tau)}\,,
\tag{3.2.105a}
$$

$$
s_{123}^{(\sigma)}\zeta_2^{(\sigma)} = -2^{-1}(1 - U_1)T_{123}^{(\sigma)}\xi_2^{(\sigma)}(1 - U_3) + 2^{-1/2}U_{1,23}T_2^{(\sigma)}\xi_2^{(\sigma)}\,.
\tag{3.2.105b}
$$

According to the results of the previous subsection, these expressions obey

$$\sum_{\sigma,\tau} \langle \zeta_1^{(\sigma)}, \zeta_2^{(\tau)} \rangle = U_{12},$$

$$\sum_{\sigma,\tau} (s_{143}^{(\sigma)} \langle \zeta_4^{(\sigma)}, \zeta_2^{(\tau)} \rangle + s_{243}^{(\sigma)} \langle \zeta_1^{(\tau)}, \zeta_4^{(\sigma)} \rangle) = U_{12,3},$$

$$\sum_{\sigma,\tau,\pi} \langle \zeta_1^{(\sigma)}, \zeta_2^{(\tau)}, \zeta_3^{(\pi)} \rangle = U_{123}. \tag{3.2.106}$$

We now denote

$$U_{12,3}^- = \sum_{\sigma,\tau} s_{143}^{(\sigma)} \langle \zeta_4^{(\sigma)}, \zeta_2^{(\tau)} \rangle,$$

$$U_{12,3}^+ = \sum_{\sigma,\tau} s_{143}^{(\sigma)} \langle \zeta_2^{(\tau)}, \zeta_4^{(\sigma)} \rangle, \tag{3.2.107}$$

so that

$$U_{12,3} = U_{12,3}^- + U_{21,3}^+. \tag{3.2.108}$$

Reasoning along the same lines as in deriving (3.2.89), we find

$$U_{12,3}^\pm = -\Theta_2^\pm (U_2 U_{1,23} + p_1 p_3^{-1} U_{3,12}^{\text{t.c.}}). \tag{3.2.109}$$

Clearly, (3.2.108,109) are consistent with (3.2.89). In the case of (3.2.91), we will have to replace (3.2.109) by

$$U_{12,3}^\pm = -\Theta_2^\pm (U_2 J_2 U_{1,23} + p_1 p_3^{-1} J_1 J_2 J_3 U_{3,12}^{\text{t.c.}}). \tag{3.2.110}$$

3.2.8 Total (Unconditional) Correlators of Transmitted Waves

Using (3.2.104) and also (3.2.106,107), we can find the unconditional threefold correlator $\langle g_1^r, g_2^r, g_3^r \rangle$ for the case where g^a is a random, and even operator, quantity. By the conventional procedure, in the approximation chosen, we get

$$\begin{aligned}
\langle g_1^r, g_2^r, g_3^r \rangle =\ & U_{123} + U_3 U_{12,4} \langle g_4^a, g_3^a \rangle + U_2 (U_{13,4}^- \langle g_4^a, g_2^a \rangle \\
& + U_{31,4}^+ \langle g_2^a, g_4^a \rangle) + U_1 U_{23,4} \langle g_1^a, g_4^a \rangle \\
& + U_2 U_3 U_{1,45} \langle g_4^a, g_2^a \rangle \langle g_5^a, g_3^a \rangle + U_1 U_3 U_{2,45} \langle g_1^a, g_4^a \rangle \langle g_5^a, g_3^a \rangle \\
& + U_1 U_2 U_{3,45} \langle g_1^a, g_4^a \rangle \langle g_2^a, g_5^a \rangle + U_1 U_2 U_3 \langle g_1^a, g_2^a, g_3^a \rangle.
\end{aligned} \tag{3.2.111}$$

We must substitute here (3.2.89,100,109). To be more specific, consider the particular case where g^a corresponds to equilibrium fluctuations at T_0, when the scatterer has the temperature T. In that case, from (3.2.34) at $J_{12} = I_{12}$, we have

$$\langle g_1^a, g_2^a \rangle = k T_0 \Theta_1^+(T_0) I_{12},$$

$$\langle g_1^a, g_2^a, g_3^a \rangle = 0. \tag{3.2.112}$$

The last threefold correlator is zero because equilibrium thermal fluctuations in linear systems are Gaussian. Substituting (3.2.112) and the equations of the previous subsection into (3.2.111) yields

$$
\begin{aligned}
k^{-2}\langle g_1^r, g_2^r, g_3^r \rangle = T^2 [& \Theta_2^- \Theta_3^- (U_2 U_3 U_{1,23} - U_1 U_{1,23}^{\text{t.c.}}) \\
& + \Theta_1^+ \Theta_3^- (U_1 U_3 U_{2,13} - U_2 U_{2,13}^{\text{t.c.}}) + \Theta_1^+ \Theta_2^+ (U_1 U_2 U_{3,12} - U_3 U_{3,12}^{\text{t.c.}})] \\
& - TT_0 \{ \Theta_3^-(T_0) U_3 [\Theta_2^- U_2 U_{1,23} + \Theta_1^+ U_1 U_{2,13} + (p_1 \Theta_2 + p_2 \Theta_1) p_3^{-1} U_{3,12}^{\text{t.c.}}] \\
& + U_2 [\Theta_2^-(T_0) \Theta_3^- (U_3 U_{1,23} + p_1 p_2^{-1} U_{2,13}^{\text{t.c.}}) \\
& + \Theta_2^+(T_0) \Theta_1^+ (U_1 U_{3,12} + p_3 p_2^{-1} U_{2,13}^{\text{t.c.}})] \\
& + \Theta_1^+(T_0) U_1 [\Theta_3^- U_3 U_{2,13} + \Theta_2^+ U_2 U_{3,12} + (p_2 \Theta_3 + p_3 \Theta_2) p_1^{-1} U_{1,23}^{\text{t.c.}}] \} \\
& + T_0^2 [\Theta_2^-(T_0) \Theta_3^-(T_0) U_2 U_3 U_{1,23} + \Theta_1^+(T_0) \Theta_3^-(T_0) U_1 U_3 U_{2,13} \\
& + \Theta_1^+(T_0) \Theta_2^+(T_0) U_1 U_2 U_{3,12}] \, .
\end{aligned}
\tag{3.2.113}
$$

Those $\Theta_i^\pm$ that do not have T_0 as an argument correspond to T. To check (3.2.113) we can put $T_0 = T$. In so doing, we can readily verify that all the terms on the right-hand side cancel out. This is exactly what is to be expected since in equilibrium the fluctuations of g^r are Gaussian because the wave systems are linear and the threefold correlator of transmitted waves is zero.

Equation (3.2.113) is written in the time representation, where J_{12} coincides with I_{12}. In the spectral and other representations, the appropriate expressions will, by (3.2.91,102,110), also include the matrix J_{12}.

3.3 Cubic Kirchhoff FDRs

3.3.1 Stochastic Representation of Transmitted Waves

We will suppose that in (3.1.39) there is no quadratic nonlinearity, i.e. $U_{1,23} = 0$. (Some results for $U_{1,23} \neq 0$ can be found in Sect. A2.4). Then, according to (3.2.78), we also have $Z_{1,23} = 0$, so that, instead of [(5.6.8) v.1], we find

$$
h_1 = Z_1 J_1 + \tfrac{1}{6} Z_{1,234} J_2 J_3 J_4 + \dots .
\tag{3.3.1}
$$

Substituting (3.2.3) and solving the resultant equation for g^r gives

$$
\begin{aligned}
g_1^r = \; & U_1 g_1^a + \tfrac{1}{12} (S_1 Z_1 S_1 + 1)^{-1} S_1 Z_{1,234} \\
& \times (1 - U_2)(1 - U_3)(1 - U_4) g_2^a g_3^a g_4^a + \dots .
\end{aligned}
\tag{3.3.2}
$$

To simplify the treatment a little, we will assume that $S_{12} = I_{12}$. This assumption is purely technical, not fundamental, in nature. Comparison of (3.3.2) with (3.1.39) shows that

$$
\begin{aligned}
U_{1,234} &= \tfrac{1}{2} (Z_1 + 1)^{-1} Z_{1,234} (1 - U_2)(1 - U_3)(1 - U_4) \\
&= \tfrac{1}{4} (1 - U_1) Z_{1,234} (1 - U_2)(1 - U_3)(1 - U_4)
\end{aligned}
\tag{3.3.3}
$$

by (3.2.9). We will now take into account the random forces $\mathcal{E}$ that are engendered in the scattering body. Instead of (3.2.18), here we should take

$$
h_1 + \mathcal{E}_1 = Z_1 J_1 + \tfrac{1}{6} Z_{1,234} J_2 J_3 J_4 + \dots .
\tag{3.3.4}
$$

Substituting (3.2.3), taking into account (3.3.3) and again solving for g^r, we get

$$
\begin{aligned}
g_1^r ={}& U_1 g_1^a - 2^{-1/2}(1 - U_1)\mathcal{E}_1 \\
&+ \tfrac{1}{6} U_{1,234}(g_2^a + 2^{-1/2}\mathcal{E}_2)(g_3^a + 2^{-1/2}\mathcal{E}_3)(g_4^a + 2^{-1/2}\mathcal{E}_4) + \ldots \,. \quad (3.3.5)
\end{aligned}
$$

We now make use of the stochastic representation [(5.7.18) v.1] of random forces. It is equivalent to

$$
\mathcal{E}_1 = \sum_\sigma (\xi_1^{(\sigma)} + T_{123}^{(\sigma)}\xi_2^{(\sigma)} J_3 + \tfrac{1}{2} T_{1234}^{(\sigma)}\xi_2^{(\sigma)} J_3 J_4)\,. \qquad (3.3.6)
$$

We have put here $T_{12}^{(\sigma)} \equiv S_{12}^{(\sigma)} = I_{12}$ to simplify the equations. The same assumption has been made in Sect. 5.7.3 v.1. Substituting (3.2.3b), we find, by (3.3.5),

$$
\begin{aligned}
\mathcal{E}_1 ={}& \sum_\sigma [\xi_1^{(\sigma)} + 2^{-1/2}T_{123}^{(\sigma)}\xi_2^{(\sigma)}(1 - U_3)(g_3^a + 2^{-1/2}\mathcal{E}_3) \\
&+ \tfrac{1}{4} T_{1234}^{(\sigma)}\xi_2^{(\sigma)}(1 - U_3)(1 - U_4)(g_3^a + 2^{-1/2}\mathcal{E}_3)(g_4^a + 2^{-1/2}\mathcal{E}_4)]\,. \quad (3.3.7)
\end{aligned}
$$

Hence (3.3.5) takes the form

$$
\begin{aligned}
g_1^r ={}& U_1 g_1^a \\
&- 2^{-1/2}(1 - U_1)\sum_\sigma \left[\xi_1^{(\sigma)} + 2^{-1/2}T_{123}^{(\sigma)}\xi_2^{(\sigma)}(1 - U_3)\left(g_3^a + 2^{-1/2}\sum_\tau \xi_3^{(\tau)}\right)\right. \\
&+ 2^{-2}T_{12,34}^{(\sigma)}\xi_2^{(\sigma)}(1 - U_3)(1 - U_4)\left(g_3^a + 2^{-1/2}\sum_\tau \xi_3^{(\tau)}\right) \\
&\times \left.\left(g_4^a + 2^{-1/2}\sum_\rho \xi_4^{(\rho)}\right)\right] + \tfrac{1}{6} U_{1,234}\left(g_2^a + 2^{-1/2}\sum_\sigma \xi_2^{(\sigma)}\right) \\
&\times \left(g_3^a + 2^{-1/2}\sum_\tau \xi_3^{(\tau)}\right)\left(g_4^a + 2^{-1/2}\sum_\rho \xi_4^{(\rho)}\right) + \ldots\,. \quad (3.3.8)
\end{aligned}
$$

This expression can be employed (by using the equations from Sect. 5.7. v.1) to work out the various correlators of transmitted waves g^r. Specifically, if g^a is taken to be fixed, nonrandom and nonoperator, then we will be able to compute the conditional correlators

$$
\langle g_1^r, g_2^r\rangle_{g^a} = U_{12} + \tfrac{1}{2} U_{1234}g_3^a g_4^a \qquad (3.3.9a)
$$

$$
\langle g_1^r, g_2^r, g_3^r\rangle_{g^a} = U_{123,4}g_4^a \qquad (3.3.9b)
$$

$$
\langle g_1^r, g_2^r, g_3^r, g_4^r\rangle_{g^a} = U_{1234}\,, \qquad (3.3.9c)
$$

or, in other words, to compute the four-subscript functions that enter these expressions. Equations (3.3.9) do not contain three-subscript functions since $U_{1,23} = 0$.

3.3.2 Simplified Stochastic Representation

In addition to (3.3.8), we can write a simpler stochastic representation:

$$g_1^\tau = U_1 g_1^a + \tfrac{1}{6} U_{1,234} g_2^a g_3^a g_4^a + \sum_\sigma [\zeta_1^{(\sigma)} + s_{123}^{(\sigma)} \zeta_2 g_3^a + \tfrac{1}{2} s_{12,34}^{(\sigma)} \zeta_2^{(\sigma)} g_3^a g_4^a]. \quad (3.3.10)$$

By comparing (3.3.8) and (3.3.10), we find

$$
\begin{aligned}
\zeta_1^{(\sigma)} &= -2^{-1/2}(1 - U_1) \left[\xi_1^{(\sigma)} + 2^{-1} T_{123}^{(\sigma)} \xi_2^{(\sigma)} (1 - U_3) \sum_\tau \xi_3^{(\tau)} \right. \\
&\quad \left. + 2^{-3} T_{12,34}^{(\sigma)} \xi_2^{(\sigma)} (1 - U_3)(1 - U_4) \sum_{\tau,\rho} \xi_3^{(\tau)} \xi_4^{(\rho)} \right] \\
&\quad + \tfrac{1}{6} \cdot 2^{-3/2} U_{1,234} \xi_2^{(\sigma)} \sum_{\tau,\rho} \xi_3^{(\tau)} \xi_4^{(\rho)} , \quad (3.3.11)
\end{aligned}
$$

$$
\begin{aligned}
s_{123}^{(\sigma)} \zeta_2^{(\sigma)} &= -2^{-1}(1 - U_1) \left[T_{123}^{(\sigma)} \xi_2^{(\sigma)} (1 - U_3) \right. \\
&\quad \left. + 2^{-1} T_{12,34}^{(\sigma)} (1 - U_3)(1 - U_4) \xi_2^{(\sigma)} \sum_\tau \xi_4^{(\tau)} \right] \\
&\quad + 2^{-2} U_{1,234} \xi_2^{(\sigma)} \sum_\tau \xi_4^{(\tau)} , \quad (3.3.12)
\end{aligned}
$$

$$s_{1234}^{(\sigma)} \zeta_2^{(\sigma)} = -2^{-3/2}(1 - U_1) T_{1234}^{(\sigma)} (1 - U_3)(1 - U_4) \xi_2^{(\sigma)} + 2^{-1/2} U_{1,234} \xi_2^{(\sigma)} . \quad (3.3.13)$$

If we fix g^a in (3.3.10) and find the conditional correlator $\langle g_1^\tau, g_2^\tau \rangle_{g^a}$ (see (3.3.9)), we will obtain for the quantum case

$$
\begin{aligned}
U_{12,34} &= V_{1234}^- + V_{2134}^+ + W_{1234} + W_{1243} \\
&= U_{12,34}^- + U_{21,34}^+ , \\
U_{12,34}^- &= V_{1234}^- + W_{1234} , \quad U_{12,34}^+ = V_{1234}^+ + W_{2143} , \quad (3.3.14)
\end{aligned}
$$

where

$$
V_{1234}^- = \sum_{\sigma,\tau} s_{15,34}^{(\sigma)} \langle \zeta_5^{(\sigma)} \zeta_2^{(\tau)} \rangle ,
$$

$$
V_{1234}^+ = \sum_{\sigma,\tau} s_{15,34}^{(\sigma)} \langle \zeta_2^{(\tau)}, \zeta_5^{(\sigma)} \rangle , \quad (3.3.15a)
$$

$$
W_{1234} = \sum_{\sigma,\tau} s_{153}^{(\sigma)} s_{264}^{(\tau)} \langle \zeta_5^{(\sigma)}, \zeta_6^{(\tau)} \rangle . \quad (3.3.15b)
$$

According to the chosen degree of accuracy, the correlators in (3.2.15) should be computed keeping only the terms of the order of kT. Instead of (3.3.11), we can then assume

$$\zeta_1^{(\sigma)} = -2^{-1/2}(1 - U_1) \xi_1^{(\sigma)} . \quad (3.3.16)$$

Let us now turn to the threefold correlator (see (3.3.9b)). Using (3.3.10), we obtain

$$U_{123,4} = U_{123,4}^{--} + U_{213,4}^{+-} + U_{312,4}^{++} , \quad (3.3.17)$$

where

$$U_{123,4}^{--} = \sum_{\sigma,\tau,\rho} s_{154}^{(\sigma)} \langle \zeta_5^{(\sigma)}, \zeta_2^{(\tau)}, \zeta_3^{(\rho)} \rangle, \tag{3.3.18a}$$

$$U_{123,4}^{+-} = \sum_{\sigma,\tau,\rho} s_{154}^{(\sigma)} \langle \zeta_2^{(\tau)}, \zeta_5^{(\sigma)}, \zeta_3^{(\rho)} \rangle, \tag{3.3.18b}$$

$$U_{123,4}^{++} = \sum_{\sigma,\tau,\rho} s_{154}^{(\sigma)} \langle \zeta_2^{(\tau)}, \zeta_3^{(\rho)}, \zeta_5^{(\sigma)} \rangle. \tag{3.3.18c}$$

Lastly, the fourfold correlator will be

$$U_{1234} = \sum_{\sigma,\tau,\rho,\pi} \langle \zeta_1^{(\sigma)}, \zeta_2^{(\tau)}, \zeta_3^{(\rho)}, \zeta_4^{(\pi)} \rangle. \tag{3.3.19}$$

Equations (3.3.14,15) and (3.3.17,18) resemble [(5.7.20,21) v.1] and [(5.7.23,24) v.1].

3.3.3 Relations for $U_{12,34}$

To begin with, we will consider $V_{1234}^{\pm}$. Substituting (3.3.13,16) into (3.3.15a,b), we get

$$
\begin{aligned}
V_{1234}^{\pm} &= \tfrac{1}{4}(1 - U_1)(1 - U_2)J_{1234}^{\pm}p_3^{-1}p_4^{-1}(1 - U_3)(1 - U_4) \\
&\quad - kT\Theta_2^{\pm}(1 - U_2 U_2^T)(1 - U_2^T)^{-1}U_{1,234}
\end{aligned}
\tag{3.3.20}
$$

by [(5.7.59,20a) v.1] and (3.2.84,27).

As is known from Sect. 5.7 v.1, the four-subscript functions, including $J_{1234}^{\pm}$ can be represented as the sum of dissipationally determinable and dissipationally undeterminable parts: $J_{1234}^{\pm} = J_{1234}^{\pm(1)} + J_{1234}^{\pm(2)}$. Then, by [(5.7.38,30,40) v.1], we get

$$
\begin{aligned}
J_{1234}^{\pm(1)} &= -i\hbar\Gamma_2^{\pm}Q_{1,234} \\
&= -i\hbar\Gamma_2^{\pm}Z_{1,234}p_2p_3p_4 \\
&= kT\Theta_2^{\pm}Z_{1,234}p_3p_4 \quad (kT\Theta_2^{\pm} = i\hbar p_2\Gamma_2^{\pm}),
\end{aligned}
\tag{3.3.21}
$$

where we have also used [(5.6.9) v.1]. Accordingly, (3.3.20) will read

$$
\begin{aligned}
V_{1234}^{\pm} &= kT\Theta_2^{\pm}[X_2 U_{1,234} - N_2 U_{1,234}] \\
&\quad + \tfrac{1}{4}(1 - U_1)(1 - U_2)J_{1234}^{\pm(2)}p_3^{-1}p_4^{-1}(1 - U_3)(1 - U_4).
\end{aligned}
\tag{3.3.22}
$$

Here we have taken into account (3.3.3) and (3.2.98). According to the equation $X_2 - N_2 = -U_2$, the operator X_2 on the right-hand side of (3.3.22) cancels out.

It follows that $V_{1234}^{\pm}$, like $J_{1234}^{\pm}$, can be represented as the sum

$$V_{1234}^{\pm} = V_{1234}^{\pm(1)} + V_{1234}^{\pm(2)}, \tag{3.3.23}$$

where

$$V_{1234}^{\pm(1)} = -kT\Theta_2^{\pm}U_2 U_{1,234} \tag{3.3.24}$$

is the dissipationally determinable part, and

$$V^{\pm(2)}_{1234} = \tfrac{1}{4}(1-U_1)(1-U_2)J^{\pm(2)}_{1234}p_3^{-1}p_4^{-1}(1-U_3)(1-U_4) \qquad (3.3.25)$$

is the dissipationally undeterminable part.

Let us now turn to W_{1234}. Substituting (3.3.12) into (3.3.15b) and taking into account [(5.7.20b) v.1], we arrive at

$$W_{1234} = \tfrac{1}{4}(1-U_1)(1-U_2)K_{1234}p_3^{-1}p_4^{-1}(1-U_3)(1-U_4). \qquad (3.3.26)$$

According to [(5.7.40) v.1], the dissipationally determinable part of K_{1234} is zero. Therefore, the dissipationally determinable part of W_{1234} is zero as well:

$$W^{(1)}_{1234} = 0, \qquad (3.3.27)$$

and (3.3.26) coincides with the dissipationally undeterminable part

$$W^{(2)}_{1234} = \tfrac{1}{4}(1-U_1)(1-U_2)K^{(2)}_{1234}p_3^{-1}p_4^{-1}(1-U_3)(1-U_4). \qquad (3.3.28)$$

Adding (3.3.25) and (3.3.28) together, we find, by (3.3.14) and [(5.7.30) v.1],

$$U^{\pm(2)}_{12,34} = \tfrac{1}{4}(1-U_1)(1-U_2)Z^{\pm(2)}_{12,34}(1-U_3)(1-U_4), \qquad (3.3.29)$$

where $Z^{\pm}_{12,34}p_3p_4 = Q^{\pm}_{12,34}$. Adding (3.3.23) and (3.3.26) together, from (3.3.14), we arrive at

$$U_{12,34} = -kT(\Theta_2^- U_2 U_{1,234} + \Theta_1^+ U_1 U_{2,134}) + U^{(2)}_{12,34}, \qquad (3.3.30)$$

where

$$\begin{aligned} U^{(2)}_{12,34} &= U^{-(2)}_{12,34} + U^{+(2)}_{21,34} \\ &= \tfrac{1}{4}(1-U_1)(1-U_2)Z^{(2)}_{12,34}(1-U_3)(1-U_4). \end{aligned} \qquad (3.3.31)$$

According to this and [(5.7.56a,b) v.1], the dissipationally undeterminable part $U^{(2)}_{12,34}$ obeys

$$U^{(2)}_{12,34} = U^{(2)}_{21,34}, \qquad (3.3.32a)$$

$$\Theta_3^- \Theta_4^- (U^{(2)}_{12,34})^{t.c.} = \Theta_1^+ \Theta_2^+ U^{(2)}_{34,12}. \qquad (3.3.32b)$$

The relation (3.3.32b) is derived by using (3.2.12) as well.

If (3.3.30) is solved for $U^{(2)}_{12,34}$ and the result is substituted into (3.3.32), we will obtain the equations relating $U_{12,34}$ to $U_{1,234}$.

3.3.4 Relations for $U_{123,4}$

In computing (3.3.17), i.e., the functions (3.3.18), we should take into account terms of the order of $(kT)^2$. Therefore, instead of (3.3.16) we should apply the more exact expression

$$\zeta_1^{(\sigma)} = -2^{-1/2}(1 - U_1)\left[\xi_1^{(\sigma)} + \tfrac{1}{2}T_{123}^{(\sigma)}\xi_2^{(\sigma)}(1 - U_3)\sum_\tau \xi_3^{(\tau)}\right] \tag{3.3.33}$$

which is obtained, if in (3.3.11) we consider the terms quadratic in the fluctuations. Using (3.3.12,33) we will have, from (3.3.18a),

$$U_{123,4}^{--} = -\tfrac{1}{4}(1 - U_1)(1 - U_2)(1 - U_3)$$

$$\times \sum_\sigma \left\{ T_{154}^{(\sigma)}(1 - U_4)\left[R_{523}^{(\sigma)} + \tfrac{1}{2}\sum_\tau T_{276}^{(\tau)}(1 - U_6)(R_{57}^{(\sigma)}R_{63}^{(\tau)} + R_{56}^{(\sigma)}R_{73}^{(\tau)}) \right.\right.$$

$$\left.+ \tfrac{1}{2}\sum_\tau T_{376}^{(\tau)}(1 - U_6)(R_{57}^{(\sigma)}R_{26}^{(\tau)} + R_{56}^{(\sigma)}R_{27}^{(\tau)})\right]$$

$$\left.+ \tfrac{1}{2}T_{1564}^{(\sigma)}(1 - U_6)(1 - U_4)\sum_\tau (R_{52}^{(\sigma)}R_{63}^{(\tau)} + R_{53}^{(\sigma)}R_{62}^{(\tau)})\right\}$$

$$+ \tfrac{1}{4}(1 - U_2)(1 - U_3)U_{1,564}\sum_{\sigma,\tau} R_{52}^{(\sigma)}R_{63}^{(\sigma)} . \tag{3.3.34}$$

The sums

$$\sum_\sigma T_{267}^{(\sigma)}R_{63}^{(\sigma)} , \quad \sum_\sigma T_{367}^{(\sigma)}R_{26}^{(\sigma)} , \tag{3.3.35}$$

i.e. the sums (3.2.94), which enter the last expression, vanish according to (3.2.96) and the equation $Z_{1,23} = 0$. By [(5.7.20,23) v.1] and (3.2.84), the expression (3.3.34) becomes

$$
\begin{aligned}
U_{123,4}^{--} = {}& -\tfrac{1}{4}(1 - U_1)[(1 - U_2)(1 - U_3)Q_{123,4}^{--} \\
& + (1 - U_2)K_{1246}p_6^{-1}U_{63} + (1 - U_3)K_{1346}p_6^{-1}U_{26} \\
& + (1 - U_2)J_{1264}^- p_6^{-1}U_{63} + (1 - U_3)J_{1364}^- p_6^{-1}U_{62}]p_4^{-1}(1 - U_4) \\
& + U_{1,564}(1 - U_5)^{-1}(1 - U_6)^{-1}U_{52}U_{63} .
\end{aligned} \tag{3.3.36}
$$

To begin with, we will consider the dissipationally determinable part of (3.3.36). If we take into account [(5.7.38,40,41) v.1] and (3.3.3), we obtain, after some algebra

$$U_{123,4}^{--(1)} = (kT)^2 \Theta_2^- \Theta_3^- (U_2 U_3 U_{1,234} - p_1 p_4^{-1}U_{4,123}^{\text{t.c.}}) . \tag{3.3.37}$$

The dissipationally undeterminable part is found using [(5.7.50) v.1] and the formulas (3.2.26) and (3.3.25,26) derived above. Cancelling out by using the equation $X_l - N_l = -U_l$ gives

$$U_{123,4}^{--(2)} = kT(\Theta_3^- U_{12,43}^{-(2)} + \Theta_2^- V_{1342}^{-(2)} + \Theta_2^+ W_{1342}^{(2)}) . \tag{3.3.38}$$

A similar derivation is also possible for $U_{123,4}^{+-}$ and $U_{123,4}^{++}$. In addition to (3.3.37,38), we will then find

$$U_{123,4}^{\gamma\pm(1)} = (kT)^2 \Theta_2^\gamma \Theta_3^\pm (U_2 U_3 U_{1,234} - p_1 p_4^{-1} U_{4,123}^{\text{t.c.}}), \quad \gamma = \pm, \tag{3.3.39}$$

$$U_{123,4}^{+-(2)} = kT(\Theta_3^- U_{12,43}^{+(2)} + \Theta_2^+ U_{13,42}^{-(2)}),$$
$$U_{123,4}^{++(2)} = kT(\Theta_3^+ V_{1243}^{+(2)} + \Theta_3^- W_{2134}^{(2)} + \Theta_2^+ U_{13,42}^{+(2)}). \tag{3.3.40}$$

Substituting (3.3.38–40) into (3.3.17) yields

$$U_{123,4} = U_{123,4}^{(1)} + U_{123,4}^{(2)}, \tag{3.3.41}$$

where

$$\beta^2 U_{123,4}^{(1)} = \Theta_2^- \Theta_3^- U_2 U_3 U_{1,234} + \Theta_1^+ \Theta_3^- U_1 U_3 U_{2,134} + \Theta_1^+ \Theta_2^+ U_1 U_2 U_{3,124}$$
$$- (p_1 \Theta_2^- \Theta_3^- + p_2 \Theta_1^+ \Theta_3^- + p_3 \Theta_1^+ \Theta_2^+) p_4^{-1} U_{4,123}^{\text{t.c.}} \tag{3.3.42}$$

and

$$\beta U_{123,4}^{(2)} = \Theta_3^- U_{12,34}^{(2)} + \Theta_2^- U_{13,24}^{-(2)} + \Theta_2^+ U_{31,24}^{+(2)} + \Theta_1^+ U_{23,14}^{(2)}. \tag{3.3.43}$$

The last equations are analogs of [(5.4.32,70) v.1] or rather [(5.4.106b,107c) v.1].

Using (3.3.43), the equation $U_{13,42}^{-(2)} + U_{31,42}^{+(2)} = U_{13,42}^{(2)}$, and also (3.3.41,42), we can express $U_{12,34}^{\pm(2)}$ in terms of $U_{1,234}, U_{12,34}$ and $U_{123,4}$:

$$U_{12,34}^{\pm(2)} = f_\pm [U_{1,234}, U_{12,34}, U_{123,4}]. \tag{3.3.44}$$

We will now introduce auxiliary functions similar to [(5.4.74) v.1] (see also [(5.4.76) v.1])

$$\tilde{\tilde{M}}_{1234} = (\Theta_2^\pm)^{-1} V^{\pm(2)} = (\Theta_2^- - \Theta_{24}^-)^{-1}(\Theta_{24}^+ U_{12,34}^{-(2)} - \Theta_{24}^- U_{12,34}^{+(2)}),$$
$$\tilde{\tilde{N}}_{1234} = (\Theta_{24}^-)^{-1} W_{1234}^{(2)} = (\Theta_2^- - \Theta_{24}^-)^{-1}(\Theta_2^- U_{12,34}^{+(2)} - \Theta_2^+ U_{12,34}^{-(2)}). \tag{3.3.45}$$

Using (3.3.25,26) and [(5.7.52,53), (5.4.53) v.1], we can prove that they behave in the same manner as M_{1234}, N_{1234}, viz. they obey equations similar to [(5.4.53) v.1]. Substituting (3.3.44) into the right-hand side of (3.3.45) and using equations of the type [(5.4.53) v.1], we can work out various relations for $U_{1,234}, U_{12,34}$ and $U_{123,4}$.

In the nonquantum case, we obtain from (3.3.41–43) the following FDR:

$$\beta^2 U_{123,4} = U_2 U_3 U_{1,234} + U_1 U_3 U_{2,134} + U_1 U_2 U_{3,124}$$
$$+ U_{4,123}^{\text{t.c.}} + \beta(U_{12,34}^{(2)} + U_{13,24}^{(2)} + U_{23,14}^{(2)}). \tag{3.3.46}$$

Thus, in this case $U_{123,4}$ is expressed through $U_{1,234}$ and $U_{12,34}$.

3.3.5 Relations for the Conditional Fourfold Correlator of Transmitted Waves

The conditional correlator $\langle g_1^r, g_2^r, g_3^r, g_4^r \rangle_{g^a}$, which corresponds to fixed approaching waves, i.e. the correlator (3.3.19), must be computed using the equation (3.3.11), which is more exact that (3.3.33). This derivation also takes into account the formulas [(5.7.55,38,40,41,50) v.1] as well as (3.2.26), (3.3.3,24,25,27–29). Leaving out the computation, we will only provide here the final result:

$$U_{1234} = U_{1234}^{(1)} + U_{1234}^{(2)}, \tag{3.3.47}$$

where

$$
\begin{aligned}
\beta^3 U_{1234}^{(1)} = & -\Theta_2^- \Theta_3^- \Theta_4^- (U_2 U_3 U_4 U_{1,234} + U_1 U_{1,234}^{\text{t.c.}}) \\
& -\Theta_1^+ \Theta_3^- \Theta_4^- (U_1 U_3 U_4 U_{2,134} + U_2 U_{2,134}^{\text{t.c.}}) \\
& -\Theta_1^+ \Theta_2^+ \Theta_4^- (U_1 U_2 U_4 U_{3,124} + U_3 U_{3,124}^{\text{t.c.}}) \\
& -\Theta_1^+ \Theta_2^+ \Theta_3^+ (U_1 U_2 U_3 U_{4,123} + U_4 U_{4,123}^{\text{t.c.}}),
\end{aligned} \tag{3.3.48}
$$

$$
\begin{aligned}
\beta^2 U_{1234}^{(2)} = & \; \Theta_3^- \Theta_4^- U_3 U_4 U_{12,34}^{(2)} + \Theta_4^- U_2 U_4 (\Theta_2^- U_{13,24}^{-(2)} + \Theta_2^+ U_{31,24}^{+(2)}) \\
& +U_2 U_3 (\Theta_2^- \Theta_3^- V_{1423}^{-(2)} + \Theta_2^+ \Theta_3^+ V_{4123}^{+(2)} + \Theta_2^- \Theta_3^+ W_{1423} \\
& +\Theta_2^+ \Theta_3^- W_{1432}) + \Theta_1^+ \Theta_4^- U_1 U_4 U_{23,14}^{(2)} \\
& +\Theta_1^+ U_1 U_3 (\Theta_3^- U_{24,31}^{-(2)} + \Theta_3^+ U_{42,31}^{+(2)}) + \Theta_1^+ \Theta_2^+ U_1 U_2 U_{34,12}^{(2)} . \tag{3.3.49}
\end{aligned}
$$

Since functions (3.3.45) obey equations similar to [(5.4.53) v.1], it can be readily shown that in (3.3.49) we can substitute $E_1 \Theta_2^+ \Theta_3^+ U_{14,23}^{(2)}$ for $\Theta_2^- \Theta_3^- V_{1423}^{-(2)} + \Theta_2^+ \Theta_3^+ V_{4123}^{+(2)} + \Theta_2^- \Theta_3^+ W_{1423} + \Theta_2^+ \Theta_3^- W_{1432}$. Consequently, U_{1234} is expressed through $U_{1,234}, U_{12,34}$ and $U_{123,4}$. In the nonquantum case, from (3.3.47–49), we have

$$
\begin{aligned}
U_{1234} = & -(kT)^3 P_{(1234)} (U_2 U_3 U_4 U_{1,234} + U_1 U_{1,234}^{\text{t.c.}}) \\
& +P_{(234)}[P_{14}(U_3 U_4 U_{12,34}^{(2)})], \tag{3.3.50}
\end{aligned}
$$

so that U_{1234} is expressed through $U_{1,234}$ and $U_{12,34}$.

The stochastic representation (3.3.10) can also be used to find unconditional correlators for random and operator incident waves. Specifically, in the nonquantum case we can easily derive the expression

$$
\begin{aligned}
& \langle g_1^r, g_2^r, g_3^r, g_4^r \rangle \\
& = U_{1234} + P_{(1234)} (U_4 U_{123,5} R_{45}^a) + P_{(234)}[P_{14}(U_3 U_4 U_{12,56} R_{35}^a R_{46}^a)] \\
& \quad + U_1 U_2 U_3 U_4 R_{1234}^a + P_{(1234)} (U_2 U_3 U_4 U_{1,567} R_{25}^a R_{36}^a R_{47}^a), \tag{3.3.51}
\end{aligned}
$$

where

$$R_{1\dots m}^a = \langle g_1^a, \dots, g_m^a \rangle \tag{3.3.52}$$

are the correlators of incident waves. Next we should substitute (3.3.46,50) into (3.3.51). The formula giving the quantum correlator is more lengthy, and so we will not provide it here.

3.3.6 Basic Four-Subscript Relations in an Arbitrary Representation

We have thus derived various equations relating the four-subscript functions $U_{....}$. For the reasons put forward in Sect. 2.5.1 they refer to the time representation. From them we can readily obtain covariant relationships that can be used in any, in particular the spectral, representation if we introduce, where appropriate, the metric tensor $J = \| J_{12} \|$. Applying (3.3.30,42,48), we obtain for the dissipationally determinable parts the following relations:

$$\beta U^{(1)}_{12,34} = -\Theta^-_2 U_2 J_2 U_{1,234} - \Theta^+_1 U_1 J_1 U_{2,134} \,, \tag{3.3.53a}$$

$$\begin{aligned}\beta^2 U^{(1)}_{123,4} = &\ \Theta^-_2 \Theta^-_3 U_2 U_3 J_2 J_3 U_{1,234} \\ &+ \Theta^+_1 \Theta^-_3 U_1 U_3 J_1 J_3 U_{2,134} + \Theta^+_1 \Theta^+_2 U_1 U_2 J_1 J_2 U_{3,124} \\ &- (p_1 \Theta^-_2 \Theta^-_3 + p_2 \Theta^+_1 \Theta^-_3 + p_3 \Theta^+_1 \Theta^+_2) p_4^{-1} J_1 J_2 J_3 J_4 U^{\mathrm{t.c.}}_{4,123} \,, \end{aligned} \tag{3.3.53b}$$

$$\begin{aligned}\beta^3 U^{(1)}_{1234} = &-\Theta^-_2 \Theta^-_3 \Theta^-_4 \Xi_{1234} - \Theta^+_1 \Theta^-_3 \Theta^-_4 \Xi_{2134} \\ &- \Theta^+_1 \Theta^+_2 \Theta^-_4 \Xi_{3124} - \Theta^+_1 \Theta^+_2 \Theta^+_3 \Xi_{4123} \,, \end{aligned} \tag{3.3.53c}$$

where

$$\Xi_{1234} = U_2 U_3 U_4 J_2 J_3 J_4 U_{1,234} + U_1 J_2 J_3 J_4 U^{\mathrm{t.c.}}_{1,234} \,. \tag{3.3.54}$$

We can see that these equations are correct if in them we write all the indices, contravariant indices being written as superscripts. The operator J in (3.3.53) is a covariant tensor. In the spectral representation it has the form (3.1.45b).

In a similar way, we can generalize the relations (3.3.32,43) to yield

$$U^{(2)}_{12,34} = U^{(2)}_{21,34} \,,$$

$$\Theta^-_3 \Theta^-_4 U^{(2)\mathrm{t.c.}}_{12,34} J_3 J_4 = \Theta^+_1 \Theta^+_2 U^{(2)}_{34,12} J_1 J_2 \,,$$

$$\beta U^{(2)}_{123,4} = \Theta^-_3 J_3 U^{(2)}_{12,34} + \Theta^-_2 J_2 U^{-(2)}_{13,24} + \Theta^+_2 J_2 U^{+(2)}_{31,24} + \Theta^+_1 J_1 U^{(2)}_{23,14} \,. \tag{3.3.55}$$

Lastly, from (3.3.49), we get

$$\begin{aligned}\beta^2 U^{(2)}_{1234} = &\ \Theta^-_3 \Theta^-_4 U_3 U_4 J_3 J_4 U^{(2)}_{12,34} + \Theta^-_4 U_2 U_4 J_2 J_4 (\Theta^-_2 U^{-(2)}_{13,24} + \Theta^+_2 U^{+(2)}_{31,24}) \\ &+ E_1 \Theta^+_2 \Theta^+_3 U_2 U_3 J_2 J_3 U^{(2)}_{14,23} + \Theta^+_1 \Theta^-_4 U_1 U_4 J_1 J_4 U^{(2)}_{23,14} \\ &+ \Theta^+_1 U_1 U_3 J_1 J_3 (\Theta^-_3 U^{-(2)}_{24,31} + \Theta^+_3 U^{+(2)}_{42,31}) \\ &+ \Theta^+_1 \Theta^+_2 U_1 U_2 J_1 J_2 U^{(2)}_{34,12} \,. \end{aligned} \tag{3.3.56}$$

It is to be noted that for the functions

$$\bar{U}_{1...m,(m+1)...n} = U_{1...m,(m+1)...n} J_{m+1} \cdots J_n \,, \tag{3.3.57}$$

which are covariant in all their indices, instead of the above equations we will again obtain relations of the type (3.3.30,42,48) and (3.3.32,43,49). The same is true of the two- and three-subscript relations of Sect. 3.2.

From the results of this section we can conclude the validity of universal Kirchhoff relations that have a structure similar to the FDRs of the first, second and third kinds. There are some differences, however: the two-subscript FDR (3.2.27) has a somewhat different form; moreover, the relations with a larger number of subscripts include the two-subscript matrix $U_{1,2}$, which is absent in the FDRs of the first, second and third kinds.

3.4 Notes on References to Chapter 3

The ordinary Kirchhoff law has been given in many texts, for example, in [3.1, §63]. The correctness of the universal relation expressing the generalized Kirchhoff law and corresponding to the linear approximation was noticed and verified for special cases in [3.2] (see also the book [3.3]). This relation was proved in [3.4,5], in which nonlinear relations of the Kirchhoff type were obtained.

4. Method of Projection in Space of Phase Space Distributions and Irreversible Processes

This chapter differs from the others because it does not derive new results of nonequilibrium thermodynamics. It also differs in the methods it uses. Nevertheless, the contents of this chapter are closely related to the rest of the book. We consider now the question of how, from the Liouville equation, which is equivalent to the dynamic equations describing microprocesses, one can determine the irreversibility of macroprocesses, and also the Markov nature of the macroprocess in question. It is useful then to apply a projection operator defined in the space of phase space distributions. Formally, irreversibility appears as a consequence of a combination of actions of the projection operator and the time-evolution operator.

This projection operator is used further to introduce the Markov operator, whose properties are also studied. One specific procedure is considered which enables one to find the Markov operator explicitly. This procedure is the adiabatic exclusion of fast-varying variables. A specific example of a computation of the Markov operator is provided.

It is shown in Sect. 4.2 that by using the projection operator (but in another space) one can obtain linear FDRs of the first kind. Within the sphere of validity of the linear approximation they are derived as exact relations.

4.1 Markov Processes in Microdynamics

4.1.1 Markov-Like Systems

The totality of molecules, ions, etc. in a system forms a dynamical system that has a Hamiltonian $H(z)$ and which evolves in accordance with Hamilton's dynamical equations. From the point of view of this molecular dynamics, the model of the Markov process is undoubtedly approximate. Only in the case of field dynamical variables and a spatially unbounded system can the Markov model be exact. Further, by no means all discrete (i.e. many-particle) dynamical systems behave in such a way that they can be approximately described by a Markov model. Those systems that can be approximately described by a Markov model will be referred to as Markov-like. As a matter of fact, such systems are characterized by at least two time scales: the time constant of macroscopic relaxations τ_r, which characterizes the time of variation of the selected internal thermodynamic parameters $B_\alpha(z)$;

and the time constant τ_0, which is the time for establishing a local equilibrium for other variables or the correlation time of random influences of other variables on the parameters $B_\alpha(z)$. These constants must differ markedly, i.e. they must obey the inequality $\tau_r \gg \tau_0$.

We will introduce the intermediate time $\tau_i = (\tau_r \tau_0)^{1/2}$, which clearly obeys

$$\tau_r \gg \tau_i \gg \tau_0 . \tag{4.1.1}$$

4.1.2 The Operators $\hat{\Pi}$ and $\hat{\Pi}^-$

A thermodynamic treatment of a dynamical system is actually incomplete. It overlooks the behavior of an enormous number of dynamical variables z; instead it is interested in the behavior of a relatively small number of chosen thermodynamic parameters $B_\alpha(z)$, $\alpha = 1, \ldots, r$. Knowing the probability density $\rho(z)$ in phase space, we can easily find the probability density

$$w(A) = \int \delta(A - B(z))\rho(z)dz \tag{4.1.2}$$

for the thermodynamic parameters. Introducing the linear operator $\hat{\Pi}$ which maps the space of phase space distributions onto the space of probability densities w, we can write (4.1.2) as

$$w = \hat{\Pi}\rho \equiv \int \delta(A - B(z))\rho(z)dz . \tag{4.1.3}$$

This operator has the matrix elements

$$(\Pi)_{Az} = \delta(A - B(z)) . \tag{4.1.4}$$

In phase space distributions obey the Liouville equation

$$\dot{\rho}(z) = \hat{L}\rho(z) \equiv \{\mathcal{H}, \rho\} , \tag{4.1.5}$$

which follows from Hamilton's equations. Here

$$\hat{L} = \sum_{k=1}^{n} \left(\frac{\partial \mathcal{H}}{\partial q_k} \frac{\partial}{\partial p_k} - \frac{\partial \mathcal{H}}{\partial p_k} \frac{\partial}{\partial q_k} \right) \tag{4.1.6}$$

is the Liouville operator.

Since ρ varies with time, so does (4.1.2). Suppose now that $w(A)$ is specified in some independent way, and we want to find its time variation, i.e. to find its future values. This can be done as follows: restore $\rho(z)$ from $w(A)$, use the Liouville equation and then return again to w. But how can we find $\rho(z)$ obeying (4.1.2) or (4.1.3) for a given $w(A)$? The operator $\hat{\Pi}$ is highly degenerate, and thus there exists no inverse operator $\hat{\Pi}^{-1}$. However, we can introduce the generalized inverse operator $\hat{\Pi}^-$, such that $\hat{\Pi}\hat{\Pi}^-$ be the unit operator in the space of distributions $w(z)$. Then letting

$$\rho(z) = \hat{\Pi}^- w(A) \tag{4.1.7}$$

and returning from (4.1.7), according to (4.1.3), to the distribution in A, we will arrive at the earlier distribution $w(A)$. The requirement that $\hat{\Pi}\hat{\Pi}^-$ be a unit operator in the space of w is equivalent to the equation

$$\hat{\Pi}\hat{\Pi}^-\hat{\Pi} = \hat{\Pi}, \tag{4.1.8}$$

which, as is known from matrix theory, is the definition of the generalized inverse matrix $\hat{\Pi}^-$.

It is worth noting that the operator $\hat{\Pi}^-$ is not uniquely defined by (4.1.8). Its exact form is selected from additional considerations. We will now look at one of the forms possible here. Let $\rho_{eq}(z)$ be an equilibrium probability density obeying

$$\hat{L}\rho_{eq}(z) = 0. \tag{4.1.9}$$

It may be either the canonical [(2.2.5) v.1] or the microcanonical [(2.2.6) v.1] probability density. By (4.1.3), the corresponding probability density of A is

$$w_{eq}(A) = \hat{\Pi}\rho_{eq}(z). \tag{4.1.10}$$

Let the operator $\hat{\Pi}^-$ be given by

$$\hat{\Pi}^-w(A) = \rho_{eq}(z)w(B(z))/w_{eq}(B(z)). \tag{4.1.11}$$

It can be easily verified that operating on (4.1.11) with $\hat{\Pi}$ and using (4.1.2), we again will arrive at the initial probability density $w(A)$. Consequently, the necessary requirement is met.

It follows from (4.1.11) that $\hat{\Pi}^-$ has the matrix elements

$$(\hat{\Pi}^-)_{zA} = \rho_{eq}(z)\delta(B(z) - A)w_{eq}^{-1}(A). \tag{4.1.12}$$

Using (4.1.4), we can write this as

$$\hat{\Pi}^- = \hat{\rho}_{eq}\hat{\Pi}^T\hat{w}_{eq}^{-1}, \tag{4.1.13}$$

where $\hat{\rho}_{eq}$ and $\hat{w}_{eq}^{-1}$ are the operators of multiplication by $\rho_{eq}(z)$ and by $1/w_{eq}(A)$, respectively, and $\hat{\Pi}^T$ is the transposed operator.

Let us introduce the operator

$$\hat{P} = \hat{\Pi}^-\hat{\Pi}. \tag{4.1.14}$$

Then, it follows from (4.1.8) that

$$\hat{P}^2 = \hat{P}. \tag{4.1.15}$$

Hence, $\hat{P}$ is the operator of projection in the space of phase space distributions.

4.1.3 The Markov Operator $\hat{M}$

Using $\hat{\Pi}^-$ and $\hat{\Pi}$, we can rather simply write the variation of $w(A)$ in time. Let $w_{t_0}(A)$ be the probability density at the initial time t_0. The respective phase space probability density is $\hat{\Pi}^- w_{t_0}$. Solving (4.1.5) with the initial probability density $\hat{\Pi}^- w_{t_0}$, we find the time dependence of the probability density. The solution of the Liouville equation can be written as

$$\rho(z,t) = \exp[\hat{L}(t-t_0)]\hat{\Pi}^- w_{t_0}. \tag{4.1.16}$$

Using this, we can easily find, by (4.1.3), the time dependences of the probability densities of thermodynamic parameters

$$w_t(A) = \int (\hat{\Pi} \exp[\hat{L}(t-t_0)]\hat{\Pi}^-)_{AA'} w_{t_0}(A')dA'. \tag{4.1.17}$$

If we do not carry out the integration with respect to A' here, we will obtain the two-time probability density

$$w_{t t_0}(A, A') = (\hat{\Pi} \exp[\hat{L}(t-t_0)]\hat{\Pi}^-)_{AA'} w_{t_0}(A'). \tag{4.1.18}$$

On the right-hand sides of (4.1.17,18) we see the matrix elements of the operator $\hat{\Pi} \exp[\hat{L}(t-t_0)]\hat{\Pi}^-$. From (4.1.18), we find the transition probabilities, i.e. the conditional probabilities

$$w_{t t_0}(A \mid A') = (\hat{\Pi} \exp[\hat{L}(t-t_0)]\hat{\Pi}^-)_{AA'}. \tag{4.1.19}$$

If a dynamical system is Markov-like, i.e. the behavior of thermodynamic parameters is nearly Markovian, the transition probabilities (4.1.19) must approximately obey the Smolukhowski-Chapman equation [(2.3.9) v.1], i.e. we have

$$\hat{\Pi} \exp(\hat{L}t_2)\hat{P} \exp(\hat{L}t_1)\hat{\Pi}^- = \hat{\Pi} \exp[\hat{L}(t_1+t_2)]\hat{\Pi}^- + O(\mu), \tag{4.1.20}$$

where μ is a small parameter. Since the process $B(t)$ is not strictly Markovian, (4.1.20) can only hold approximately at $t_1 \gg \tau_0, t_2 \gg \tau_0$. We will assume that $t_1, t_2 \gtrsim \tau_i$, where τ_i is a quantity obeying (4.1.1).

For a real dynamical process, we should not pass to the limit $\tau \longrightarrow 0$, which is prescribed by equation [(2.3.20) v.1] defining the operator of the master equation. Instead, the operator of the master equation

$$\dot{w}(A) = \hat{M}w(A) \tag{4.1.21}$$

will be defined by

$$\hat{M} = \tau_i^{-1}\hat{\Pi}(\exp(\hat{L}\tau_i) - \hat{1})\hat{\Pi}^-, \tag{4.1.22}$$

where $\hat{1}$ is the identity operator in the space of phase space distributions. Proceeding from (4.1.20), we can prove in the normal way that equation (4.1.21) with the operator (4.1.22) is approximately valid.

The term $O(\mu)$, which describes the inaccuracy of (4.1.20), is responsible for the fact that (4.1.22) is specified to within $\tau_i^{-1}O(\mu)$. It follows that the solution

$$w(A,t) = \exp[\hat{M}(t - t_0)]w_0(A) \tag{4.1.23}$$

of equation (4.1.21) at $t - t_0 \sim \tau_r$ has a relative error of the order of $\tau_r \mu / \tau_i$. Markov-like systems must obey

$$(\tau_r/\tau_i)\mu \ll 1, \quad \text{i.e.} \quad (\tau_r/\tau_0)^{1/2}\mu \ll 1. \tag{4.1.24}$$

In (4.1.22) we can vary the time τ_i, with the result that $\hat{M}$ will vary but slightly, provided τ_i remains within the range (4.1.1). Let us take a closer look at this approximate invariance. We will equate the derivative $d\hat{M}/d\tau_i$ to zero:

$$
\begin{aligned}
0 &\approx \hat{\Pi}\frac{d}{d\tau_i}[\tau_i^{-1}(\exp(\hat{L}\tau_i) - \hat{1})]\hat{\Pi}^- \\
&= \hat{\Pi}[\tau_i^{-1}\hat{L}\exp(\hat{L}\tau_i) - \tau_i^{-2}(\exp(\hat{L}\tau_i) - \hat{1})]\hat{\Pi}^-.
\end{aligned}
\tag{4.1.25}
$$

We then have

$$\tau_i^{-1}\hat{\Pi}(\exp(\hat{L}\tau_i) - \hat{1})\hat{\Pi}^- \approx \hat{\Pi}\hat{L}\exp(\hat{L}\tau_i)\hat{\Pi}^-. \tag{4.1.26}$$

Hence, instead of (4.1.22) we can also write

$$\hat{M} = \hat{\Pi}\hat{L}\exp(\hat{L}\tau_i)\hat{\Pi}^-. \tag{4.1.27}$$

For Markov-like processes the operator $\hat{N}(\tau) = \hat{\Pi}\hat{L}\exp(\hat{L}\tau)\hat{\Pi}^-$ varies quickly at $\tau \sim \tau_0$, and then at $\tau_0 \ll \tau \ll \tau_r$ it varies slowly and is equal to (4.1.27). The right-hand side of (4.1.22) is the mean $\tau_i^{-1}\int_0^{\tau_i}\hat{N}(\tau)d\tau$ and, of course, it approximately coincides with (4.1.27).

4.1.4 Properties of the Master Equation Operator

(a) If we act with (4.1.27) on the equilibrium probability density (4.1.10), we will obtain

$$\hat{M}w_{\text{eq}}(A) = \hat{\Pi}\hat{L}\exp(\hat{L}\tau_i)\hat{\Pi}^-w_{\text{eq}}(A). \tag{4.1.28}$$

From (4.1.11) we get

$$\hat{\Pi}^-w_{\text{eq}}(A) = \rho_{\text{eq}}(z). \tag{4.1.29}$$

Therefore, (4.1.28) becomes

$$\hat{M}w_{\text{eq}}(A) = \hat{\Pi}\hat{L}\exp(\hat{L}\tau_i)\rho_{\text{eq}}(z). \tag{4.1.30}$$

But, by virtue of (4.1.9), the expression $\hat{L}\exp(\hat{L}\tau_i)\rho_{\text{eq}}$ is zero. Accordingly,

$$\hat{M}w_{\text{eq}}(A) = 0. \tag{4.1.31}$$

This equation is absolutely natural: an equilibrium distribution must be stable in time. Thus this equation is evidence of the consistency of the theory.

(b) Let us now turn to time reversibility, which, as noted in Sect. 3.2 v.1, requires

$$\mathcal{H}(q, -p) = \mathcal{H}(q, p) \quad \text{or} \quad \mathcal{H}(\varepsilon z) = \mathcal{H}(z) \,. \tag{4.1.32}$$

Next we introduce the ε-conjugation operation to be denoted as $(\ldots)_\varepsilon$. It transforms the operator $\hat{R}$ with matrix elements $R_{zz'}$ or R_{zB}, etc., into the operator $(\hat{R})_\varepsilon$, which has the matrix elements $R_{\varepsilon z, \varepsilon z'}$ or $R_{\varepsilon z, \varepsilon B}$, and so on. Here, as before, $\varepsilon z = \varepsilon(q, p) = (q, -p)$, $\varepsilon B = \{\varepsilon_\alpha B_\alpha\}$.

Let us now apply this operation to the Liouville operator (4.1.6), which has the matrix elements

$$L_{zz'} = \sum_{k=1}^{n} \left(\frac{\partial \mathcal{H}(q,p)}{\partial q_k} \frac{\partial}{\partial p_k} - \frac{\partial \mathcal{H}(q,p)}{\partial p_k} \frac{\partial}{\partial q_k} \right) \delta(q - q')\delta(p - p') \,. \tag{4.1.33}$$

With this definition we obtain

$$(\hat{L})_\varepsilon = \left\| \sum_{k=1}^{n} \left(-\frac{\partial \mathcal{H}(q,-p)}{\partial q_k} \frac{\partial}{\partial p_k} + \frac{\partial \mathcal{H}(q,-p)}{\partial p_k} \frac{\partial}{\partial q_k} \right) \delta(z - z') \right\| \,. \tag{4.1.34}$$

We thus see that, subject to (4.1.32), the right-hand side of (4.1.34) only differs in sign from (4.1.33). Consequently, if the time reversal condition (4.1.32) is met, then

$$(\hat{L})_\varepsilon = -\hat{L} \,. \tag{4.1.35}$$

Owing to this, the Liouville equation $\dot{\rho} = \hat{L}\rho$ remains invariant under time reversal. Yet (4.1.21) is not invariant under the interchange $t \longrightarrow -t$, since the equation $(\hat{M})_\varepsilon = -\hat{M}$ of the type (4.1.35) does not hold for the operator (4.1.22) or (4.1.27). This is because, unlike (4.1.5), equation (4.1.21) can describe irreversible relaxation processes. It is important, however, that the operator $\hat{M}$, which is responsible for the fact that the processes $A(t)$ are irreversible due to dissipation, satisfies a time reversibility condition of another type. It implies the invariance of many-time stationary probability densities, but not the processes themselves, under time reversal. Let us prove this invariance.

Substituting (4.1.13) into (4.1.27) yields

$$\hat{M} = \hat{\Pi}\hat{L} \exp(\hat{L}\tau_m)\hat{\rho}_{eq}\hat{\Pi}^T \hat{w}_{eq}^{-1} \,. \tag{4.1.36}$$

Next we will operate on each side of this equation with the operation of ε-conjugation. Since this operation possesses the property of distributivity, we obtain

$$(\hat{M})_\varepsilon = (\hat{\Pi})_\varepsilon (\hat{L})_\varepsilon (\exp(\hat{L}\tau_m))_\varepsilon (\hat{\rho}_{eq})_\varepsilon (\hat{\Pi}^T)_\varepsilon (\hat{w}^{-1})_\varepsilon \,. \tag{4.1.37}$$

It can be easily verified, for example by expanding the exponential function in a series, that

$$(\exp(\hat{L}\tau_m))_\varepsilon = \exp[(\hat{L})_\varepsilon \tau_m] \,. \tag{4.1.38}$$

From (4.1.35) it follows that

$$(\hat{M})_\varepsilon = -(\hat{\Pi})_\varepsilon \hat{L} \exp(-\hat{L}\tau_m)(\hat{\rho}_{eq})_\varepsilon (\hat{\Pi}^T)_\varepsilon (\hat{w}^{-1})_\varepsilon \,. \tag{4.1.39}$$

Further, by (4.1.4),

$$(\hat{\Pi})_{\varepsilon A, \varepsilon z} = \delta(\varepsilon A - B(\varepsilon z)) \,. \tag{4.1.40}$$

The right-hand side is equal to $\delta(\varepsilon A - \varepsilon B(z))$, since $B(\varepsilon z) = \varepsilon B(z)$. Hence

$$(\hat{\Pi})_\varepsilon = \hat{\Pi} \,. \tag{4.1.41}$$

The equilibrium probability density $\rho_{\mathrm{eq}}(z)$ is expressed in terms of the Hamiltonian $\mathcal{H}(z)$. Therefore, it follows from (4.1.32) that $\rho_{\mathrm{eq}}(\varepsilon z) = \rho_{\mathrm{eq}}(z)$. This means that $(\hat{\rho}_{\mathrm{eq}}(z))_\varepsilon = \hat{\rho}_{\mathrm{eq}}(z)$. Lastly, from (4.1.10) we have

$$
\begin{aligned}
w_{\mathrm{eq}}(\varepsilon A) &= (\hat{\Pi})_\varepsilon \rho_{\mathrm{eq}}(\varepsilon z) \\
&= \hat{\Pi} \rho_{\mathrm{eq}}(z) \,,
\end{aligned} \tag{4.1.42}
$$

so that $w_{\mathrm{eq}}(\varepsilon B) = w_{\mathrm{eq}}(B)$ and $(\hat{w}_{\mathrm{eq}}^{-1})_\varepsilon = \hat{w}_{\mathrm{eq}}^{-1}$. Using the above relationships equation (4.1.39) becomes

$$(\hat{M})_\varepsilon = -\hat{\Pi}\hat{L}\exp(-\hat{L}\tau_i)\hat{\rho}_{\mathrm{eq}}\hat{\Pi}^T \hat{w}_{\mathrm{eq}}^{-1} \,. \tag{4.1.43}$$

Transposing operators on either side of the equation, we get

$$(\hat{M})_\varepsilon^T = \hat{w}_{\mathrm{eq}}^{-1}\hat{\Pi}\hat{\rho}_{\mathrm{eq}}\exp(\hat{L}\tau_i)\hat{L}\hat{\Pi}^T \,. \tag{4.1.44}$$

We have used here the easily verifiable property $\hat{L}^T = -\hat{L}$ of the operator (4.1.6).

The operator $\hat{L}$ commutes with $\hat{\rho}_{\mathrm{eq}}$. As a matter of fact, the equation $\hat{L}\hat{\rho}_{\mathrm{eq}} - \hat{\rho}_{\mathrm{eq}}\hat{L} = 0$ is simply another way of representing (4.1.9). Using this commutativity, (4.1.44) can be written as $(\hat{M})_\varepsilon^T = \hat{w}_{\mathrm{eq}}^{-1}\hat{\Pi}\hat{L}\exp(\hat{L}\tau_i)\hat{\rho}_{\mathrm{eq}}\hat{\Pi}^T$. Comparing this with (4.1.36) we have

$$(\hat{M})_\varepsilon^T = \hat{w}_{\mathrm{eq}}^{-1}\hat{M}\hat{w}_{\mathrm{eq}} \,. \tag{4.1.45}$$

This is the desired condition related to the invariance of the probability densities. It coincides with the equation [(3.2.30) v.1] derived in Sect. 3.2. v.1 by purely Markov methods. It is obtained here by another method for Markov-like dynamical processes.

4.1.5 Transformation of the Markov Operator to Generalized Fokker-Planck Operator Form

The operator (4.1.22) [or (4.1.27)] can be reduced to various special forms. We now want to derive one such form. Using (4.1.4,12), we will represent (4.1.27) in more detail

$$(\hat{M})_{AA'} = \int \delta(A - B(z))[\hat{L}\exp(\hat{L}\tau_i)\rho_{\mathrm{eq}}(z)\delta(A' - B(z))]dz w_{\mathrm{eq}}^{-1}(A') \,. \tag{4.1.46}$$

We will now change the order of the factors in the integrand and refer $\hat{L}$ to the other factor, this operation being associated with transposition of $\hat{L}$. Then, from (4.1.46), we obtain

$$
\begin{aligned}
(\hat{M})_{AA'} &= \int \delta(A' - B(z))\rho_{\mathrm{eq}}(z) \\
&\quad \times [(-\hat{L})\exp(-\hat{L}\tau_i)\delta(A - B(z))]dz w_{\mathrm{eq}}^{-1}(A') \,.
\end{aligned} \tag{4.1.47}
$$

We have utilized here the property $\hat{L}^T = -\hat{L}$ mentioned above. Consider the expression

$$-\hat{L}\exp(-\hat{L}\tau)\delta(A-B(z)) = \frac{\partial}{\partial\tau}[\exp(-\hat{L}\tau)\delta(A-B(z))]\,. \qquad (4.1.48)$$

For any $f, F_1, \ldots, F_n$, we have the following relation:

$$\exp(-\hat{L}\tau)f(F_1(z),\ldots,F_n(z))$$
$$= f(\exp(-\hat{L}\tau)F_1(z),\ldots,\exp(-\hat{L}\tau)F_n(z))\,. \qquad (4.1.49)$$

Using this, we reduce (4.1.48) to the form

$$-\hat{L}\exp(-\hat{L}\tau)\delta(A-B(z)) = \frac{\partial}{\partial\tau}\delta(A-\exp(-\hat{L}\tau)B(z))\,. \qquad (4.1.50)$$

The right-hand side is differentiated as a composite function: first with respect to the argument of the function δ, or A, and then the argument is differentiated with respect to τ. We arrive at

$$-\hat{L}\exp(-\hat{L}\tau)\delta(A-B(z))$$
$$= \frac{\partial}{\partial A}\{\delta(A-\exp(-\hat{L}\tau)B(z))[\exp(-\hat{L}\tau)\hat{L}B(z)]\}$$
$$= \frac{\partial}{\partial A}\{[\exp(-\hat{L}\tau)\delta(A-B(z))][\exp(-\hat{L}\tau)\hat{L}B(z)]\}\,. \qquad (4.1.51)$$

We can rewrite this as

$$-\hat{L}\exp(-\hat{L}\tau)\delta(A-B(z)) = \frac{\partial}{\partial A}\{\delta(A-B(z))[\exp(-\hat{L}\tau)\hat{L}B(z)]\}$$
$$-\frac{\partial}{\partial A}\left\{\left[\frac{\exp(\hat{L}\tau)-\hat{1}}{\hat{L}}\hat{L}\exp(-\hat{L}\tau)\delta(A-B(z))\right]\right.$$
$$\times[\exp(-\hat{L}\tau)\hat{L}B(z)]\Big\}\,. \qquad (4.1.52)$$

Using (4.1.52), we reduce (4.1.47) to

$$(\hat{M})_{AA'} = -\frac{\partial}{\partial A}\int\delta(A'-B(z))\rho_{\mathrm{eq}}(z)\delta(A-B(z))v(z,\tau_i)dzw_{\mathrm{eq}}^{-1}(A')$$
$$-\frac{\partial}{\partial A}\int\delta(A'-B(z))\rho_{\mathrm{eq}}(z)v(z,\tau_i)$$
$$\times\left[\frac{\exp(\hat{L}\tau_i)-\hat{1}}{\hat{L}}(-\hat{L})\exp(-\hat{L}\tau_i)\delta(A-B(z))\right]dzw_{\mathrm{eq}}^{-1}(A')\,,$$
$$\qquad (4.1.53)$$

where $v(z,\tau) = -\hat{L}\exp(-\hat{L}\tau)B(z) = d[\exp(-\hat{L}\tau)B(z)]/d\tau$. Since the second term includes the expression $-\hat{L}\exp(-\hat{L}\tau)\delta(A-B(z))$, we can apply the transformation (4.1.51) or (4.1.52) to it.

Using (4.1.51), we reduce the expression in brackets in (4.1.53) to

$$\hat{L}^{-1}[\exp(\hat{L}\tau_i)-\hat{1}]\frac{\partial}{\partial A}\{[\exp(-\hat{L}\tau_i)\delta(A-B(z))](\exp(-\hat{L}\tau_i)\hat{L}B(z))\} \qquad (4.1.54)$$

or, by (4.1.49), to

$$\frac{\partial}{\partial A}\{\hat{L}^{-1}[\exp(\hat{L}\tau_i) - \hat{1}]\exp(-\hat{L}\tau_i)[\delta(A - B(z))(\hat{L}B(z))]\}$$
$$= -\frac{\partial}{\partial A}\{\hat{L}^{-1}(\hat{1} - \exp(-\hat{L}\tau_i))[\delta(A - B(z))v(z,0)]\}. \qquad (4.1.55)$$

Therefore, (4.1.53) becomes

$$(\hat{M})_{AA'} = -\frac{\partial}{\partial A_\alpha}\int \delta(A - B(z))v_\alpha(z,\tau_i)\rho_{eq}(z)\delta(A' - B(z))dz$$
$$\times w_{eq}^{-1}(A') + \frac{\partial^2}{\partial A_\alpha \partial A_\beta}\int \delta(A' - B(z))\rho_{eq}(z)v_\alpha(z,\tau_i)$$
$$\times \{L^{-1}(1 - \exp(-\hat{L}\tau_i))[\delta(A - B(z))v_\beta(z,0)]\}dz w_{eq}^{-1}(A'). \qquad (4.1.56)$$

The form of the master equation (4.1.21) corresponding to the matrix elements of operator $\hat{M}$ is given by

$$\dot{w}(A) = -\frac{\partial}{\partial A_\alpha}[K_\alpha(A)w(A)] + \frac{\partial^2}{\partial A_\alpha \partial A_\beta}\left[\int D_{\alpha\beta}(A, A')w(A')dA'\right], \quad (4.1.57)$$

where

$$K_\alpha(A) = \int \delta(A - B(z))v_\alpha(z,\tau_i)\rho_{eq}(z)dz w_{eq}^{-1}(A), \qquad (4.1.58a)$$

$$D_{\alpha\beta}(A, A') = \int \delta(A - B(z))v_\beta(z,0)\{\hat{L}^{-1}[\exp(\hat{L}\tau_i) - \hat{1}]$$
$$\times [v_\alpha(z,\tau_i)\rho_{eq}(z)\delta(A' - B(z))]\}dz w_{eq}^{-1}(A'). \qquad (4.1.58b)$$

In the second integral here we have changed the order of the operators. This integral can be written in the shorter form

$$D_{\alpha\beta}(A, A') = (\hat{\Pi}\hat{v}_\beta(0)\hat{L}^{-1}[\exp(\hat{L}\tau_i) - 1]\hat{v}_\alpha(\tau_i)\hat{\Pi}^-)_{AA'}. \qquad (4.1.59)$$

Equation (4.1.57) has the form of the generalized Fokker-Planck equation. It differs from the latter in that it has an integral with respect to A' in the last term. If $D_{\alpha\beta}(A, A')$ had the form $D'_{\alpha\beta}(A)\delta(A - A')$, then (4.1.57) would turn into a conventional Fokker-Planck equation. Significantly, the formula (4.1.52) can be applied any number of times to obtain an equation with any number of derivatives with respect to A.

4.1.6 Coefficients of the Master Equation $K_{\alpha_1\ldots\alpha_j}(A)$

To arrive at a master equation of the form [(2.3.24) v.1] we should instead of (4.1.52) use somewhat different formulas. Let us denote

$$\varphi_\tau(z) = \exp(-\hat{L}\tau)\delta(A - B(z)). \qquad (4.1.60)$$

This can be rewritten as

$$
\begin{aligned}
\varphi_\tau(z) &= \delta(A - B(z)) - [1 - \exp(-\hat{L}\tau)]\delta(A - B(z)) \\
&= \delta(A - B(z)) - \int_0^\tau \hat{L}\exp(-\hat{L}\sigma)\delta(A - B(z))d\sigma\,.
\end{aligned}
\tag{4.1.61}
$$

We can substitute (4.1.51) with τ replaced by σ into the integrand of the right-hand side of (4.1.61) and obtain

$$
\varphi_\tau(z) = \delta(A - B(z)) - \frac{\partial}{\partial A_\alpha}\int_0^\tau v_\alpha(z,\sigma)\varphi_\sigma(z)d\sigma\,.
\tag{4.1.62}
$$

Substituting this equation with σ taken instead of τ into its integrand yields

$$
\begin{aligned}
\varphi_\tau(z) &= \delta(A - B(z)) - \frac{\partial}{\partial A_\alpha}\int_0^\tau d\sigma v_\alpha(z,\sigma)\delta(A - B(z)) \\
&\quad + \frac{\partial^2}{\partial A_\alpha \partial A_\beta}\int_0^\tau d\sigma \int_0^\sigma d\pi v_\alpha(z,\sigma)v_\beta(z,\pi)\varphi_\pi(z)\,.
\end{aligned}
\tag{4.1.63}
$$

Multiple application of (4.1.62) gives

$$
\begin{aligned}
\varphi_\tau(z) &= \sum_{j=0}^{n-1}(-1)^j \frac{\partial^j}{\partial A_{\alpha_1}\dots\partial A_{\alpha_j}}\int_0^\tau d\sigma_1 v_{\alpha_1}(z,\sigma_1)\int_0^{\sigma_1} d\sigma_2 v_{\alpha_2}(z,\sigma_2) \\
&\quad \times \dots \times \int_0^{\sigma_{j-1}} d\sigma_j v_{\alpha_j}(z,\sigma_j)\delta(A - B(z)) \\
&\quad + (-1)^n \frac{\partial^n}{\partial A_{\alpha_1}\dots\partial A_{\alpha_n}}\int_0^\tau d\sigma_1 v_{\alpha_1}(z,\sigma_1)\int_0^{\sigma_1} d\sigma_2 v_{\alpha_2}(z,\sigma_2) \\
&\quad \times \dots \times \int_0^{\sigma_{n-1}} d\sigma_n v_{\alpha_n}(z,\sigma_n)\varphi_{\sigma_n}(z)\,.
\end{aligned}
\tag{4.1.64}
$$

Now, using (4.1.51,60), we transform (4.1.47) into

$$
(\hat{M})_{AA'} = -\frac{\partial}{\partial A_\alpha}\int \delta(A' - B(z))\rho_{\mathrm{eq}}(z)v_\alpha(z,\tau_i)\varphi_{\tau_i}(z)dz w_{\mathrm{eq}}^{-1}(A')
\tag{4.1.65}
$$

and apply (4.1.64) to arrive at

$$
\begin{aligned}
(\hat{M})_{AA'} &= \sum_{m=1}^n \frac{(-1)^m}{m!}\frac{\partial^m}{\partial A_{\alpha_1}\dots\partial A_{\alpha_m}}[K_{\alpha_1\dots\alpha_m}(A)\delta(A - A')] \\
&\quad + (-1)^{n+1}\frac{\partial^{n+1}}{\partial A_{\alpha_1}\dots\partial A_{\alpha_{n+1}}}D_{\alpha_1\dots\alpha_{n+1}}(A, A')\,,
\end{aligned}
\tag{4.1.66}
$$

where

$$
\begin{aligned}
K_{\alpha_1\dots\alpha_m}(A) &= P_{\alpha_1\dots\alpha_m}\int dz v_{\alpha_1}(z,\tau_i)\int_0^{\tau_i} d\sigma_1 v_{\alpha_2}(z,\sigma_1)\int_0^{\sigma_1} d\sigma_2 v_{\alpha_3}(z,\sigma_2) \\
&\quad \times \dots \times \int_0^{\sigma_{m-2}} d\sigma_{m-1} v_{\alpha_m}(z,\sigma_{m-1})\rho_{\mathrm{eq}}(z)\delta(A - B(z))w_{\mathrm{eq}}^{-1}(A)\,, \quad (4.1.67a)
\end{aligned}
$$

$$
\begin{aligned}
D_{\alpha_1\dots\alpha_{n+1}} &= \int dz \delta(A' - B(z))v_{\alpha_1}(z,\tau_i)\int_0^{\tau_i} d\sigma_1 v_{\alpha_2}(z,\sigma_1)\int_0^{\sigma_1} d\sigma_2 v_{\alpha_3}(z,\sigma_2) \\
&\quad \times \dots \times \int_0^{\sigma_{n-1}} d\sigma_n v_{\alpha_{n+1}}(z,\sigma_n)[\exp(-\hat{L}\sigma_n)\delta(A - B(z))]\rho_{\mathrm{eq}}(z)w_{\mathrm{eq}}^{-1}(A')\,.
\end{aligned}
$$
$$
\tag{4.1.67b}
$$

In (4.1.67a) $P_{\alpha_1 \ldots \alpha_m}$ is the $m!$-term symmetrizing sum over the subscripts $\alpha_1, \ldots, \alpha_m$. In (4.1.67a), the integral of the type

$$\int \ldots \rho_{eq}(z)\delta(A - B(z))dz / w_{eq}(A)$$

$$= \int \ldots \rho_{eq}(z)\delta(A - B(z))dz \bigg/ \int \int \rho_{eq}(z)\delta(A - B(z))dz \qquad (4.1.68)$$

is nothing but the conditional equilibrium mean $\langle \ldots \mid A \rangle_{eq} = \langle \ldots \rangle_A$. Therefore, we can write (4.1.67a) as

$$K_{\alpha_1 \ldots \alpha_m}(A) = P_{\alpha_1 \ldots \alpha_m} \int \ldots \int_{0 < \sigma_{m-1} < \ldots < \sigma_1 < \tau_i} \langle v_{\alpha_1}(z, \tau_i) v_{\alpha_2}(z, \sigma_1) $$
$$\times \ldots \times v_{\alpha_m}(z, \sigma_{m-1}) \rangle_A d\sigma_1 \ldots d\sigma_{m-1} . \qquad (4.1.69)$$

Since (4.1.69) contains a symmetrizing sum, the integration region can be simplified:

$$K_{\alpha_1 \ldots \alpha_m}(A)$$
$$= P_{(\alpha_1 \ldots \alpha_m)} \int_0^{\tau_i} \ldots \int_0^{\tau_i} \langle v_{\alpha_1}(z, \tau_i) v_{\alpha_2}(z, \sigma_1) $$
$$\times \ldots \times v_{\alpha_m}(z, \sigma_{m-1}) \rangle_A d\sigma_1 \ldots d\sigma_{m-1}$$
$$= P_{(\alpha_1 \ldots \alpha_m)} \langle v_{\alpha_1}(z, \tau_i) \Delta B_{\alpha_2}(z) \ldots \Delta B_{\alpha_m}(z) \rangle_A , \qquad (4.1.70)$$

where

$$\Delta B_\alpha(z) = \int_0^{\tau_i} v_\alpha(z, \sigma)d\sigma$$
$$= \exp(-\hat{L}\tau_i)B_\alpha(z) - B_\alpha(z)$$
$$= B_\alpha(\exp(-\hat{L}\tau_i)z) - B_\alpha(z) , \qquad (4.1.71)$$

and $P_{(\alpha_1 \ldots \alpha_m)}$ is the m-term sum over cyclic permutations of the subscripts. It is easily seen that (4.1.70) can be represented as a derivative

$$K_{\alpha_1 \ldots \alpha_m}(A) = \frac{d}{d\tau_i} \langle \Delta B_{\alpha_1}(z) \ldots \Delta B_{\alpha_m}(z) \rangle_A , \qquad (4.1.72)$$

As n in (4.1.66) tends to infinity, we will obtain an operator whose corresponding master equation is of the form

$$\dot{w} = \sum_{m=1}^{\infty} \frac{1}{m!}(-1)^m \frac{\partial^m}{\partial A_{\alpha_1} \ldots \partial A_{\alpha_m}}[K_{\alpha_1 \ldots \alpha_m}(A)w] \qquad (4.1.73)$$

that is common for the Markov theory.

If we take (4.1.22) as $\hat{M}$, we can go over the same procedure. In so doing, instead of (4.1.70) we will have

$$K_{\alpha_1 \ldots \alpha_m}(A) = \tau_i^{-1} P_{\alpha_1 \ldots \alpha_m} \int \ldots \int_{0 < \sigma_m < \ldots < \sigma_1 < \sigma_i} \langle v_{\alpha_1}(z, \sigma_1) v_{\alpha_2}(z, \sigma_2) $$
$$\times \ldots \times v_{\alpha_m}(z, \sigma_m) \rangle_A \times d\sigma_1 \ldots d\sigma_m$$
$$= \tau_i^{-1} \langle \Delta B_{\alpha_1} \ldots \Delta B_{\alpha_m} \rangle_A . \qquad (4.1.74)$$

It is easily seen that (4.1.72) and (4.1.74) are approximately equal, provided that the conditional mean

$$\langle v_{\alpha_1}(z, \sigma_1) \ldots v_{\alpha_m}(z, \sigma_m) \rangle_A \qquad (4.1.75)$$

has a small correlation time $\tau_{\text{cor}} \ll \tau_i$ and is approximately stationary, i.e. approximately invariant under shifts $\sigma_1 \longrightarrow \sigma_1 + a, \ldots, \sigma_m \longrightarrow \sigma_m + a$ by $a \sim \tau_i$. These properties are accounted for by the fact that the correlation time τ_{cor} is about τ_0, and the nonstationarity due to the condition $B(z) = A$ has no time to manifest itself at time intervals of the order of τ_i. It takes a time τ_r for the nonstationarity to show up.

The above formula

$$K_{\alpha_1 \ldots \alpha_m}(A) = \tau_i^{-1} \langle \Delta B_{\alpha_1} \ldots \Delta B_{\alpha_m} \rangle_A \qquad (4.1.76)$$

is quite natural: for Markov-like processes it is a counterpart of [(2.3.25) v.1] in the theory of exact Markov processes.

4.1.7 The Zwanzig Equation

If a process $B(z(t))$ is not Markov-like, equation (4.1.21) does not hold for it. In that case, we have a more complicated equation, which we now proceed to derive.

Let $\rho(z, t_0)$ be some initial distribution in a phase space. We will replace it by $\rho_0(z) = \hat{P}\rho(z, t_0)$, a distribution obtained by using the projection operator $\hat{P} = \hat{\Pi}^- \hat{\Pi}$. Besides the replacement $\rho(z, t_0) \longrightarrow \rho_0(z)$, no other restrictions of generality will be introduced. If we take $\rho_0(z)$ as an initial distribution, then using the Liouville equation we can find a distribution for future times

$$\rho(z, t) = \exp[\hat{L}(t - t_0)]\rho_0(z) . \qquad (4.1.77)$$

To these distributions in phase space there correspond the following probability densities for internal thermodynamic parameters

$$\begin{aligned} w_t &= \hat{\Pi}\rho(z, t) \\ &= \hat{\Pi}\exp[\hat{L}(t - t_0)]\rho_0 . \end{aligned} \qquad (4.1.78)$$

We will now introduce the notation $\rho_1 = \hat{\Pi}^- w_t$. Substituting this into (4.1.78) gives

$$\rho_1 = \hat{P}\exp[\hat{L}(t - t_0)]\rho_0 . \qquad (4.1.79)$$

Differentiating (4.1.78) with respect to time yields

$$\begin{aligned} \dot{w}_t &= \hat{\Pi}\hat{L}\exp[\hat{L}(t - t_0)]\rho_0 \\ &= \hat{\Pi}\hat{L}(\rho_1 + \rho_2) , \end{aligned} \qquad (4.1.80)$$

where

$$\begin{aligned} \rho_2 &= \rho - \rho_1 \\ &= (\hat{1} - \hat{P})\exp[\hat{L}(t - t_0)]\rho_0 . \end{aligned} \qquad (4.1.81)$$

Here we have used (4.1.77,79).

If we now differentiate (4.1.81) with respect to time, we find

$$
\begin{aligned}
\dot{\rho}_2 &= (\hat{1} - \hat{P})\hat{L}\exp[\hat{L}(t - t_0)]\rho_0 \\
&= (\hat{1} - \hat{P})\hat{L}(\rho_1 + \rho_2)\,.
\end{aligned}
\tag{4.1.82}
$$

We will write this as

$$
\dot{\rho}_2 - (\hat{1} - \hat{P})L\rho_2 = (\hat{1} - \hat{P})L\rho_1
\tag{4.1.83}
$$

and will treat it as an equation with respect to an unknown function $\rho_2(t)$ assuming that $\rho_1(t)$ is known. Integrating (4.1.83) with the initial condition

$$
[\rho_2]_{t=t_0} = (\hat{1} - \hat{P})\rho_0 = 0\,,
\tag{4.1.84}
$$

we will have

$$
\rho_2(z, t) = \int_{t_0}^{t} \exp[(\hat{1} - \hat{P})\hat{L}(t - t')](\hat{1} - \hat{P})\hat{L}\rho_1(z, t')dt'\,.
\tag{4.1.85}
$$

Substituting (4.1.85) into (4.1.80) gives

$$
\dot{w} = \hat{\Pi}\hat{L}\left[\rho_1 + \int_{t_0}^{t} \exp[(\hat{1} - \hat{P})\hat{L}(t - t')](\hat{1} - \hat{P})\hat{L}\rho_1(z, t')dt'\right]\,.
\tag{4.1.86}
$$

If we insert the expression $\rho_1 = \hat{\Pi}^- w_t$, which was a definition of ρ_1, we will arrive at

$$
\dot{w}_t = \hat{\Pi}\hat{L}\hat{\Pi}^- w_t + \hat{\Pi}\hat{L}\int_{t_0}^{t} dt' \exp[(\hat{1} - \hat{P})\hat{L}(t - t')](\hat{1} - \hat{P})\hat{L}\hat{\Pi}^- w_{t'}\,.
\tag{4.1.87}
$$

In the formula (4.1.87) we can write $\hat{Q}\exp(\hat{Q}\hat{L}s)\hat{Q}$ instead of $\exp(\hat{Q}\hat{L}s)\hat{Q}$ ($\hat{Q} = \hat{1} - \hat{P}$) as is easy to check by expanding the exponent function into a series. The formula (4.1.87) is the desired equation for the probability density of an arbitrary process. We thus see that, unlike (4.1.21), it additionally includes a time integral. This equation was first derived by *Zwanzig* [4.1].

If the operators $\hat{\Pi}$ and $\hat{\Pi}^-$ in (4.1.87) are written using (4.1.4) and (4.1.12), then we have

$$
\begin{aligned}
\dot{w}_t(A) &= \int\int \delta(A - B(z))\hat{L}\rho_{\mathrm{eq}}(z)\delta(A' - B(z))dz \cdot w_{\mathrm{eq}}^{-1}(A')w_{t'}(A')dA' \\
&+ \int_{t_0}^{t} dt' \int\int \delta(A - B(z))\hat{L}(\hat{1} - \hat{P})\exp[(\hat{1} - \hat{P})\hat{L}(t - t')] \\
&\times \rho_{\mathrm{eq}}(z)(\hat{1} - \hat{P})\hat{L}\delta(A' - B(z))dz w_{\mathrm{eq}}^{-1}(A')w_{t'}(A')dA'\,.
\end{aligned}
\tag{4.1.88}
$$

Here we have placed the additional operator $\hat{1} - \hat{P}$ before the exponential function. This is possible because, just like $\hat{P}$, it is idempotent: $(\hat{1} - \hat{P})^2 = \hat{1} - \hat{P}$. Considering the form of the Liouville operator (4.1.6), we can readily verify the validity of the relationship

$$
\hat{L}\delta(A - B(z)) = -\frac{\partial}{\partial A_\alpha}[\delta(A - B(z))(\hat{L}B_\alpha(z))]\,.
\tag{4.1.89}
$$

Owing to this relation we can write the combination in (4.1.88) as

$$(\hat{1} - \hat{P})\hat{L}\delta(A - B(z)) = -\frac{\partial}{\partial A_\alpha}X_\alpha(z, A)\,, \qquad (4.1.90)$$

where

$$X_\alpha(z, A) = (\hat{1} - \hat{P})[\delta(A - B(z))(\hat{L}B_\alpha(z))]\,. \qquad (4.1.91)$$

Using (4.1.89) and (4.1.90), we reduce (4.1.88) to

$$
\begin{aligned}
\dot{w}_t(A) &= -\frac{\partial}{\partial A_\alpha}[K_\alpha(A)w_t] \\
&+ \frac{\partial}{\partial A_\alpha}\int_{t_0}^t dt' \int D_{\alpha\beta}(A, A', t - t')\frac{\partial}{\partial A'}\frac{w_{t'}(A')}{w_{\mathrm{eq}}(A')}dA'\,,
\end{aligned}
\qquad (4.1.92)
$$

where

$$K_\alpha(A) = -\int(\hat{L}B(z))\rho_{\mathrm{eq}}(z)\delta(A - B(z))dz\,w_{\mathrm{eq}}^{-1}(A) \equiv \langle v(z, 0)\rangle_A\,,$$

$$D_{\alpha\beta}(A, A', t - t') = \int X_\alpha(z, A)\exp[(\hat{1} - \hat{P})\hat{L}(t - t')]\rho_{\mathrm{eq}}(z)X_\beta(z, A')dz\,.$$

$$(4.1.93)$$

It is seen from the derivation of the Zwanzig equation (4.1.92) that it is exact and valid for all systems. In real situations, however, the functions $D_{\alpha\beta}(A, A', t-t')$ differ markedly from zero only at $t - t'$ of the order of a characteristic time τ_{char} of a large system. At $t - t' \gg \tau_{\mathrm{char}}$ these functions vanish, and thus the exact value of t_0 in (4.1.87) and (4.1.92) becomes of no significance if $t - t_0 \gg \tau_{\mathrm{char}}$. Specifically, we may let $t_0 = -\infty$.

4.1.8 One Method of Using a Large Parameter to Derive the Markov Equation

When applying equations (4.1.87) and (4.1.92) to a Markov-like process, one should bear in mind that the functions $D_{\alpha\beta}(A, A', t - t')$ are substantially nonzero only at $t - t' \sim \tau_0 \ll \tau_i$, and that w_t within $t - t_0 \sim \tau_i$ has had no time yet to change. This enables the Markov equation to be derived from (4.1.87) and (4.1.92) at $t - t_0 \sim \tau_i$ by substituting w_t for $w_{t'}$. The Markov operator will now assume another form

$$\hat{M} = \hat{\Pi}\hat{L}\left[\hat{1} + \int_0^{\tau_i}ds\exp(\hat{Q}\hat{L}s)\hat{Q}\hat{L}\right]\hat{\Pi}^- = \hat{\Pi}\hat{L}\exp(\hat{Q}\hat{L}\tau_i)\hat{\Pi}^- \qquad (4.1.94)$$

in addition to (4.1.22) and (4.1.27) ($\hat{Q} = \hat{1} - \hat{P}$).

Let us now consider the transition to the Markov process in more detail. Suppose that the parameters $B_\alpha(z)$ are dynamic variables Q_j, P_j of the subsystem S that interacts with a thermostat having a temperature T. Let the complete Liouville operator consist of three parts

$$\hat{L} = \hat{L}_1 + \hat{L}_2 + \hat{L}_3\,, \qquad (4.1.95)$$

where $\hat{L}_3$ corresponds to the Hamilton function $\mathcal{H}_0(Q, P)$ for the subsystem S; $\hat{L}_1$ describes the time variation of the thermostat variables at fixed Q and P; and $\hat{L}_2$ describes the interaction of the subsystem with the thermostat. We also assume that Q and P are Markov-like variables, because the operators $\hat{L}_1$ and $\hat{L}_2$ depend in a suitable way on the large parameter γ. So, we assume

$$\hat{L}_1 = \gamma^2 \hat{L}_1', \tag{4.1.96a}$$
$$\hat{L}_2 = \gamma \hat{L}_2', \tag{4.1.96b}$$

where $\hat{L}_1', \hat{L}_2'$ are now independent of γ. According to (4.1.96a), the thermostat variables vary over a time-scale of about γ^{-2}, so that $\tau_0 = a/\gamma^2$, where a is independent of γ. It follows that the inequality $\tau_r \gg \tau_0$ results from the large value of γ and that $\tau_i = (\tau_r \tau_0)^{1/2} = b/\gamma$, where $b = (\tau_r a)^{1/2}$ is independent of γ.

In many cases we have

$$\hat{\Pi} \hat{L}_1' = 0,$$
$$\hat{L}_1' \hat{\Pi}^- = 0,$$
$$\hat{\Pi} \hat{L}_2' \hat{\Pi}^- = 0 \tag{4.1.97}$$

Hence, $\hat{P} \hat{L}_1' = 0$, $\hat{P} \hat{L}_2' \hat{\Pi}^- = 0$, etc. Substituting (4.1.95,96) into (4.1.87), where we put $t - t_0 = \tau_i$, and using (4.1.97), we have

$$\begin{aligned} \dot{w}_t &= \hat{\Pi} \hat{L}_3 \hat{\Pi}^- w_t + \hat{\Pi} (\gamma \hat{L}_2' + \hat{L}_3 \hat{Q}) \\ &\times \int_{t-b/\gamma}^{t} dt' \exp[(\gamma^2 \hat{L}_1' + \gamma \hat{Q} \hat{L}_2' + \hat{Q} \hat{L}_3)(t - t')](\gamma \hat{L}_2' + \hat{Q} \hat{L}_3) \hat{\Pi}^- w_t, \end{aligned} \tag{4.1.98}$$

$(\hat{Q} = \hat{1} - \hat{P})$. Introducing the new integration variables $s = \gamma^2(t - t')$, we find

$$\begin{aligned} \dot{w}_t &= \hat{\Pi} \hat{L}_3 \hat{\Pi}^- w_t + \hat{\Pi} (\hat{L}_2' + \gamma^{-1} \hat{L}_3 \hat{Q}) \\ &\times \int_0^{b\gamma} ds \exp[(\hat{L}_1' + \gamma^{-1} \hat{Q} \hat{L}_2' + \gamma^{-2} \hat{Q} \hat{L}_3)s](\hat{L}_2' + \gamma^{-1} \hat{Q} \hat{L}_3) \hat{\Pi}^- w_{t-s/\gamma^2}. \end{aligned} \tag{4.1.99}$$

Considering that γ is large (or letting $\gamma \longrightarrow \infty$), we obtain the Markov master equation

$$\dot{w}_t = \hat{\Pi} \hat{L}_3 \hat{\Pi}^- w_t + \hat{\Pi} \hat{L}_2' \int_0^{\infty} ds \exp(\hat{L}_1' s) \hat{L}_2' \hat{\Pi}^- w_t. \tag{4.1.100}$$

We thus see that it does not explicitly depend on γ, although this by no means implies that it is independent of γ – its terms may depend on γ implicitly. In fact, by (4.1.3,11), the expression $\hat{\Pi} \dots \hat{\Pi}^-$ contains the integral that averages over the thermostat variables z' with the weight

$$\rho_{\text{eq}}(z)/w_{\text{eq}}(Q, P) = \rho_{\text{eq}}(z' \mid Q, P). \tag{4.1.101}$$

Using the Hamilton function $\mathcal{H}_1(z') = \gamma^2 \mathcal{H}_1'(z)$ that corresponds to the operator $\hat{L}_1 = \gamma^2 \hat{L}_1'$, the equilibrium conditional probability density (4.1.101) can be represented in the form of the Gibbs distribution

$$\rho_{eq}(z' \mid Q, P) = \text{const} \cdot \exp[-\gamma^2 \mathcal{H}_1'(z'; Q, P)/kT]. \tag{4.1.102}$$

Therefore, due to the averaging, equation (4.1.100) will, generally speaking, contain a dependence on γ. Note that the result (4.1.100) can also be obtained by the same method from the equation $\dot{w} = \hat{M}w$ and the formula

$$\hat{M} = \hat{\Pi}\hat{L}\hat{\Pi}^- + \hat{\Pi}\hat{L} \int_0^{\tau_i} ds \exp(\hat{L}\hat{s})\hat{L}\hat{\Pi}^-,$$

which is equivalent to (4.1.27).

If we return to $\hat{L}_1, \hat{L}_2$ from $\hat{L}_1', \hat{L}_2'$, the equation (4.1.100) will read

$$\dot{w}_t = \hat{\Pi}\left[\hat{L}_3 + \hat{L}_2 \int_0^\infty d\sigma e^{\hat{L}_1\sigma} \hat{L}_2\right] \hat{\Pi}^- w_t \tag{4.1.103}$$

$(\sigma = s/\gamma^2)$. The procedure used to derive (4.1.100) is called adiabatic exclusion. It can also be used (see, e.g. [4.2]) in the quantum case.

4.1.9 Example of Derivation of the Master Equation

(a) Formulation of the problem. Consider a simple mechanical system S with variables Q and P, attached to a chain of coupled linear oscillators, which plays the role of a thermostat. Let the system have the Hamiltonian $\mathcal{H}_0(Q, P)$, and the complete Hamiltonian have the form

$$\mathcal{H}(z) = \mathcal{H}_0(Q, P) + \frac{1}{2m} \sum_{j=1}^N p_j^2 + \frac{\kappa}{2} \sum_{j=1}^N (q_j - q_{j-1})^2, \tag{4.1.104}$$

where $q_0 = Q$.

The small time τ_0 here is $(m/\kappa)^{1/2}$, and the relaxation (or variation) time of the subsystem S can be estimated as

$$\tau_r \sim \left(\frac{\partial^2 \mathcal{H}_0}{\partial Q^2} \frac{\partial^2 \mathcal{H}_0}{\partial P^2}\right)^{-1/2}. \tag{4.1.105}$$

Consequently, the process $\{Q(t), P(t)\}$ will be Markov-like if

$$(m/\kappa)^{1/2} \ll \tau_r. \tag{4.1.106}$$

This condition will be met with ever more ease as the chain of coupled linear oscillators transforms into the continuous wave system having the Lagrange function

$$\hat{L}_1 = \tfrac{1}{2} \int \{\rho[\dot{q}(x)]^2 - \kappa_0[\partial q(x)/\partial x]^2\}dx. \tag{4.1.107}$$

If the neighboring links of the chain are separated by Δx, then for such a limiting process we should put

$$m = \rho\Delta x, \quad \kappa = \kappa_0/\Delta x.$$

Then $\tau_0 \equiv (m/\kappa)^{1/2} = (\rho/\kappa_0)^{1/2}\Delta x$, and the condition (4.1.106) will be ensured by the small value of Δx. The large parameter γ used in the previous subsection can conveniently be introduced by the relation $\Delta x = \gamma^{-2}$. Then,

$$m = \rho/\gamma^2, \quad \kappa = \gamma^2 \kappa_0.$$ (4.1.108)

Corresponding to the Hamiltonian (4.1.104), the Liouville operator is

$$\hat{L} = \hat{L}_1 + \hat{L}_2 + \hat{L}_3,$$

where

$$\hat{L}_3 = \frac{\partial \mathcal{H}_0}{\partial Q} \frac{\partial}{\partial P} - \frac{\partial \mathcal{H}_0}{\partial P} \frac{\partial}{\partial Q},$$ (4.1.109)

$$\hat{L}_2 = -\kappa(q_1 - Q)\frac{\partial}{\partial P},$$ (4.1.110)

$$\dot{\hat{L}}_1 = -\sum_{j=1}^{N} \frac{p_j}{m} \frac{\partial}{\partial q_j}$$
$$-\kappa \left[\sum_{j=1}^{N-1} (q_{j+1} - 2q_j + q_{j-1})\frac{\partial}{\partial p_j} + (q_{N-1} - q_N)\frac{\partial}{\partial p_N} \right].$$ (4.1.111)

In this case, instead of (4.1.96), we have, by (4.1.108), $\hat{L}_1 = \gamma^2 \hat{L}'_1$, $\hat{L}_2 = \gamma^2 \hat{L}'_2$. It is because of this difference that the final equation loses γ when the thermostat temperature T is independent of γ.

The equilibrium probability density $\rho_{eq}(z)$ is given by the Gibbs formula $\rho_{eq} = \text{const} \cdot \exp(-\mathcal{H}/kT)$, into which we must substitute (4.1.104). It is easily seen that, according to this probability density, the random variables $p_1, \ldots, p_N, \xi_1 = q_1 - Q$, $\xi_2 = q_2 - q_1, \ldots, \xi_N = q_N - q_{N-1}$ are statistically independent of one another and of the pair (Q, P). Consequently, integrating with respect to these variables readily gives

$$w_{eq}(Q, P) = \text{const} \cdot \exp[-\mathcal{H}_0(Q, P)/kT].$$ (4.1.112)

Further, we can easily find the equilibrium conditional probability density defined by (4.1.101):

$$\rho_{eq}(q, p \mid Q, P) = \text{const} \cdot \exp[-\mathcal{H}_1(q, p; Q)/kT],$$ (4.1.113)

where

$$\mathcal{H}_1(q, p; Q) = \frac{1}{2m} \sum_{j=1}^{N} p_j^2 + \frac{\kappa}{2}(q_1 - Q)^2 + \frac{\kappa}{2} \sum_{j=2}^{N} (q_j - q_{j-1})^2,$$ (4.1.114)

$q = (q_1, \ldots, q_N)$, $p = (p_1, \ldots, p_N)$. This Hamiltonian describes the above oscillator chain with one end, namely Q, fixed. Note that the quantities $\xi_j = q_j - q_{j-1}$ are, by (4.1.113), distributed independently and have zero mean. We then readily find

$$\langle q_j \rangle_Q = Q, \quad j = 1, \ldots, N,$$ (4.1.115)

where $\langle \ldots \rangle_Q$ is the mean with the conditional probability density (4.1.113).

Equations (4.1.3,11) for $\hat{\Pi}$ and $\hat{\Pi}^-$ now take the form

$$\hat{\Pi}\rho = \int \rho(Q,P,q,p)dqdp,$$

$$\hat{\Pi}^-w = \rho_{\text{eq}}(q,p\mid Q)w(Q,P).\tag{4.1.116}$$

Using them we can easily see that in the case of (4.1.109–111) the conditions (4.1.97) are met, the third one being valid by (4.1.115). This example is thus in many ways similar to the case discussed in the previous subsection. The only difference is that now instead of $\hat{L}_2 = \gamma\hat{L}_2'$ we have $\hat{L}_2 = \gamma^2\hat{L}_2'$. This difference, however, is significant since the exponential function in (4.1.99) will contain the sum $\hat{L}_1' + \hat{Q}\hat{L}_2' + \gamma^{-2}\hat{Q}\hat{L}_3$, which already cannot be replaced by $\hat{L}_1'$. In this connection, the master equation will have to be derived by another method, namely by using the equations (4.1.67a) or (4.1.74). To begin with, however, we will derive some useful results.

(b) Dynamic equations and statistical properties of waves in a chain of coupled oscillators. The following Hamilton equations correspond to the Hamiltonian (4.1.104):

$$\dot{Q} = \partial\mathcal{H}_0/\partial P,\tag{4.1.117a}$$

$$\dot{P} = -\partial\mathcal{H}_0/\partial Q + \kappa(q_1 - Q),\tag{4.1.117b}$$

$$\left.\begin{array}{l}\dot{q}_j = p_j/m,\quad j=1,\ldots,N,\\[4pt]\dot{p}_j = \kappa(q_{j+1} - 2q_j + q_{j-1}),\quad j=1,\ldots,N-1,\\[4pt]\dot{p}_N = \kappa(q_{N-1} - q_N).\end{array}\right\}\tag{4.1.118}$$

Equations (4.1.118) yield

$$\tau_0^2\ddot{q}_j = q_{j+1} - 2q_j + q_{j-1}.\tag{4.1.119}$$

If the solution of these equations is sought in the form $q_j = C\cdot\exp(\pm i\omega t \pm i\lambda j)$, then from (4.1.119) we will obtain the dispersion equation

$$\begin{aligned}\tau_0^2\omega^2 &= 2(1-\cos\lambda)\\ &= 4\sin^2(\lambda/2).\end{aligned}\tag{4.1.120}$$

We will assume that $\omega \geq 0, \lambda \geq 0$, and then the last equation can be replaced by

$$\tau_0\omega = 2\sin(\lambda/2).\tag{4.1.121}$$

Let us initially regard the chain as infinite in both directions. Using unitary transformations we can go over from the variables q_j to the complex variables

$$\tilde{q}(\lambda) = (2\pi)^{-1/2}\sum_{j=-\infty}^{\infty}\exp(-i\lambda j)q_j,\quad -\pi < \lambda \leq \pi.\tag{4.1.122}$$

The reverse transformation has the form

$$\begin{aligned}q_j &= (2\pi)^{-1/2}\int_{-\pi}^{\pi}d\lambda\exp(i\lambda j)\tilde{q}(\lambda)\\ &= (2\pi)^{-1/2}\int_0^{\pi}d\lambda\exp(i\lambda j)\tilde{q}(\lambda) + \text{c.c.}\end{aligned}\tag{4.1.123}$$

We have here used the relation $\tilde{q}(-\lambda) = \tilde{q}^*(\lambda)$, which holds because the q_j are real. The transformation is unitary since either of the following equations is valid

$$\frac{1}{2\pi} \int_{-\pi}^{\pi} d\lambda \exp[i\lambda(j - l)] = \delta_{jl}, \qquad (4.1.124a)$$

$$\frac{1}{2\pi} \sum_{j=-\infty}^{\infty} \exp[-ij(\lambda - \lambda')] = \delta(\lambda - \lambda'). \qquad (4.1.124b)$$

If now we use (4.1.123) and take into account the exponential form of the solution of (4.1.119), we can write q_j as a function of time as follows:

$$q_j(t) = (2\pi)^{-1/2} \int_0^{\pi} d\lambda e^{i\lambda j}[r_-(\lambda)e^{i\omega t} + r_+(\lambda)e^{-i\omega t}] + \text{c.c.}, \qquad (4.1.125)$$

where ω, λ are related by (4.1.121) and c.c. stands for complex conjugate. If (4.1.123) refers to time $t = 0$, it is then clear that $r_-(\lambda) + r_+(\lambda) = \tilde{q}(\lambda)$. The values of $r_\pm(\lambda)$ are determined by $\tilde{q}(\lambda), \tilde{p}(\lambda)$ at $t = 0$.

We can express the Hamilton function

$$\mathcal{H}_1 = \frac{1}{2} \sum [m^{-1}p_j^2 + \kappa(q_{j+1} - q_j)^2] \qquad (4.1.126)$$

in terms of r_+ and r_-.

For this purpose, we will have to use the equations

$$p_j(t) = m\dot{q}_j(t)$$
$$= m(2\pi)^{-1/2} \int_0^{\pi} d\lambda i\omega e^{i\lambda j}[r_-(\lambda)e^{i\omega t} - r_+(\lambda)e^{-i\omega t}] + \text{c.c.}, \qquad (4.1.127a)$$

$$q_{j+1} - q_j = (2\pi)^{-1/2} \int_0^{\pi} d\lambda (e^{i\lambda} - 1)e^{i\lambda j}[r_-(\lambda)e^{i\omega t} + r_+(\lambda)e^{-i\omega t}] + \text{c.c.} \qquad (4.1.127b)$$

which follow from (4.1.125).

Squaring the last expressions, substituting the result into (4.1.126) and summing up over j, we obtain

$$\mathcal{H}_1 = 2m \int_0^{\pi} d\lambda \omega^2(|r_-(\lambda)|^2 + |r_+(\lambda)|^2) \qquad (4.1.128)$$

by (4.1.124b,121). Expressing $\mathcal{H}_1$ in terms of the functions $y_l(\lambda)$ defined by $r_-(\lambda) = y_1(\lambda) + iy_2(\lambda)$, $r_+ = y_3 + iy_4$, we will have

$$\mathcal{H}_1 = 2m \int_0^{\pi} d\lambda \omega^2 \sum_{l=1}^{4} y_l^2(\lambda). \qquad (4.1.129)$$

Let us now write the Gibbs distribution $w(y) = \text{const} \cdot \exp(-\mathcal{H}_1/kT)$ with this Hamiltonian. This distribution is Gaussian. Applying the conventional formulas, we can easily find the correlators of the functions $y_l(\lambda)$:

$$\langle y_l(\lambda), y_m(\lambda') \rangle = kT(4m\omega_\lambda^2)^{-1}\delta_{lm}\delta(\lambda - \lambda'), \quad 0 < \lambda, \lambda' \leq \pi. \qquad (4.1.130)$$

This equation, together with the relation $\langle y_l(\lambda) \rangle = 0$, completely determines the statistics of waves travelling through an infinite chain.

(c) Allowance for reflection at end. In order that we may ignore the reflection of waves at the distant end, we will pass to the limit $N \longrightarrow \infty$. This limiting process is of crucial importance, since it is only due to it that the system achieves complete irreversibility and the characteristic frequency spectrum corresponding to the Hamiltonian (4.1.114) becomes continuous.

It now remains only to consider the influence of the near end $j = 0$. At $N = \infty$ all the equations (4.1.119) will hold if (4.1.125) is valid for all $q_0, q_1, \ldots$, i.e. for $q_0 = Q$ among them. The appropriate equation for $j = 0$ is of the form

$$Q(t) = u_-(t) + u_+(t) , \tag{4.1.131}$$

where

$$u_\pm(t) = (2\pi)^{-1/2} \int_0^\pi d\lambda \exp(\mp i\omega t) r_\pm(\lambda) + \text{c.c.} \tag{4.1.132}$$

Since $Q(t)$ is given by (4.1.117), equation (4.1.131) defines not $Q(t)$, but $u_+ = Q - u_-$, i.e. the reflected wave.

We will now consider the force $f(t) = \kappa(q_1 - Q)$ that enters into (4.1.117b). Putting in (4.1.127b) $j = 0$, we obtain

$$\begin{aligned} f &\equiv \kappa(q_1 - Q) \\ &= \kappa(2\pi)^{-1/2} \int_0^\pi d\lambda (e^{i\lambda} - 1)[e^{i\omega t} r_-(\lambda) + e^{-i\omega t} r_+(\lambda)] + \text{c.c.} \end{aligned} \tag{4.1.133}$$

Equation (4.1.121) enables the identity

$$e^{i\lambda} - 1 = \left(\cos\frac{\lambda}{2} + i\sin\frac{\lambda}{2} \right) 2i \sin\frac{\lambda}{2} \tag{4.1.134}$$

to be transformed to

$$\exp(i\lambda) - 1 = F(i\omega) \cdot i\omega\tau_0 , \tag{4.1.135}$$

where

$$F(x) = [1 + (\tfrac{1}{2}\tau_0 x)^2]^{1/2} + \tfrac{1}{2}\tau_0 x . \tag{4.1.136}$$

Substituting (4.1.135) into (4.1.133) we readily see that in it we can replace $i\omega$ by $\partial/\partial t$ (the term with r_-) and by $-\partial/\partial t$ (the term with r_+). If we then factor appropriate operators outside the integral sign and take into account (4.1.132), we get

$$f(t) = (\kappa m)^{1/2} \frac{\partial}{\partial t} \left[F\left(\frac{\partial}{\partial t} \right) u_-(t) - F\left(-\frac{\partial}{\partial t} \right) u_+(t) \right]. \tag{4.1.137}$$

Hence, using (4.1.131), we find

$$f(t) = -(\kappa m)^{1/2} \frac{\partial}{\partial t} F\left(-\frac{\partial}{\partial t} \right) Q(t) + \eta(t) , \tag{4.1.138}$$

where

$$
\begin{aligned}
\eta(t) &= (\kappa m)^{1/2} \frac{\partial}{\partial t} \left[F\left(\frac{\partial}{\partial t}\right) + F\left(-\frac{\partial}{\partial t}\right) \right] u_- \\
&= 2(\kappa m)^{1/2} (2\pi)^{-1/2} \int_0^\pi d\lambda(i\omega) \left(1 - \tfrac{1}{4}\tau_0^2 \omega^2\right)^{1/2} e^{i\omega t} r_-(\lambda) + \text{c.c.}
\end{aligned}
$$

(4.1.139)

by (4.1.136,132). It is convenient to expand the function (4.1.136), which enters into (4.1.138), into a Taylor series to obtain

$$
f(t) = -(\kappa m)^{1/2}[\dot{Q}(t) - \tfrac{1}{2}\tau_0 \ddot{Q}(t) + \tfrac{1}{8}\tau_0^2 \, \dddot{Q}\,(t) + \ldots] + \eta(t)\,.
$$

(4.1.140)

The expression (4.1.138) can also be represented in integral form as

$$
f(t) = -(\kappa m)^{1/2} \frac{\partial}{\partial t} \int_{-\infty}^t \varphi(t - s) Q(s)\,ds + \eta(t)\,,
$$

(4.1.141)

where $\varphi(t)$ is the function whose Laplace transform is

$$
\mathcal{L}[\varphi(t)] = \sqrt{1 + (\tfrac{1}{2}\tau_0 p)^2} - \tfrac{1}{2}\tau_0 p\,.
$$

(4.1.142)

Using item 29.3.58 of the handbook [4.3], we have

$$
\varphi(t) = \tfrac{1}{t} J_1(2\tfrac{t}{t_0})\vartheta(t)\,, \quad t \geq 0
$$

(4.1.143)

with $2\vartheta(t) = 1 + \operatorname{sign} t$. Therefore, (4.1.141) can be written as

$$
f(t) = -(\kappa m)^{1/2} \int_{-\infty}^{t+\varepsilon} G(t - s) Q(s)\,ds + \eta(t)\,,
$$

(4.1.144)

where

$$
\begin{aligned}
G(\tau) &= \frac{d}{d\tau}[\varphi(\tau)\vartheta(\tau)] \\
&= \delta(\tau) + \frac{d\varphi(\tau)}{d\tau}\vartheta(\tau)\,.
\end{aligned}
$$

(4.1.145)

The random function $\eta(t)$, which describes the action of the incident wave, is statistically independent of $Q(\cdot)$ and has zero mean and correlator

$$
\langle \eta(t + \tau)\eta(t)\rangle = kT\kappa\pi^{-1} \int_{-\pi}^\pi e^{i\omega\tau}(1 - \tfrac{1}{4}\tau_0^2 \omega^2)\,d\lambda\,.
$$

(4.1.146)

This has been found using (4.1.139) and the relationships

$$
\begin{aligned}
&m\omega_\lambda^2\langle r_-(\lambda)r_-^*(\lambda')\rangle = \tfrac{1}{2}kT\delta(\lambda - \lambda')\,, \\
&\langle r_-(\lambda)r_-(\lambda')\rangle = 0\,, \\
&\langle r_-^*(\lambda)r_-^*(\lambda')\rangle = 0\,,
\end{aligned}
$$

(4.1.147)

which follow from (4.1.130). If now we use (4.1.121) to change the integration variable in (4.1.146), we will have

$$\langle\eta(t+\tau)\eta(t)\rangle = kT(\kappa m)^{1/2}\frac{1}{\pi}\int_{-2/\tau_0}^{2/\tau_0} e^{i\omega\tau}(1-\tfrac{1}{4}\tau_0^2\omega^2)^{1/2}d\omega. \tag{4.1.148}$$

It follows that the correlation time of the process $\eta(t)$ is τ_0. Applying the Fourier transformation, which is inverse to the transformation given by the formula 11.4.25 of [4.3] at $n = 1$, from (4.1.148) we obtain

$$\langle\eta(t+\tau)\eta(t)\rangle = kT(\kappa m)^{1/2}\tau^{-1}J_1(2\tau/\tau_0). \tag{4.1.149}$$

(d) Coefficients of the master equation. The results (4.1.144,149) hold at any τ_0. At $\tau_0 \ll \tau_r$ we can with their help readily obtain the coefficients $K_\alpha(A), K_{\alpha\beta}(A)$ (recall that $A_1 = Q, A_2 = P$). For this purpose we use (4.1.72), i.e. the formulas

$$K_\alpha(A^0) = \frac{\partial}{\partial\tau_i}\langle\Delta A_\alpha\rangle_{A^0}, \tag{4.1.150a}$$

$$K_{\alpha\beta}(A^0) = \frac{\partial}{\partial\tau_i}\langle\Delta A_\alpha\Delta A_\beta\rangle_{A^0} \tag{4.1.150b}$$

with $\Delta A_\alpha \equiv \Delta A_\alpha(t_0 + \tau_i) = A_\alpha(t_0 + \tau_i) - A_\alpha^0$, $A_\alpha^0 = A_\alpha(t_0)$. According to (4.1.117,144) Q and P are governed by the equations

$$\dot{Q} = \frac{\partial\mathcal{H}_0}{\partial P}, \tag{4.1.151a}$$

$$\dot{P} = -\frac{\partial\mathcal{H}_0}{\partial Q} - (\kappa m)^{1/2}\int_{-\infty}^{t} G(t-s)Q(s)ds + \eta(t). \tag{4.1.151b}$$

The conditional averaging $\langle\ldots\rangle_{A^0} = \int dqdp\ldots\rho_{\mathrm{eq}}(q,p\mid A^0)$ in (4.1.150) means that the conditions

$$Q(t) = Q^0 \quad \text{at} \quad t \le t_0, \tag{4.1.152}$$

$$P(t) = P^0 \quad \text{at} \quad t \le t_0 \tag{4.1.153}$$

must be fulfilled. In fact, the equilibrium described by the probability density (4.1.113) is maintained when the evolution is governed by L_1, which leaves Q, P unchanged. Therefore, (4.1.151b) must be changed into

$$\begin{aligned}\dot{P} &= -\frac{\partial\mathcal{H}_0}{\partial Q} - (\kappa m)^{1/2}\left\{\int_{t_0}^{t+\varepsilon} G(t-s)Q(s)ds + \int_{-\infty}^{t_0} G(t-s)Q^0ds\right\} \\ &= -\frac{\partial\mathcal{H}_0}{\partial Q} - (\kappa m)^{1/2}\int_{t_0}^{t+\varepsilon} G(t-s)[Q(s) - Q^0]ds \end{aligned} \tag{4.1.154}$$

$(\varepsilon > 0)$. Here the property

$$\int_{-\infty}^{\varepsilon} G(-\sigma)d\sigma = 0$$

of the function (4.1.145) has been used. Applying (4.1.145) we have

$$\begin{aligned}G(t-s)ds &= -\frac{d}{ds}[\varphi(t-s)\eta(t-s)]ds \\ &= -d[\varphi(t-s)\eta(t-s)].\end{aligned}$$

Integration by parts gives

$$\int_{t_0}^{t+\varepsilon} G(t-s)\Delta Q(s)ds = -\Delta Q(s)\varphi(t-s)\vartheta(t-s)\Big|_{t_0}^{t+\varepsilon}$$
$$+ \int_{t_0}^{t} \varphi(t-s)\dot{Q}(s)ds \,.$$

Therefore (4.1.154) takes the form

$$\dot{P} = -\frac{\partial \mathcal{H}_0}{\partial Q} - (\kappa m)^{1/2} \int_{t_0}^{t} \varphi(t-s)\frac{\partial \mathcal{H}_0}{\partial P}(A(s))ds + \eta(t) \,. \qquad (4.1.155)$$

Equations (4.1.151a) and (4.1.154) can be written as

$$\dot{A}_\alpha = F_\alpha(A) + \int_{t_0}^{t} ds\varphi_\alpha(t-s)F_\alpha'(A(s)) + \eta_\alpha(t) \qquad (4.1.156)$$

with $F_1 = F_2' = \partial \mathcal{H}_0/\partial P$, $F_2 = -\partial \mathcal{H}_0/\partial Q$, $\varphi_1(\tau) = 0$, $\varphi_2(\tau) = -(\kappa m)^{1/2}\varphi(\tau)$, $\eta_1(t) = 0$, $\eta_2(t) = \eta(t)$.
If we integrate both sides of (4.1.155) from t_0 to $t_0 + \tau_i$, we get

$$\Delta A_\alpha(t_0+\tau_i) = \int_{t_0}^{t_0+\tau_i} dt \left\{ F_\alpha(A^0) + \frac{\partial F_\alpha}{\partial A_\beta}(A^0)\Delta A_\beta(t) \right.$$
$$+ \frac{1}{2}\frac{\partial^2 F_\alpha}{\partial A_\beta \partial A_\gamma}(A^0)\Delta A_\alpha(t)\Delta A_\beta(t)$$
$$+ \ldots + \int_{t_0}^{t} ds\varphi_\alpha(t-s)\Big[F_\alpha'(A^0)$$
$$\left. + \frac{\partial F_\alpha'}{\partial A_\beta}(A^0)\Delta A_\beta(s) + \ldots] + \eta(t) \right\} \,. \qquad (4.1.157)$$

Here the expansion of $F_\alpha(A) = F_\alpha(A^0 + \Delta A)$ and $F_\alpha'(A) = F_\alpha'(A^0 + \Delta A)$ into a Taylor series is used. By the iterative substitutions of the equation (4.1.157) (with $t_0 + \tau_i$ changed to $t, s, \ldots$) into its right-hand side we can find $\Delta A_\alpha(t_0 + \tau_i)$ to any degree of accuracy. Thus we obtain

$$\Delta A_\alpha(t_0+\tau_i)$$
$$= F_\alpha(A^0)\tau_i + \frac{1}{2}\frac{\partial F_\alpha}{\partial A_\beta}(A^0)F_\beta(A^0)\tau_i^2 + F_\alpha'(A^0)\int_{t_0}^{t_0+\tau_i} dt$$
$$\times \int_{t_0}^{t} ds\varphi_\alpha(t-s) + \int_{t_0}^{t_0+\tau_i} \eta_\alpha(t)dt + \frac{\partial F_\alpha}{\partial A_\beta}(A^0)$$
$$\times \int_{t_0}^{t_0+\tau_i} (t_0+\tau_i-t)\eta_\beta(t)dt + \ldots \,, \qquad (4.1.158)$$

where the terms with triple and higher integrals are omitted. In (4.1.158) the integral with $\varphi_\alpha(t-b)$ can be simplified

$$\int_{t_0}^{t_0+\tau_i} dt \int_{t_0}^{t} ds\varphi_\alpha(t-s) = \int_0^{\tau_i} (\tau_i - t')\varphi_\alpha(t')dt' \,.$$

According to (4.1.158) we have

$$\Delta Q(t_0 + \tau_i)$$

$$= \left(\frac{\partial \mathcal{H}_0}{\partial P}\right)_0 \tau_i + \frac{1}{2}\left(\frac{\partial^2 \mathcal{H}_0}{\partial Q \partial P}\frac{\partial \mathcal{H}_0}{\partial P} - \frac{\partial^2 \mathcal{H}_0}{\partial P^2}\frac{\partial \mathcal{H}_0}{\partial Q}\right)_0 \tau_i^2$$

$$+ \left(\frac{\partial^2 \mathcal{H}_0}{\partial P^2}\right)_0 \int_{t_0}^{t_0+\tau_i}(t_0 + \tau - t)\eta(t)dt + \dots ,$$

$$\Delta P(t_0 + \tau_i)$$

$$= -\left(\frac{\partial \mathcal{H}_0}{\partial Q}\right)_0 \tau_i + \frac{1}{2}\left(\frac{\partial^2 \mathcal{H}_0}{\partial Q^2}\frac{\partial \mathcal{H}_0}{\partial P} - \frac{\partial^2 \mathcal{H}_0}{\partial Q \partial P}\frac{\partial \mathcal{H}_0}{\partial Q}\right)_0 \tau_i^2$$

$$- (\kappa m)^{1/2}\left(\frac{\partial \mathcal{H}_0}{\partial P}\right)_0 \int_0^{\tau_i}(\tau_i - s)\varphi(s)ds + \int_{t_0}^{t_0+\tau_i}\eta(t)dt$$

$$- \left(\frac{\partial^2 \mathcal{H}_0}{\partial P \partial Q}\right)_0 \int_{t_0}^{t_0+\tau_i}(t_0 + \tau - t)\eta(t)dt + \dots . \tag{4.1.159}$$

Here $(\dots)_0$ denotes that functions are taken at $A = A^0$. Hence, by (4.1.150a) we obtain

$$K_1(A^0) = \frac{\partial \mathcal{H}_0}{\partial P}(A^0)\left[1 + O\left(\frac{\tau_i}{\tau_r}\right)\right], \tag{4.1.160a}$$

$$K_2(A^0) = \left[-\frac{\partial \mathcal{H}_0}{\partial Q}(A^0) - (\kappa m)^{1/2}\frac{\partial \mathcal{H}_0}{\partial P}(A^0)\right]\left[1 + O\left(\frac{\tau_i}{\tau_r}\right)\right]. \tag{4.1.159b}$$

Terms with $\eta(t)$ have been dropped out because of averaging $\langle \dots \rangle_{A^0}$. In (4.1.159b) we have used that

$$\int_0^{\tau_i}\varphi(s)ds \approx \int_0^\infty \varphi(s)ds \quad \text{for} \quad \tau_i \gg \tau_0 \tag{4.1.160}$$

and that

$$\int_0^\infty \varphi(s)ds = 1. \tag{4.1.161}$$

The latter equation follows from (4.1.142) if the formula

$$\int_0^\infty \varphi(t)dt = \mathcal{L}[\varphi(t)]\Big|_{p=0}$$

is taken into account. A more exact formula than (4.1.160) can be obtained with the help of the asymptotic expansion of the function $J_1(x)$ for large x (see 9.2.5,9,10 from [4.3]). Namely, we have

$$\int_0^{\tau_i}\varphi(s)ds - 1 = -\int_{\tau_i}^\infty J_1\left(2\frac{s}{\tau_0}\right)s^{-1}ds$$

$$= \left(\frac{2}{\pi}\right)^{1/2}\left[\left(\frac{\tau_0}{2\tau_i}\right)^{3/2}\sin\chi\right.$$

$$\left. -\frac{3}{2}\left(\frac{\tau_0}{2\tau_i}\right)^{5/2}\cos\chi + \dots\right] \tag{4.1.162}$$

with $\chi = 2\tau_i/\tau_0 - 3\pi/4$.

Further we substitute (4.1.158) into (4.1.150b) and obtain

$$K_{11}(A^0) = \dots , \tag{4.1.163a}$$

$$K_{12}(A^0) = K_{21}(A^0) = \dots , \tag{4.1.163b}$$

$$K_{22}(A^0) = 2 \int_0^{\tau_i} \langle \eta(t+\tau)\eta(t)\rangle d\tau + \dots \tag{4.1.163c}$$

since

$$\left\langle \left\{ \int_{t_0}^{t+\tau_i} \eta(t)dt \right\}^2 \right\rangle = 2 \int_0^{\tau_i} (\tau_i - \tau)\langle \eta(t+\tau)\eta(t)\rangle d\tau . \tag{4.1.164}$$

The terms not written in (4.1.163) and denoted by dots are proportional to τ_i and smaller. Subsequently, they play no essential part owing to $\tau_i \ll \tau_r$. Substituting (4.1.149) into (4.1.163c) gives

$$\begin{aligned} K_{22} &= 2kT(\kappa m)^{1/2} \int_0^{\tau_i} J_1\left(2\frac{\tau}{\tau_0}\right)\tau^{-1}d\tau + \dots \\ &= 2kT(\kappa m)^{1/2} + \dots . \end{aligned} \tag{4.1.165}$$

More exact estimation of this integral is given by (4.1.162). Considering (4.1.160, 163a,163b,165) and the fact that higher coefficients $K_{\alpha\beta\dots}$ are proportional to $(\tau_i)^m, m \geq 2$, we can write out the master equation (4.1.73) for our example. It will have the form

$$\begin{aligned} \dot{w}(Q,P) &= \left\{ \hat{L}_3 w + (\kappa m)^{1/2}\frac{\partial}{\partial P}\left[\frac{\partial \mathcal{H}_0}{\partial P}w\right] + kT(\kappa m)^{1/2}\frac{\partial^2 w}{\partial P^2} \right\} \\ &\quad \times \left[1 + O\left(\frac{\tau_i}{\tau_r}\right)\right] . \end{aligned} \tag{4.1.166}$$

Thus we see that the operator (4.1.27) gives, with due accuracy, the ordinary Fokker-Planck equation having the frictional and fluctuational terms for the present example. It can be easily verified that the probability density (4.1.112) is its stationary solution.

Note that the application of the formula (4.1.103) to our case gives the equation

$$\dot{w}(Q,P) = \hat{L}_3 w + kT(\kappa m)^{1/2}\frac{\partial^2 w}{\partial P^2} , \tag{4.1.167}$$

where the second term – the frictional term – in braces on the right-hand side of (4.1.166) is absent. Therefore, the fluctuational input caused by the approaching wave is not balanced by dissipation and the stationary solution of (4.1.167) does not exist.

4.1.10 More Exact Formulas for the Markov Operator

The definition (4.1.22) or (4.1.27) of the Markov operator $\hat{M}$ can be improved. To do this we must take into account that the transformation from $w_t(A)$ to

$w_{t+\tau_i}(A) = \exp(\hat{M}\tau_i)w_t(A)$ corresponds to the transformation from $\rho_t(z)$ to $\rho_{t+\tau_i}(z) = \exp(\hat{L}\tau_i)\rho_t(z)$. Therefore we have

$$
\begin{aligned}
(e^{\hat{M}\tau_i} - \hat{I})w_t(A) &= \hat{\Pi}(e^{\hat{L}\tau_i} - \hat{1})\rho_t(z) \\
&= \hat{\Pi}(e^{\hat{L}\tau_i} - \hat{1})\hat{\Pi}^- w_t(A)
\end{aligned}
$$

and

$$
e^{\hat{M}\tau_i} - \hat{I} = \hat{\Pi}(e^{\hat{L}\tau_i} - \hat{1})\hat{\Pi}^- . \tag{4.1.168}
$$

This definition is more exact than (4.1.22), i.e. the dependence of $\hat{M}$ on $\tau_i \gg \tau_0$ has been made still weaker. To solve (4.1.168) for $\hat{M}$, we use the expansion of $\exp(\hat{M}\tau_i)$ into the series. This gives

$$
\sum_{n=1}^{\infty} \frac{1}{n!}(\hat{M}\tau_i)^n w_t = \hat{\Pi}(e^{\hat{L}\tau_i} - \hat{1})\hat{\Pi}^- w_t . \tag{4.1.169}
$$

Since $\tau_i \ll \tau_r$ we have

$$
\left| \hat{M}\tau_i w_t \right| \ll w_t , \tag{4.1.170}
$$

so that the series on the left-hand side of (4.1.169) rapidly converges. From the dependence

$$
y = x + \tfrac{1}{2}x^2 + \tfrac{1}{6}x^3 + \dots \tag{4.1.171}
$$

the following dependence ensues

$$
x = y - \tfrac{1}{2}y^2 + \tfrac{1}{3}y^3 + \dots . \tag{4.1.172}
$$

This can be easily checked.

Applying (4.1.172), from (4.1.169) we obtain

$$
\begin{aligned}
\hat{M} &= \tau_i^{-1}\hat{\Pi}(e^{\hat{L}\tau_i} - \hat{1})\hat{\Pi}^- - \tfrac{1}{2}\tau_i^{-1}\hat{\Pi}(e^{\hat{L}\tau_i} - \hat{1})\hat{P}(e^{\hat{L}\tau_i} - \hat{1})\hat{\Pi}^- \\
&\quad + \tfrac{1}{3}\tau_i^{-1}\left[\hat{\Pi}(e^{\hat{L}\tau_i} - \hat{1})\hat{\Pi}^-\right]^3 + \dots .
\end{aligned} \tag{4.1.173}
$$

We see that additional terms appear in (4.1.173) in comparison with (4.1.22). Analogously we can find the operator that is more exact than (4.1.27).

Now we turn to the Zwanzig equation (4.1.87) and obtain the operator more exact than (4.1.94) for the Markov-like processes. Setting $t_0 = t - \tau_i$ in (4.1.87), we have the equation

$$
\dot{w}_t = \hat{\Pi}\hat{L}\hat{\Pi}^- w_t + \hat{\Pi}\hat{L}\int_0^{\tau_i} ds \, \exp(\hat{Q}\hat{L}s)\hat{Q}\hat{L}w_{t-s} . \tag{4.1.174}
$$

But

$$
w_{t-s} = e^{-\hat{M}s}w_t , \tag{4.1.175}
$$

so that (4.1.174) takes the form of (4.1.21) for

$$\hat{M} = \hat{\Pi}\hat{L}\hat{\Pi}^- + \hat{\Pi}\hat{L}\int_0^{\tau_i} ds\, \exp(\hat{Q}\hat{L}s)\hat{Q}\hat{L}\hat{\Pi}^- \exp(-\hat{M}s)\,. \qquad (4.1.176)$$

Exponent $\exp(-\hat{M}s)$ is understood as a series expansion

$$\hat{M} = \hat{\Pi}\hat{L}\hat{\Pi}^- + \hat{\Pi}\hat{L}\sum_{n=0}^{\infty}\frac{(-1)^n}{n!}\int_0^{\tau_i}\exp(\hat{Q}\hat{L}s)\hat{Q}\hat{L}s^n ds\,\hat{\Pi}^-\hat{M}^n\,. \qquad (4.1.177)$$

Integrating gives

$$\hat{M} = \hat{\Pi}\hat{L}e^{\hat{Q}\hat{L}\tau_i}\hat{\Pi}^- + \hat{\Pi}\hat{L}\left[(\hat{Q}\hat{L})^{-1}(e^{\hat{Q}\hat{L}\tau_i} - \hat{1} - \hat{Q}\hat{L}\tau_i)\right]\hat{\Pi}^-\hat{M} + \ldots \quad (4.1.178)$$

We may solve this equation for $\hat{M}$ by iterations and finally arrive at

$$\hat{M} = \hat{\Pi}\hat{L}e^{\hat{Q}\hat{L}\tau_i}\hat{\Pi}^- + \hat{\Pi}\hat{L}\left[(\hat{Q}\hat{L})^{-1}(e^{\hat{Q}\hat{L}\tau_i} - \hat{1} - \hat{Q}\hat{L}\tau_i)\right]\hat{P}\hat{L}e^{\hat{Q}\hat{L}\tau_i}\hat{\Pi}^- + \ldots\,.$$
$$(4.1.179)$$

Thus, the terms additional to the expression on the right-hand side of (4.1.94) can be obtained. Note that the function

$$x^{-1}(e^{x\tau_i} - 1 - x\tau_i) = \sum_{n=2}^{\infty}\frac{1}{n!}x^{n-1}\tau_i^n$$

(just as the functions for higher terms) is entire. So there is no need to calculate $(\hat{Q}\hat{L})^{-1}$, which is impossible because the projection operator $\hat{Q} = \hat{1} - \hat{P}$ is degenerate.

According to (4.1.173,179) the operator $\hat{M}$ or rather the function $\hat{M}w$ can be calculated essentially as a series in τ_i/τ_r.

4.2 Derivation of Linear FDR of the First Kind and Reciprocal Relation Using Projection Operator

4.2.1 Modified Projection Operator

In Sect. 4.1 we have thought that the probability density in phase space, $\rho(z)$, evolves following the law $\rho_t = \exp(\hat{L}t)\rho_0$, and the dynamic variables z or their functions are constant. In analogy with quantum theory this picture can be called "the Schrödinger representation". Another interpretation is possible, however. In it, no distribution varies in time, whereas the dynamical variables or their functions $C(z)$ vary as $C_t = \exp(-\hat{L}t)C_0$. This interpretation is the nonquantum analog of the Heisenberg representation. It can be easily seen that these interpretations are equivalent since they yield the same mean

$$\begin{aligned} m(t) &= \int C(z)\rho_t(z)dz \\ &= \int C_t(z)\rho(z)dz\,, \end{aligned} \qquad (4.2.1)$$

where $C(z)$ is arbitrary.

Let us go over from the Schrödinger representation to the Heisenberg representation. Clearly, now, instead of the projection operation $\hat{P}\rho = \hat{\Pi}^-\hat{\Pi}\rho$ (see (4.1.14)) we should consider the operation $\hat{P}^T C(z) = \hat{\Pi}^T(\hat{\Pi}^-)^T C(z)$ applied to the functions of dynamical variables, so that the mean $\int C\hat{P}\rho\,dz$ should be the same in both representations.

Using (4.1.4) and (4.1.12), we can readily verify that the operator $\hat{P}^T = \hat{\Pi}^T(\hat{\Pi}^-)^T = \hat{\Pi}^T w_{eq}^{-1}\hat{\Pi}\rho_{eq}$ leaves unchanged any function $F(B(z))$ of the variables $B_1(z),\ldots,B_r(z)$:

$$\hat{P}^T F(B(z)) = F(B(z))\,. \tag{4.2.2}$$

Linear relations of the first kind can conveniently be derived using a "stronger" projection (projection onto a narrower space), which only leaves invariant linear functions of $B_\alpha(z)$, i.e. the functions of the form $F(B(z)) = C_1 B_1(z)+\ldots+C_r B_r(z)$. We can readily verify that this property is inherent in the operator given by

$$\hat{P}C(z) = \sum_{\gamma,\beta} B_\gamma(z)\langle BB\rangle_{\gamma\beta}^{-1}\langle B_\beta C\rangle\,. \tag{4.2.3}$$

where we write $\hat{P}$ instead of $\hat{P}^T$ only to simplify the presentation; $\langle BB\rangle_{\gamma\beta}^{-1}$ is the matrix inverse of $\langle B_\beta(z)B_\gamma(z)\rangle$; the mean is taken with the equilibrium distribution ρ_{eq}.

4.2.2 Derivation of the Mori Formula

In what follows we will use the identity

$$\exp(-\hat{L}t) = \exp(-\hat{Q}\hat{L}t) - \int_0^t d\tau \exp(-\hat{L}\tau)\hat{P}\hat{L}\exp[-\hat{Q}\hat{L}(t-\tau)]\,, \tag{4.2.4}$$

where $\hat{Q} = \hat{1} - \hat{P}$. We can easily see that it is valid by using on both side the Laplace transformation, which gives

$$(s\hat{1} + L)^{-1} = (s\hat{1} + QL)^{-1} - (s\hat{1} + L)^{-1}PL(s\hat{1} + QL)^{-1}\,. \tag{4.2.5}$$

The last equation can be verified by multiplying both sides by $s\hat{1} + \hat{L}$ on the left, and by $s\hat{1} + \hat{Q}\hat{L}$ on the right.

Differentiating $B_\alpha(z,t) = \exp(-\hat{L}t)B_\alpha(z)$ with respect to time yields

$$\begin{aligned}
\dot{B}_\alpha(t) &= -\exp(-\hat{L}t)\hat{L}B_\alpha(0) \\
&= -\exp(-\hat{L}t)(\hat{P} + \hat{Q})\hat{L}B_\alpha(0) \\
&= T_1 + T_2\,.
\end{aligned} \tag{4.2.6}$$

Using (4.2.3), we can write $T_1 = -\exp(-\hat{L}t)\hat{P}\hat{L}B_\alpha(0)$ as

$$T_1 = -\exp(-\hat{L}t)B_\gamma(0)\langle BB\rangle_{\gamma\beta}^{-1}\langle B_\beta(0)[\hat{L}B_\alpha(0)]\rangle\,. \tag{4.2.7}$$

Denoting

$$\langle[\hat{L}B_\alpha(0)]B_\beta(0)\rangle\langle BB\rangle_{\beta\gamma}^{-1} = d_{\alpha\gamma}\,, \tag{4.2.8}$$

we obtain

$$T_1 = -d_{\alpha\gamma} B_\gamma(t) \,. \tag{4.2.9}$$

We will now proceed to transform T_2. By (4.2.4),

$$
\begin{aligned}
T_2 &\equiv -\exp(-\hat{L}t)\hat{Q}\hat{L}B_\alpha(0) \\
&= -\exp(-\hat{Q}\hat{L}t)\hat{Q}\hat{L}B_\alpha(0) \\
&\quad + \int_0^t d\tau \exp(-\hat{L}\tau)\hat{P}\hat{L}\exp[-\hat{Q}\hat{L}(t-\tau)]\hat{Q}\hat{L}B_\alpha(0)
\end{aligned}
\tag{4.2.10}
$$

or

$$T_2 = F_\alpha(t) - \int_0^t d\tau \exp(-\hat{L}\tau)\hat{P}\hat{L}F_\alpha(t-\tau) \,, \tag{4.2.11}$$

where

$$F_\alpha(t) = -\exp(-\hat{Q}\hat{L}t)\hat{Q}\hat{L}B_\alpha(0) \,. \tag{4.2.12}$$

From (4.2.3,11) we have

$$T_2 = F_\alpha(t) - \int_0^t d\tau \exp(-\hat{L}\tau)B_\gamma(0)\langle BB\rangle^{-1}_{\gamma\beta}\langle B_\beta(0)[\hat{L}F_\alpha(t-\tau)]\rangle \,. \tag{4.2.13}$$

Here we can substitute $B_\gamma(\tau)$ for $\exp(-\hat{L}\tau)B_\gamma(0)$. By (4.2.12) and $\hat{Q}^2 = \hat{Q}$, $F_\alpha(t)$ has the property $F_\alpha(t) = \hat{Q}F_\alpha(t)$. Therefore, we easily have $\langle B_\beta(0)[\hat{L}F_\alpha(t-\tau)]\rangle = \langle B_\beta(0)[\hat{L}\hat{Q}F_\alpha(t-\tau)]\rangle$. Further, using the property $\hat{L}^T = -\hat{L}$ of the Liouville operator, we get

$$
\begin{aligned}
&\langle[\hat{L}\hat{Q}F_\alpha(t-\tau)]B_\beta(0)\rangle \\
&= \int [\hat{Q}F_\alpha(z,t-\tau)]\{\hat{L}^T[\rho_{\rm eq}(z)B_\beta(z,0)]\}dz \\
&= -\int [\hat{Q}F_\alpha(z,t-\tau)]\{\hat{L}[\rho_{\rm eq}(z)B_\beta(z,0)]\}dz \,.
\end{aligned}
\tag{4.2.14}
$$

Since $\hat{L}\rho_{\rm eq} = 0$, the right-hand side of (4.2.14) is

$$-\langle[\hat{Q}F_\alpha(t-\tau)][\hat{L}B_\beta(0)]\rangle \,, \quad \text{i.e.}$$
$$\langle B_\beta(0)[\hat{L}F_\alpha(t-\tau)]\rangle = -\langle[\hat{Q}F_\alpha(t-\tau)][\hat{L}B_\beta(0)]\rangle \,. \tag{4.2.15}$$

Using (4.2.3), we can readily obtain

$$
\begin{aligned}
\langle(\hat{P}f)g\rangle &= \langle B_\alpha g\rangle\langle BB\rangle^{-1}_{\alpha\beta}\langle B_\beta f\rangle = \langle f(\hat{P}g)\rangle \,, \\
\langle(\hat{Q}f)g\rangle &= \langle fg\rangle - \langle(\hat{P}f)g\rangle = \langle f(\hat{Q}g)\rangle \,.
\end{aligned}
\tag{4.2.16}
$$

Therefore, (4.2.15) gives

$$\langle B_\beta(0)[\hat{L}F_\alpha(t-\tau)]\rangle = -\langle F_\alpha(t-\tau)[\hat{Q}\hat{L}B_\beta(0)]\rangle \,. \tag{4.2.17}$$

But, by (4.2.12), $-\hat{Q}\hat{L}B_\beta(0) = F_\beta(0)$, hence (4.2.17) can be written as

$$\langle B_\beta(0)[\hat{L}F_\alpha(t-\tau)]\rangle = \langle F_\alpha(t-\tau)F_\beta(0)\rangle \,. \tag{4.2.18}$$

Substituting this into (4.2.13) gives

$$T_2 = F_\alpha(t) - \int_0^t d\tau \langle F_\alpha(t-\tau)F_\beta(0)\rangle \langle BB\rangle_{\beta\gamma}^{-1} B_\gamma(\tau)\,. \qquad (4.2.19)$$

By (4.2.9,19), equation (4.2.6) becomes

$$\dot{B}_\alpha(t) = -d_{\alpha\beta}B_\beta(t) - \int_0^t d\tau \langle F_\alpha(t-\tau)F_\beta(0)\rangle \langle BB\rangle_{\beta\gamma}^{-1} B_\gamma(\tau) + F_\alpha(t)\,. \quad (4.2.20)$$

This equation was derived by *Mori* [4.4] in 1965). The functions $F_\alpha(t)$ are treated as random Langevin forces acting on the variables B_α. Their mean is zero:

$$
\begin{aligned}
\langle F_\alpha(t)\rangle &= -\langle \hat{Q}\hat{L}\exp(-\hat{Q}\hat{L}t)B_\alpha(0)\rangle \\
&= \int [\exp(-\hat{Q}\hat{L}t)B_\alpha(0)][(\hat{L}-\hat{L}\hat{P}^T)\rho_{eq}]dz \\
&= 0\,, \qquad\qquad\qquad\qquad\qquad\qquad\qquad (4.2.21)
\end{aligned}
$$

since $\hat{L}\rho_{eq} = 0$, $\hat{P}^T\rho_{eq} = \rho_{eq}B_\beta(z)\langle BB\rangle_{\beta\gamma}^{-1}\langle B_\gamma(z)\rangle = 0$, if all the equilibrium means $\langle B_\alpha\rangle$ are assumed to be zero.

4.2.3 Linear FDR of the First Kind

Introducing the function

$$
\begin{aligned}
f_{\alpha\beta}(t_1,t_2) &= d_{\alpha\beta}\delta(t_1-t_2-0) + \langle F_\alpha(t_1-t_2)F_\gamma(0)\rangle \langle BB\rangle_{\gamma\beta}^{-1}\eta(t_1-t_2) \\
&\quad (2\eta(\tau) = 1 + \operatorname{sign}\tau)\,, \qquad\qquad\qquad\qquad (4.2.22)
\end{aligned}
$$

we can write (4.2.20) in the form

$$\dot{B}_\alpha(t) = -\int f_{\alpha\beta}(t_1,t_2)B_\beta(t_2)dt_2 + F_\alpha(t)\,. \qquad (4.2.23)$$

This equation should be compared with

$$\dot{B}_\alpha(t) = -\int \Phi_{\alpha,\beta}(t_1;t_2)x_\beta(t_2)dt_2 + F_\alpha(t)\,, \qquad (4.2.24)$$

i.e. with the equation [(5.1.72) v.1] taken in the linear approximation. The only difference is that in (4.2.24) the variables $x(B)$ are used. In the linear approximation, x is related to B quite simply: $x_\alpha = u_{\alpha\beta}B_\beta$ (see [(5.5.10) v.1]), where $u_{\alpha\beta} = \partial^2 F(B)/\partial B_\alpha \partial B_\beta$ at $B = 0$. In the linear approximation the distribution [(2.2.61) v.1] has the form

$$
\begin{aligned}
w(B) &= \text{const} \cdot \exp(-F(B)/kT) \\
&= \text{const} \cdot \exp[-(2kT)^{-1}u_{\alpha\beta}B_\alpha B_\beta]\,. \qquad (4.2.25)
\end{aligned}
$$

It follows that $\langle B_\alpha B_\beta\rangle = kT u_{\alpha\beta}^{-1}$, and hence this relationship $x_\alpha = u_{\alpha\beta}B_\beta$ can be represented as

$$x_\alpha = kT\langle BB\rangle_{\alpha\beta}^{-1}B_\beta\,. \qquad (4.2.26)$$

Substituting (4.2.26) into (4.2.24) and comparing with (4.2.23) gives

$$kT\Phi_{\alpha,\gamma}(t_1; t_2)\langle BB\rangle^{-1}_{\gamma\beta} = f_{\alpha\beta}(t_1, t_2),\tag{4.2.27}$$

i.e. by (4.2.8,22),

$$kT\Phi_{\alpha,\beta}(t_1; t_2) = \langle[\hat{L}B_\alpha(0)]B_\beta(0)\rangle\delta(t_{12} - 0) + \langle F_\alpha(t_{12})F_\beta(0)\rangle\eta(t_{12}).\tag{4.2.28}$$

If $\{F_\alpha(t)\}$ is a stationary process in the sense that the condition

$$\langle F_\alpha(t + a)F_\beta(t' + a)\rangle = \langle F_\alpha(t)F_\beta(t')\rangle\tag{4.2.29}$$

is met at arbitrary a (to be proved later), then in (4.2.28) $\langle F_\alpha(t_{12})F_\beta(0)\rangle$ conincides with $\langle F_\alpha(t_1)F_\beta(t_2)\rangle$. Then, from (4.2.28) we can obtain the formula

$$\Phi_{\alpha\beta}(t_1, t_2) \equiv \langle F_\alpha(t_1)F_\beta(t_2)\rangle = kT\Phi_{\alpha,\beta}(t_1; t_2) \quad \text{for} \quad t_1 - t_2 > \varepsilon,\tag{4.2.30}$$

where ε is a small positive quantity.

To prove the FDR [(5.1.53) v.1], which is valid at any $t_1 - t_2$, we will form the sum

$$\begin{aligned}
kT[\Phi_{\alpha,\beta}(t_1; t_2) &+ \Phi_{\beta,\alpha}(t_2; t_1)]\\
&= \{\langle[\hat{L}B_\alpha(0)]B_\beta(0)\rangle + \langle[\hat{L}B_\beta(0)]B_\alpha(0)\rangle\}\delta(t_{12})\\
&\quad + \langle F_\alpha(t_1)F_\beta(t_2)\rangle[\eta(t_{12}) + \eta(t_{21})].
\end{aligned}\tag{4.2.31}$$

We have used here (4.2.28,29). The difference between $\delta(t_{12} - 0)$ and $\delta(t_{21} - 0)$ in this relationship has become insignificant, and so in their place we have taken $\delta(t_{12})$. By allowing the operator $\hat{L}$ to be transposed from $B_\alpha(0)$ to $B_\beta(0)$ (such an operation has been performed in (4.2.15)), we can easily verify that the tensor $\langle[\hat{L}B_\alpha(0)]B_\beta(0)\rangle$ is antisymmetric, so that the sum in braces in (4.2.31) vanishes. Since $\eta(t_{12}) + \eta(t_{21}) = 1$ and $\langle F_\alpha(t_1)F_\beta(t_2)\rangle \equiv \Phi_{\alpha\beta}(t_1, t_2)$, we obtain from (4.2.31) the desired FDR: $\Phi_{12} = kT(\Phi_{1,2} + \Phi_{2,1})$ (see [(5.1.53) and (5.5.27) v.1] at $\hbar = 0$).

4.2.4 Proof of Stationarity of Fluctuational Inputs $F_\alpha(t)$

Consider the correlator $\langle F_\alpha(t_1)F_\beta(t_2)\rangle$. According to (4.2.12), we have

$$\langle F_\alpha(t_1)F_\beta(t_2)\rangle = \langle[\exp(-\hat{Q}\hat{L}t_1)F_\alpha(0)]F_\beta(t_2)\rangle.\tag{4.2.32}$$

The right-hand side of this can be transformed as follows:

$$\begin{aligned}
\langle[\exp(-\hat{Q}\hat{L}t_1)\hat{Q}F_\alpha(0)]F_\beta(t_2)\rangle\\
\equiv \int dz[\exp(-\hat{Q}\hat{L}t)\hat{Q}F_\alpha(z, 0)]\rho_{eq}(z)F_\beta(z, t_2)\\
= \int dzF_\alpha(0)\{\hat{Q}^T[\exp(-\hat{Q}\hat{L}t_1)]^T\rho_{eq}F_\beta(t_2)\}\\
= \int dzF_\alpha(0)\{\hat{Q}^T\exp(\hat{L}\hat{Q}^Tt_1)\rho_{eq}F_\beta(t_2)\}
\end{aligned}\tag{4.2.33}$$

or, by expanding the exponential function into a series,

$$\langle [\exp(-\hat{Q}\hat{L}t_1)F_\alpha(0)]F_\beta(t_2)\rangle$$
$$= \int dz F_\alpha(0) \left\{ \sum_{n=0}^{\infty} \frac{t_1^n}{n!} (\hat{Q}^T \hat{L})^n \hat{Q}^T \rho_{eq} F_\beta(t_2) \right\}. \tag{4.2.34}$$

Let us now consider how $\hat{Q}^T = \hat{1} - \hat{P}^T$ operates on $\rho_{eq}(z)F_\beta(z, t_2)$. According to (4.2.3), the operator $\hat{P}^T$ has the following matrix elements

$$(\hat{P}^T)_{zz'} = (\hat{P})_{z'z} = \rho_{eq}(z)B_\beta(z)\langle BB\rangle_{\beta\delta}^{-1} B_\delta(z'). \tag{4.2.35}$$

We see that they have the property

$$(\hat{P}^T)_{zz'}\rho_{eq}(z') = \rho_{eq}(z)P_{zz'}, \tag{4.2.36}$$

hence,

$$(\hat{Q}^T)_{zz'}\rho_{eq}(z') = \rho_{eq}(z)Q_{zz'}. \tag{4.2.37}$$

Since $\hat{L}\rho_{eq} = 0$, it follows that

$$(\hat{Q}^T \hat{L})_{zz'}\rho_{eq}(z') = \rho_{eq}(z)(\hat{Q}\hat{L})_{zz'}; \tag{4.2.38}$$

therefore

$$[(\hat{Q}^T \hat{L})^n]_{zz'}\rho_{eq}(z') = \rho_{eq}(z)[(\hat{Q}\hat{L})^n]_{zz'}. \tag{4.2.39}$$

We thus arrive at

$$\sum_{n=0}^{\infty} \frac{t_1^n}{n!} (\hat{Q}^T \hat{L})^n \hat{Q}^T \rho_{eq} F_\beta(t_2) = \rho_{eq}[\exp(\hat{Q}\hat{L}t_1)F_\beta(t_2)] \tag{4.2.40}$$

$(\hat{Q}F_\beta(t_2) = F_\beta(t_2))$. Thus, from (4.2.32,34,40), we have

$$\langle F_\alpha(t_1)F_\beta(t_2)\rangle = \langle F_\alpha(0)[\exp(\hat{Q}\hat{L}t_1)F_\beta(t_2)]\rangle. \tag{4.2.41}$$

To arrive at the final result

$$\langle F_\alpha(t_1)F_\beta(t_2)\rangle = \langle F_\alpha(0)F_\beta(t_{21})\rangle, \tag{4.2.42}$$

we simply have to take account of (4.2.12). This proves the stationarity condition (4.2.29), and concludes the proof of the FDR [(5.1.53) v.1].

4.2.5 Derivation of Reciprocal Relation for $\Phi_{1,2}$

This relation follows from the invariance of the Hamiltonian under the change $(q, p) \longrightarrow (q, -p)$, or $z \longrightarrow \varepsilon z$, which corresponds to time reversal. It has been pointed out earlier that this invariance has the form $\mathcal{H}(\varepsilon z) = \mathcal{H}(z)$. It gives rise to the invariance of the equilibrium distribution

$$\rho_{eq}(\varepsilon z) = \rho_{eq}(z). \tag{4.2.43}$$

Specifically, this suggests the relationship

$$\varepsilon_\alpha \varepsilon_\beta \langle B_\alpha(z,0) B_\beta(z,0) \rangle = \langle B_\alpha(z,0) B_\beta(z,0) \rangle , \qquad (4.2.44)$$

which can readily be proved using the equation

$$\varepsilon_\alpha B_\alpha(z,0) = B_\alpha(\varepsilon z,0) \qquad (4.2.45)$$

and changing the integration variable to $\varepsilon z = z'$ in the averaging integral. We can easily verify the invariance

$$P_{\varepsilon z, \varepsilon z'} = P_{zz'} \qquad (4.2.46)$$

of the projection operator $P_{zz'} = B_\beta(z,0) \langle BB \rangle_{\beta\alpha}^{-1} B_\alpha(z',0) \rho_{\text{eq}}(z')$. To this end, it is sufficient to take into account (4.2.43–45).

Let us consider the Liouville operator. It follows from its explicit form that it possesses the property

$$L_{\varepsilon z, \varepsilon z'} = -L_{zz'} . \qquad (4.2.47)$$

We will now proceed to prove the reciprocal relation, a procedure that will include two stages.
(a) To begin with, we will prove the relation

$$\varepsilon_\alpha \varepsilon_\beta \langle [\hat{L} B_\beta(0)] B_\alpha(0) \rangle = \langle [\hat{L} B_\alpha(0)] B_\beta(0) \rangle . \qquad (4.2.48)$$

For this purpose, we will expand the left-hand side

$$\varepsilon_\alpha \varepsilon_\beta \langle [\hat{L} B_\beta(0)] B_\alpha(0) \rangle = \varepsilon_\alpha \varepsilon_\beta \int L_{zz'} B_\beta(z',0) B_\alpha(z,0) \rho_{\text{eq}}(z) dz dz' , \qquad (4.2.49)$$

make allowance for (4.2.45), introduce the new integration variables $\varepsilon z = y$, $\varepsilon z' = y'$, and use (4.2.47). The resultant formula will be

$$\varepsilon_\alpha \varepsilon_\beta \langle [\hat{L} B_\beta(0)] B_\alpha(0) \rangle = - \int L_{yy'} B_\beta(y',0) B_\alpha(y,0) \rho_{\text{eq}}(y) dy dy' . \qquad (4.2.50)$$

To justify (4.2.48), we will only have to use the property $\hat{L}^T = -\hat{L}$ of the Liouville operator.
(b) It is also a straightforward exercise to prove the relation

$$\varepsilon_\alpha \varepsilon_\beta \langle F_\alpha(t_2) F_\beta(t_1) \rangle = \langle F_\alpha(t_1) F_\beta(t_2) \rangle . \qquad (4.2.51)$$

By (4.2.29), it is equivalent to the formula

$$\varepsilon_\alpha \varepsilon_\beta \langle F_\alpha(t_{21}) F_\beta(0) \rangle = \langle F_\alpha(t_{12}) F_\beta(0) \rangle \qquad (4.2.52)$$

or, by (4.2.12),

$$\varepsilon_\alpha \varepsilon_\beta \langle [\exp(-\hat{Q}\hat{L}_{21}) \hat{Q}\hat{L} B_\alpha(0)][\hat{Q}\hat{L} B_\beta(0)] \rangle$$
$$= \langle [\exp(\hat{Q}\hat{L} t_{21}) \hat{Q}\hat{L} B_\alpha(0)][\hat{Q}\hat{L} B_\beta(0)] \rangle . \qquad (4.2.53)$$

Let us write the left-hand side of this in more detail

$$\varepsilon_\alpha \varepsilon_\beta \langle F_\alpha(t_{21}) F_\beta(0) \rangle$$
$$= \varepsilon_\alpha \varepsilon_\beta \int [\exp(-\hat{Q}\hat{L}t_{21})\hat{Q}\hat{L}]_{zz'} B_\alpha(z',0)(\hat{Q}\hat{L})_{zz''} B_\beta(z'',0)\rho_{\mathrm{eq}}(z) dz dz' dz'' .$$

$$(4.2.54)$$

According to (4.2.46,47), it holds that

$$(\hat{Q}\hat{L})_{\varepsilon y, \varepsilon y'} = -(\hat{Q}\hat{L})_{yy'} ,$$
$$(\exp(-\hat{Q}\hat{L}t_{21})\hat{Q}\hat{L})_{\varepsilon y, \varepsilon y'} = -(\exp(-\hat{Q}\hat{L}t_{12})\hat{Q}\hat{L})_{yy'} .$$

$$(4.2.55)$$

Therefore, after we have introduced the new integration variables $y = \varepsilon z$, $y' = \varepsilon z'$, $y'' = \varepsilon z''$ and used (4.2.45), the integral in (4.2.54) will be equal to the mean on the right-hand side of (4.2.53). This proves (4.2.51).

If now we take into account (4.2.28), then by (4.2.48,51), we immediately obtain the reciprocal relation $\varepsilon_\alpha \varepsilon_\beta \Phi_{\beta,\alpha}(t_1, t_2) = \Phi_{\alpha,\beta}(t_1, t_2)$ (see [(5.1.15) and (5.5.19) v.1]).

In conclusion we note that the works cited in this chapter do not exhaust literature on the subject (see, for example, [4.5]).

Appendices

A1. Derivation of Equation (1.1.59)

A1.1 Identity for $\Phi(p_1, \ldots, p_m)$

Let us denote the functions (1.1.51) by $\Phi_{1\ldots m}$. The following identity is valid:

$$\sum_{k=1}^{m} \Phi_{k(k+1)\ldots m12\ldots(k-1)} \exp[-i\hbar\beta(p_1 + p_2 + \ldots + p_{k-1})] = \Phi_{1\ldots(m-1)} \quad (A1.1)$$

if $p_1 + p_2 + \ldots + p_m = 0$.

Proof. By (1.1.51), we have

$$\beta^m \Phi_{k\ldots m12\ldots(k-1)} = \int_0^\beta d\lambda_k \int_0^{\lambda_k} d\lambda_{k+1} \ldots \int_0^{\lambda_{m-1}} d\lambda_m \int_0^{\lambda_m} d\lambda_1$$
$$\times \ldots \times \int_0^{\lambda_{k-2}} d\lambda_{k-1} \exp\left(-i\hbar \sum_{j=1}^m \lambda_j p_j\right). \quad (A1.2)$$

Changing the sequence of integration, we can write this as

$$\beta^m \Phi_{k\ldots m12\ldots(k-1)} = \int_0^\beta d\lambda_m \int_{\lambda_m}^\beta d\lambda_{m-1} \int_{\lambda_{m-1}}^\beta d\lambda_{m-2} \ldots \int_{\lambda_{k+1}}^\beta d\lambda_k \int_0^{\lambda_m} d\lambda_1$$
$$\times \ldots \times \int_0^{\lambda_{k-2}} d\lambda_{k-1} \exp\left(-i\hbar \sum_{j=1}^m \lambda_j p_j\right). \quad (A1.3)$$

We will now multiply (A1.3) by $\exp(-i\hbar\beta \sum_{j=1}^{k-1} p_j)$, denote $\lambda_j + \beta = \lambda_j'$, at $j = 1, \ldots, k-1$, and sum up over k. Then the left-hand side of (A1.1), which we will denote by Z, will become

$$Z = \beta^{-m} \sum_{k=1}^m \int_0^\beta d\lambda_m \int_{\lambda_m}^\beta d\lambda_{m-1} \ldots \int_{\lambda_{k+1}}^\beta d\lambda_k \int_\beta^{\beta+\lambda_m} d\lambda_1' \int_\beta^{\lambda_1'} d\lambda_2'$$
$$\times \ldots \times \int_\beta^{\lambda_{k-2}'} d\lambda_{k-1}' \exp\left(-i\hbar \sum_{j=1}^{k-1} \lambda_j' p_j - i\hbar \sum_{j=k}^m \lambda_j p_j\right). \quad (A1.4)$$

Using the equation $p_1 + \ldots + p_m = 0$, we can write the expression in the exponential function as

$$-i\hbar \sum_{j=1}^{k-1}(\lambda_j' - \lambda_m)p_j - i\hbar \sum_{j=k}^{m}(\lambda_j - \lambda_m)p_j \,.$$

Introducing the new integration variables $\mu_j = \lambda_j' - \lambda_m$ at $j \leq k-1$, $\mu_j = \lambda_j - \lambda_m$ at $m > j \geq k$, we get from (A1.4)

$$Z = \frac{1}{\beta^m} \int_0^\beta d\lambda_m Y \,, \tag{A1.5}$$

where

$$
\begin{aligned}
Y &= \sum_{k=1}^{m} \int_0^{\beta-\lambda_m} d\mu_{m-1} \int_{\mu_{m-1}}^{\beta-\lambda_m} d\mu_{m-2} \cdots \int_{\mu_{k+1}}^{\beta-\lambda_m} d\mu_k \int_{\beta-\lambda_m}^{\beta} d\mu_1 \int_{\beta-\lambda_m}^{\mu_1} d\mu_2 \\
&\quad \times \cdots \times \int_{\beta-\lambda_m}^{\mu_{k-2}} d\mu_{k-1} \exp\left(-i\hbar \sum_{j=1}^{m-1} \mu_j p_j\right) \\
&= \sum_{k=1}^{m} \int_{\beta-\lambda_m}^{\beta} d\mu_1 \int_{\beta-\lambda_m}^{\mu_1} d\mu_2 \cdots \int_{\beta-\lambda_m}^{\mu_{k-2}} d\mu_{k-1} \int_0^{\beta-\lambda_m} d\mu_k \int_0^{\mu_k} d\mu_{k+1} \\
&\quad \times \cdots \times \int_0^{\mu_{m-2}} d\mu_{m-1} \exp\left(-i\hbar \sum_{j=1}^{m-1} \mu_j p_j\right) \,. \tag{A1.6}
\end{aligned}
$$

Let E_k be the integration region for the kth term in (A1.6). It is easily seen that E_k includes all the points for which among the numbers $\mu_1, \mu_2, \ldots, \mu_{m-1}$ there are $m-k$ numbers less than $\beta - \lambda_m$ and $k-1$ numbers larger than $\beta - \lambda_m$. The region E_k is the part of the region E defined by

$$\beta > \mu_1 > \mu_2 > \ldots > \mu_{m-2} > \mu_{m-1} > 0 \,. \tag{A1.7}$$

Each point in E must belong to one of the subregions $E_1, E_2, \ldots, E_m$. Accordingly, the regions E_k add up to E and we get from (A1.6)

$$Y = \int_E \exp(-i\hbar\mu_1 p_1 - \ldots - i\hbar\mu_{m-1}p_{m-1})d\mu_1 \ldots d\mu_{m-1} \,. \tag{A1.8}$$

It follows from (1.1.51) that $Y = \beta^{m-1}\Phi_{m-1}(p_1, \ldots, p_{m-1})$. Therefore, a trivial integration in (A1.5) yields (A1.1).

A1.2 Relation for Moments

The sum on the left-hand side of (1.1.59) can be written as

$$P_{1\ldots n}(\Phi_{1\ldots n}\langle B_1 \ldots B_n\rangle_0) = P_{1\ldots(n-1)}[P_{(1\ldots n)}(\Phi_{1\ldots n}\langle B_1 \ldots B_n\rangle_0)] \,. \tag{A1.9}$$

Next, each term

$$\Phi_{k(k+1)\ldots n12\ldots(k-1)}\langle B_k B_{k+1} \ldots B_n B_1 B_2 \ldots B_{k-1}\rangle_0$$

of the sum $P_{(1\ldots n)}$ can, by the use of [(5.2.71) v.1], be reduced to the form

$$\Phi_{k(k+1)...n12...(k-1)}\exp[-i\hbar\beta(p_1+p_2+\ldots+p_{k-1})]\langle B_1 B_2 \ldots B_n\rangle_0\,. \qquad (A1.10)$$

Using (A1.1) gives

$$P_{(1...n)}\Phi_{1...n}\langle B_1 \ldots B_n\rangle_0 = \Phi_{1...(n-1)}\langle B_1 \ldots B_n\rangle_0\,. \qquad (A1.11)$$

To arrive at (1.1.59), we simply have to carry out the summation $P_{1...(n-1)}$ in (A1.11).

Instead of $B_1, \ldots, B_n$ in (A1.11) we can take any other operators $D_1, \ldots, D_n$. Lastly, note that, according to (1.1.36,51), the appropriate equation (with D_j) can also be represented in the form

$$\begin{aligned}
P_{(1...n)}&\int_0^\beta d\lambda_1 \int_0^{\lambda_1} d\lambda_2 \ldots \int_0^{\lambda_{n-1}} d\lambda_n \langle [\exp(\lambda_1\hat{\mathcal{H}}_0)\hat{D}_1\exp(-\lambda_1\hat{\mathcal{H}}_0)] \\
&\times \ldots \times [\exp(\lambda_n\hat{\mathcal{H}}_0)\hat{D}_n\exp(-\lambda_n\hat{\mathcal{H}}_0)]\rangle_0 \\
= \beta&\int_0^\beta d\lambda_1 \int_0^{\lambda_1} d\lambda_2 \ldots \int_0^{\lambda_{n-2}} d\lambda_{n-1} \langle [\exp(\lambda_1\hat{\mathcal{H}}_0)\hat{D}_1\exp(-\lambda_1\hat{\mathcal{H}}_0)] \\
&\times \ldots \times [\exp(\lambda_{n-1}\hat{\mathcal{H}}_0)\hat{D}_{n-1}\exp(-\lambda_{n-1}\hat{\mathcal{H}}_0)]\hat{D}_n\rangle_0\,. \qquad (A1.12)
\end{aligned}$$

Here, as in (1.2.1), $\hat{H}_0$ is a nonperturbed Hamiltonian.

A2. Some Additional Results Concerning Nonlinear Fluctuation-Dissipation Relations

A2.1 The Nonquantum Cubic FDRs of the First Kind

Here we derive the cubic FDRs of the first kind by using the alternative method (see Sect. 6.1.3, v.1) of inclusion of the external forces into the phenomenological equation. In this respect the present theory differs from the theory of Sect. 5.5, v.1.

Before the inclusion of the forces, the linear-cubic approximation reads

$$\dot{B}_1 = -\Phi_{1,2}x_2 - \tfrac{1}{6}\Phi_{1,234}x_2 x_3 x_4 - F_1[x,\xi] \qquad (A2.1)$$

instead of [(5.5.44) v.1], where

$$\begin{aligned}
F_1[x,\xi] &= -\sum_\sigma [S_{12}^{(\sigma)}\xi_2^{(\sigma)} + S_{123}^{(\sigma)}\xi_2^{(\sigma)}x_3 \\
&\quad +\tfrac{1}{2}S_{12,34}^{(\sigma)}\xi_2^{(\sigma)}x_3 x_4] \qquad (A2.2)
\end{aligned}$$

in analogy with [(5.7.18) v.1]. In (A2.1) internal forces x_l are defined by

$$\begin{aligned}
x_\alpha &= \frac{\partial F(B)}{\partial B_\alpha} \\
&= u_{\alpha\beta}B_\beta + \tfrac{1}{6}s_{\alpha\beta\gamma\delta}B_\beta B_\gamma B_\delta \qquad (A2.3)
\end{aligned}$$

$(u_{\alpha\beta} = r_{\alpha\beta}^{-1})$ according to [(6.1.78) v.1]. If new variables

$$B_\alpha' = u_{\alpha\beta}B_\beta + \tfrac{1}{18}s_{\alpha\beta\gamma\delta}B_\beta B_\gamma B_\delta \qquad (A2.4)$$

(see [(6.1.79) v.1]) are introduced, equation (A2.1) assumes the form

$$
\dot{B}'_1 = -(U_{12} + \tfrac{1}{6}S_{1342}B_3B_4)\{\Phi_{2,3}x_3
$$

$$
+ \tfrac{1}{6}\Phi_{2,345}x_3x_4x_5 + F_2[x,\xi]\}
\tag{A2.5}
$$

with $U_{12} = u_{\alpha_1\alpha_2}\delta(t_1 - t_2)$, $S_{1234} = s_{\alpha_1\alpha_2\alpha_3\alpha_4}\delta$.

If the fluctuational source $F[x,\xi]$ is absent, from (A2.5) we have the phenomenological equation

$$
\dot{B}'_1 = -(U_{12} + \tfrac{1}{6}S_{1342}B_3B_4)[\Phi_{2,3}U_{34}B_4 + \tfrac{1}{6}(\Phi_2 S_{2345}
$$

$$
+ \Phi_{2,345}U_3U_4U_5)B_3B_4B_5],
\tag{A2.6}
$$

where (A2.3) has been used. The external forces are introduced in the same manner as in [(6.1.61) v.1], i.e. we obtain

$$
\dot{B}'_1 - \tilde{h}_1 = -(U_{12} + \tfrac{1}{6}S_{1342}B_3B_4)[\Phi_{2,3}U_{34}B_4 + \tfrac{1}{6}(\Phi_2 S_{2345}
$$

$$
+ \Phi_{2,345}U_3U_4U_5)B_3B_4B_5].
\tag{A2.7}
$$

Substituting (A2.4) into the left-hand side of (A2.7) and solving this equation for $\tilde{h}_1$ gives

$$
\tilde{h}_1 = (U_{12} + \tfrac{1}{6}S_{1342}B_3B_4)[\dot{B}_2 + \Phi_2 U_2 B_2 + \tfrac{1}{6}(\Phi_2 S_{2345}
$$

$$
+ \Phi_{2,345}U_3U_4U_5)B_3B_4B_5].
\tag{A2.8}
$$

Comparing (A2.8) with the formula

$$
\tilde{h}_1 = \tilde{Z}_{1,2}\tilde{J}_2 + \tfrac{1}{6}Z_{1,234}J_2J_3J_4
\tag{A2.9}
$$

(see [(5.6.8) v.1]) and taking into account that $\tilde{J} = \dot{\tilde{A}} = B$ we obtain

$$
\tilde{Z}_{1,2} = U_{12}p_2 + U_1\Phi_{1,2}U_2,
\tag{A2.10}
$$

$$
\tilde{Z}_{1,234} = \tfrac{1}{3}S_{1234}(p_2 + p_3 + p_4 + \Phi_2 U_2 + \Phi_3 U_3 + \Phi_4 U_4)
$$

$$
+ U_1\Phi_1 S_{1234} + U_1\Phi_{1,234}U_2U_3U_4
$$

$$
= \tfrac{1}{3}(p_1 + 3U_1\Phi_1 + U_2\Phi_2^T + U_3\Phi_3^T + U_4\Phi_4^T)S_{1234}
$$

$$
+ U_1\Phi_{1,234}U_2U_3U_4.
\tag{A2.11}
$$

Now we return to the case when the fluctuational source $F[x,\xi]$ is present (see (A2.5)). Then instead of (A2.8) we have

$$
\tilde{h}_1 = \tilde{Z}_{1,2}B_2 + \tfrac{1}{6}\tilde{Z}_{1,234}B_2B_3B_4 + (U_{12} + \tfrac{1}{6}S_{1342}B_3B_4)
$$

$$
\times F_2[UB + \tfrac{1}{6}SBBB, \xi].
\tag{A2.12}
$$

Comparison of this formula with the equation

$$
\tilde{h}_1 = \tilde{Z}_{1,2}\tilde{J}_2 + \tfrac{1}{6}\tilde{Z}_{1,234}\tilde{J}_2\tilde{J}_3\tilde{J}_4 - \tilde{\mathcal{E}}_1,
\tag{A2.13}
$$

which is equivalent to [(5.6.46) v.1] in the linear-cubic approximation, gives

$$\tilde{\mathcal{E}}_1 = -(U_{12} + \tfrac{1}{6}S_{1342}B_3B_4)F_2[UB + \tfrac{1}{6}SBBB, \xi]. \tag{A2.14}$$

Using this formula and [(5.6.99), (5.1.92,93) v.1], we obtain

$$\begin{aligned}
\tilde{Z}_{12,34} &= U_1U_2\Phi_{12,34}U_3U_4 + \tfrac{1}{3}[U_2S_{1345}\Phi_{52} \\
&\quad +U_1S_{1345}\Phi_{15}],
\end{aligned} \tag{A2.15}$$

$$\tilde{Z}_{123,4} = -U_1U_2U_3\Phi_{123,4}U_4, \tag{A2.16}$$

$$\tilde{Z}_{1234} = U_1U_2U_3U_4\Phi_{1234}. \tag{A2.17}$$

The cubic relations of the third kind [(5.7.55,56) v.1] can be used for deriving the cubic FDRs of the first kind. In the non-quantum case, substituting (A2.11,15–17) into these relations and considering [(5.5.27) v.1] yields

$$\begin{aligned}
\beta\Phi_{12,34}^{(1)} &= P_{12}\Phi_{1,234}, \\
\beta^2\Phi_{123,4}^{(1)} &= P_{(123)}\Phi_{1,234} + \Phi_{4,123}^{\text{t.c.}}, \\
\beta^3\Phi_{1234}^{(1)} &= P_{(1234)}[\Phi_{1,234} + \Phi_{1,234}^{\text{t.c.}}], \\
\Phi_{12,34}^{(2)} &= \Phi_{34,12}^{(2)\text{t.c.}}, \\
\beta\Phi_{123,4}^{(2)} &= P_{(123)}\Phi_{12,34}^{(2)}, \\
\beta^2\Phi_{1234}^{(2)} &= P_{(123)}P_{14}\Phi_{12,34}^{(2)}
\end{aligned} \tag{A2.18}$$

if we set

$$\begin{aligned}
\Phi_{12,34}^{(2)} &= U_1^{-1}U_2^{-1}U_3^{-1}U_4^{-1}\left\{ \tilde{Z}_{12,34}^{(2)} + \left[\tfrac{1}{3}(p_1 + p_2)\right.\right. \\
&\quad \left.\left. + \tfrac{2}{3}(U_1\Phi_1 + U_2\Phi_2 + U_3\Phi_3^T + U_4\Phi_4^T)\right] S_{1234}\right\}.
\end{aligned} \tag{A2.19}$$

We see that these FDRs are similar to the nonquantum cubic FDRs of the second and third kinds.

A2.2 FDRs of the Second Kind and Transformation of the Stable Point

If the stable point A^0 is not placed at the origin of coordinates, the formula [(5.2.4) v.1] has the form

$$\begin{aligned}
A_1 &= A_1^0 + G_{1,2}h_2 + \tfrac{1}{2}G_{1,23}h_2h_3 \\
&\quad + \tfrac{1}{6}G_{1,234}h_2h_3h_4 + \dots,
\end{aligned} \tag{A2.20}$$

where $G_{1,23\dots} \equiv G_{\alpha_1,\alpha_2\alpha_3\dots}(t_1; t_2, t_3, \dots)$ are the admittances, $h_l \equiv h_{\alpha_l}(t_l)$ are generally nonconstant external forces conjugate to A_{α_l}, and $A_1^0 \equiv A_{\alpha_1}^0(t_1) = A_{\alpha_1}^0$ are independent of t_1.

We suppose for a moment that $h_\alpha(t) = h_\alpha^0$, i.e. that $h_\alpha(t)$ are independent of t. Then from (A2.20) we will obtain the new stable point

$$
\begin{aligned}
\bar{A}_\alpha^0 &= A_\alpha^0 + h_\beta^0 \int G_{\alpha,\beta}(t_1; t_2) dt_2 \\
&\quad + \tfrac{1}{2} h_\beta^0 h_\gamma^0 \int G_{\alpha,\beta\gamma}(t_1; t_2, t_3) dt_2 dt_3 + \dots .
\end{aligned}
\tag{A2.21}
$$

Let us return to the general case where the forces $h_\alpha(t)$ are nonconstant. Defining $\bar{h}_\alpha(t) = h_\alpha(t) - h_\alpha^0$, where h_α^0 are the constant forces, we can write (A2.20) in the form

$$
\begin{aligned}
A_1 &= A_1^0 + G_{1,2}(h_2^0 + \bar{h}_2) \\
&\quad + \tfrac{1}{2} G_{1,23}(h_2^0 + \bar{h}_2)(h_3^0 + \bar{h}_3) + \dots .
\end{aligned}
\tag{A2.22}
$$

After regrouping the terms and using (A2.21), we obtain

$$
\begin{aligned}
A_1 &= \bar{A}_1^0 + \bar{G}_{1,2}\bar{h}_2 + \tfrac{1}{2}\bar{G}_{1,23}\bar{h}_2\bar{h}_3 \\
&\quad + \tfrac{1}{6}\bar{G}_{1,234}\bar{h}_2\bar{h}_3\bar{h}_4 + \dots ,
\end{aligned}
\tag{A2.23}
$$

with

$$
\bar{G}_{1,2} = G_{1,2} + G_{1,23}h_3^0 + \tfrac{1}{2}G_{1,234}h_3^0 h_4^0 + \dots ,
\tag{A2.24a}
$$

$$
\bar{G}_{1,23} = G_{1,23} + G_{1,234}h^0 + \dots ,
\tag{A2.24b}
$$

$$
\bar{G}_{1,234} = G_{1,234} + \dots .
\tag{A2.24c}
$$

Here $h_l^0 \equiv h_{\alpha_l}^0(t) = h_{\alpha_l}^0$ and $\bar{A}_1^0$ in (A2.23) is determined by (A2.21). Thus two interpretations of one reality can be given: we can say that the stable state A^0 is perturbed by the forces $h_\alpha(t)$ (see (A2.20)), or we can say that the stable state $\bar{A}^0$ is perturbed by the forces $\bar{h}_\alpha(t) = h_\alpha(t) - h_\alpha^0$ (see (A2.23)). These two different interpretations can be applied to fluctuations as well. Applying the notations introduced in Sects. 5.3,4, v.1, the nonequilibrium correlators can be written in the form

$$
\langle B_1, B_2 \rangle = G_{12} + G_{12,3}h_3 + \tfrac{1}{2}G_{12,34}h_3 h_4 + \dots ,
\tag{A2.25a}
$$

$$
\langle B_1, B_2, B_3 \rangle = G_{123} + G_{123,4}h_4 + \dots ,
\tag{A2.25b}
$$

$$
\langle B_1, B_2, B_3, B_4 \rangle = G_{1234} + \dots .
\tag{A2.25c}
$$

Substituting $h_l = h_l^0 + \bar{h}_l$ and regrouping the terms gives

$$
\langle B_1, B_2 \rangle = \bar{G}_{12} + \bar{G}_{12,3}\bar{h}_3 + \tfrac{1}{2}\bar{G}_{12,34}\bar{h}_3 \bar{h}_4 + \dots ,
\tag{A2.26a}
$$

$$
\langle B_1, B_2, B_3 \rangle = \bar{G}_{123} + \bar{G}_{123,4}\bar{h}_4 + \dots ,
\tag{A2.26b}
$$

$$
\langle B_1, B_2, B_3, B_4 \rangle = \bar{G}_{1234} + \dots ,
\tag{A2.26c}
$$

$$\bar{G}_{12} = G_{12} + G_{12,3}h_3^0 + \tfrac{1}{2}G_{12,34}h_3^0 h_4^0 + \dots , \tag{A2.27a}$$

$$\bar{G}_{12,3} = G_{12,3} + G_{12,34}h_4^0 + \dots , \tag{A2.27b}$$

$$\bar{G}_{123} = G_{123} + G_{123,4}h_4^0 + \dots , \tag{A2.27c}$$

$$\bar{G}_{12,34} = G_{12,34} + \dots , \tag{A2.27d}$$

$$\bar{G}_{123,4} = G_{123,4} + \dots , \tag{A2.27e}$$

$$\bar{G}_{1234} = G_{1234} + \dots . \tag{A2.27f}$$

From (A2.24,27) we see that in the linear-quadratic-cubic approximation the four-subscript functions $\bar{G}_{...}$ coincide with $G_{...}$, the three-subscript functions $\bar{G}_{...}$ differ from $G_{...}$ only by one term and the two-subscript functions $\bar{G}_{...}$ differ from $G_{...}$ by two terms.

It is important that the same FDRs are valid in both interpretations, i.e. the functions $\bar{G}_{...}$ are related to each other by the same FDRs as $G_{...}$. Let us take, for instance, the FDR [(5.3.89) v.1]. We have

$$G_{12,3} = i\hbar[\Gamma_2^- G_{1,23} + \Gamma_1^+ G_{2,13} - (\Gamma_2^- + \Gamma_1^+)G_{3,12}^{\text{t.c.}}] . \tag{A2.28}$$

Furthermore, the analogous FDR

$$\bar{G}_{12,3} = i\hbar[\Gamma_2^- \bar{G}_{1,23} + \Gamma_1^+ \bar{G}_{2,13} - (\Gamma_2^- + \Gamma_1^+)\bar{G}_{3,12}^{\text{t.c.}}] . \tag{A2.29}$$

must hold for $\bar{G}_{...}$. We will now find the condition for this to be true. If we substitute (A2.24b,27b) into (A2.29) and use (A2.28), we obtain

$$\begin{aligned}
h_{\alpha_4}^0 \int dt_4 G_{\alpha_1\alpha_2,\alpha_3\alpha_4}(t_1, t_2; t_3, t_4) \\
= i\hbar h_{\alpha_4}^0 \int dt_4 [\Gamma_2^- G_{\alpha_1,\alpha_2\alpha_3\alpha_4}(t_1; t_2, t_3, t_4) \\
+ \Gamma_1^+ G_{\alpha_2,\alpha_1\alpha_3\alpha_4}(t_2; t_1, t_3, t_4) \\
- (\Gamma_2^- + \Gamma_1^+)\varepsilon_{\alpha_1}\varepsilon_{\alpha_2}\varepsilon_{\alpha_3}G_{\alpha_3,\alpha_1\alpha_2\alpha_4}^*(-t_3; -t_1, -t_2, t_4)] .
\end{aligned} \tag{A2.30}$$

This equation may be written in the shorter form

$$\begin{aligned}
h_{\alpha_4}^0 \int dt_4 G_{12,34} &= i\hbar h_{\alpha_4}^0 \int dt_4 [\Gamma_2^- G_{1,234} + \Gamma_1^+ G_{2,134} \\
&\quad -(\Gamma_2^- + \Gamma_1^+)\varepsilon_{\alpha_4}G_{3,124}^{\text{t.c.}}] .
\end{aligned} \tag{A2.31}$$

Since all $h_{\alpha_4}^0$ are arbitrary, we will have

$$\begin{aligned}
\int dt_4 G_{12,34} &= i\hbar \int dt_4 [\Gamma_2^- G_{1,234} + \Gamma_1^+ G_{2,134} \\
&\quad -(\Gamma_2^- + \Gamma_1^+)\varepsilon_{\alpha_4}G_{3,124}^{\text{t.c}}] .
\end{aligned} \tag{A2.32}$$

Using [(5.4.45,48) v.1], and after cancellations, we find from (A2.32)

$$\int dt_4 G_{12,34}^{(2)} = -i\hbar(\Gamma_2^- + \Gamma_1^+)\varepsilon_{\alpha_4} \int dt_4 G_{3,124}^{\text{t.c.}} . \tag{A2.33}$$

This new FDR is a necessary and sufficient condition for invariance of the FDR (A2.28) under the stable point transformations. Note that integrating both sides of (A2.33) with respect to t_3 from $-\infty$ to ∞ yields

$$\int G^{(2)}_{12,34} dt_3 dt_4 = 0 \tag{A2.34}$$

because, after integration with respect to t_3, t_4, we have $(p_1 + p_2)J = 0$ and $(\Gamma_2^- + \Gamma_1^+)J = 0$ according to [(5.2.103) v.1]. Here J is the double integral $\int\int dt_3 dt_4 G^{\text{t.c.}}_{3,124}$.

We now pass to another quadratic FDR [(5.3.65) v.1]. An identical relation must be valid for $\bar{G}_{...}$:

$$
\begin{aligned}
\bar{G}_{123} = {} & -\hbar^2 [\Gamma_2^- \Gamma_3^- (\bar{G}_{1,23} + \bar{G}^{\text{t.c.}}_{1,23}) + \Gamma_1^+ \Gamma_3^- (\bar{G}_{2,13} + \bar{\Gamma}^{\text{t.c.}}_{2,13}) \\
& + \Gamma_1^+ \Gamma_2^+ (\bar{G}_{3,12} + \bar{G}^{\text{t.c.}}_{3,12})].
\end{aligned}
\tag{A2.35}
$$

Substituting (A2.24b,27c) into this equation and using [(5.3.65) v.1], we have

$$
\begin{aligned}
h^0_{\alpha_4} & \int dt_4 G_{123,4} \\
& = -\hbar^2 h^0_{\alpha_4} \int dt_4 [\Gamma_2^- \Gamma_3^- (G_{1,234} + \varepsilon_{\alpha_4} G^{\text{t.c.}}_{1,234}) \\
& \quad + \Gamma_1^+ \Gamma_3^- (G_{2,134} + \varepsilon_{\alpha_4} G^{\text{t.c.}}_{2,134}) \\
& \quad + \Gamma_1^+ \Gamma_2^+ (G_{3,124} + \varepsilon_{\alpha_4} G^{\text{t.c.}}_{3,124})].
\end{aligned}
\tag{A2.36}
$$

Inserting equations [(5.4.28,32,70) v.1] into the left-hand side of (A2.36) and using the arbitrariness of $h^0_{\alpha_4}$ yields

$$
\begin{aligned}
i\hbar & \int dt_4 (\Gamma_3^- G^{(2)}_{12,34} + \Gamma_2^- G^{-(2)}_{13,24} + \Gamma_2^+ G^{+(2)}_{31,24} + \Gamma_1^+ G_{23,14}) \\
& - \hbar^2 (\Gamma_2^- \Gamma_3^- + \Gamma_1^+ \Gamma_3^- + \Gamma_1^+ \Gamma_2^+) \int dt_4 G^{\text{t.c.}}_{4,\bar{1}23} \\
& = -\hbar^2 \varepsilon_{\alpha_4} \int dt_4 (\Gamma_2^- \Gamma_3^- G^{\text{t.c.}}_{1,234} + \Gamma_1^+ \Gamma_3^- G^{\text{t.c.}}_{2,134} \\
& \quad + \Gamma_1^+ \Gamma_2^+ G^{\text{t.c.}}_{3,124}).
\end{aligned}
\tag{A2.37}
$$

But after integration with respect to t_4 we will have $p_1 + p_2 + p_3 = 0$, so that $\Gamma^- 2\Gamma_3^- + \Gamma_1^+ \Gamma_3^- + \Gamma_1^+ \Gamma_2^+ = 0$ according to [(5.2.104) v.1]. Consequently, we can write (A2.37) in the form

$$
\begin{aligned}
\int & dt_4 (\Gamma_3^- G^{(2)}_{12,34} + \Gamma_2^- G^{-(2)}_{13,24} + \Gamma_2^+ G^{+(2)}_{31,24} + \Gamma_1^+ G^{(2)}_{23,14}) \\
& = -i\hbar \varepsilon_{\alpha_4} [\Gamma_1^+ (\Gamma_3^- + \Gamma_2^+) \int dt_4 G^{\text{t.c.}}_{1,234} + (\Gamma_2^- \Gamma_3^- \\
& \quad + \Gamma_1^+ \Gamma_2^+) \int dt_4 G^{\text{t.c.}}_{2,134} + \Gamma_3^- (\Gamma_2^- \\
& \quad + \Gamma_1^+) \int dt_4 G^{\text{t.c.}}_{3,124}].
\end{aligned}
\tag{A2.38}
$$

Equation (A2.33) obtained earlier allows one to cancel some terms in (A2.38) and to get

$$
\begin{aligned}
\int & dt_4 (\Gamma_2^- G^{-(2)}_{13,24} + \Gamma_2^+ G^{+(2)}_{31,24}) \\
& = -i\hbar (\Gamma_2^- \Gamma_3^- + \Gamma_1^+ \Gamma_2^+) \varepsilon_{\alpha_4} \int dt_4 G^{\text{t.c.}}_{2,134}
\end{aligned}
\tag{A2.39}
$$

or after subscript permutation,

$$\Gamma_3^- \int dt_4 G_{12,34}^{-(2)} + \Gamma_3^+ \int dt_4 G_{21,34}^{+(2)}$$
$$= -i\hbar(\Gamma_2^- \Gamma_3^- + \Gamma_1^+ \Gamma_3^+)\varepsilon_{\alpha_4} \int dt_4 G_{3,\bar{1}24}^{\text{t.c.}} . \qquad (A2.40)$$

According to [(5.4.69) v.1], we can write (A2.33) in the form

$$\int dt_4 G_{12,34}^{-(2)} + \int dt_4 G_{21,34}^{+(2)} = -i\hbar(\Gamma_2^- + \Gamma_1^+)\varepsilon_{\alpha_4} \int dt_4 G_{3,\bar{1}24}^{\text{t.c.}} . \qquad (A2.41)$$

Formulas (A2.40,41) can be regarded as a system of equations for the unknown quantities $\int dt_4 G_{12,34}^{-(2)}$ and $\int dt_4 G_{21,34}^{+(2)}$. The system is nondegenerate in the quantum case. The solution of this system gives

$$\int dt_4 G_{12,34}^{\pm(2)} = -i\hbar \Gamma_2^\pm \varepsilon_{\alpha_4} \int dt_4 G_{3,\bar{1}24}^{\text{t.c.}} . \qquad (A2.42)$$

This result is stronger than (A2.33).

Consider now the reciprocal relation [(5.3.46) v.1]. Its invariance under the transformation of the stable point means that an identical relation

$$\bar{G}_{2,1}^{\text{t.c}} = \bar{G}_{1,2} \qquad (A2.43)$$

holds for every h^0. Substitution of (A2.24a) into this equation and use of [(5.3.46) v.1] gives

$$h_{\alpha_3}^0 \varepsilon_{\alpha_3} \int dt_3 G_{2,\bar{1}3}^{\text{t.c.}} + \frac{1}{2} h_{\alpha_3}^0 h_{\alpha_4}^0 \varepsilon_{\alpha_3} \varepsilon_{\alpha_4} \int dt_3 dt_4 G_{2,\bar{1}34}^{\text{t.c.}}$$
$$= h_{\alpha_3}^0 \int dt_3 G_{1,23} + \frac{1}{2} h_{\alpha_3}^0 h_{\alpha_4}^0 \int dt_3 dt_4 G_{1,234} . \qquad (A2.44)$$

Differentiating both sides of (A2.44) with respect to $h_{\alpha_3}^0$ and setting $h^0 = 0$, we get

$$\varepsilon_{\alpha_3} \int G_{2,\bar{1}3}^{\text{t.c.}} dt_3 = \int G_{1,23} dt_3 . \qquad (A2.45)$$

If we differentiate both sides of (A2.44) with respect to $h_{\alpha_3}^0$ and $h_{\alpha_4}^0$, we obtain

$$\varepsilon_{\alpha_3} \varepsilon_{\alpha_4} \int dt_3 dt_4 G_{2,\bar{1}34}^{\text{t.c.}} = \int dt_3 dt_4 G_{1,234} . \qquad (A2.46)$$

The formula (A2.45) can be regarded as a quadratic reciprocal relation, and (A2.46) as a cubic one.

As is easily checked, the invariance of the linear FDR [(5.3.6) v.1] under the stable point transformation needs no additional condition. The foregoing investigation of the FDR invariance under the stable point transformation shows the considerable degree of concordance in the form of the FDRs for the various numbers of subscripts.

The final relations (A2.33,42,45,46) can be written in the spectral form:

$$G_{12,34}^{(2)}\Big|_{\omega_4=0} = -i\hbar(\Gamma_1 + \Gamma_2)\varepsilon_{\alpha_4}G_{3,124}^{\text{t.c.}}\Big|_{\omega_4=0}\,,\tag{A2.47a}$$

$$G_{12,34}^{\pm(2)}\Big|_{\omega_4=0} = -i\hbar\Gamma_2^{\pm}\varepsilon_{\alpha_4}G_{3,124}^{\text{t.c.}}\Big|_{\omega_4=0}\,,\tag{A2.47b}$$

$$\varepsilon_{\alpha_3}G_{2,13}^{\text{t.c.}}\Big|_{\omega_3=0} = G_{1,23}\big|_{\omega_3=0}\,,\tag{A2.47c}$$

$$\varepsilon_{\alpha_3}\varepsilon_{\alpha_4}G_{2,134}^{\text{t.c.}}\Big|_{\omega_3=\omega_4=0} = G_{1,234}\big|_{\omega_3=\omega_4=0}\,.\tag{A2.47d}$$

Let us consider some consequences of (A2.42). Integrating the second equation [(5.4.74) v.1] with respect to t_4 and substituting (A2.42), we get

$$\int dt_4 N_{1234} = 0\,.\tag{A2.48}$$

Further, after integrating both sides of [(5.4.68) v.1] with respect to t_4 from $-\infty$ to ∞ and using (A2.42,48) we have

$$\int dt_4 M_{1234} = -i\hbar^3\varepsilon_{\alpha_4}\int dt_4 G_{3,124}^{\text{t.c.}}\,.\tag{A2.49}$$

Note that $\Gamma^{\pm}$ do not enter into equations (A2.48,49).

The consequences of another type are obtained from (A2.45,46) by integrating with respect to t_1. This gives

$$\varepsilon_{\alpha_1}\varepsilon_{\alpha_2}\int G_{2,13}dt_1 dt_3 = \int G_{1,23}dt_2 dt_3\,,$$

$$\varepsilon_{\alpha_1}\varepsilon_{\alpha_2}\int G_{2,134}dt_1 dt_3 dt_4 = \int G_{1,234}dt_2 dt_3 dt_4\tag{A2.50}$$

since $\int_{-\infty}^{\infty} f(t_1 - t_2)dt_1 = \int_{-\infty}^{\infty} f(t_1 - t_2)dt_2$.

According to (A2.21) these equations can be written as

$$\varepsilon_{\alpha}\varepsilon_{\beta}\frac{\partial^2 \bar{A}_{\beta}^0}{\partial h_{\alpha}^0 \partial h_{\gamma}^0}\bigg|_{h^0=0} = \frac{\partial^2 \bar{A}_{\alpha}^0}{\partial h_{\beta}^0 \partial h_{\gamma}^0}\bigg|_{h^0=0}\,,$$

$$\varepsilon_{\alpha}\varepsilon_{\beta}\frac{\partial^3 \bar{A}_{\beta}^0}{\partial h_{\alpha}^0 \partial h_{\gamma}^0 \partial h_{\delta}^0}\bigg|_{h^0=0} = \frac{\partial^3 \bar{A}_{\alpha}^0}{\partial h_{\beta}^0 \partial h_{\gamma}^0 \partial h_{\delta}^0}\bigg|_{h^0=0}\,.\tag{A2.51}$$

Applying [(4.2.6) v.1] the dependence $\bar{A}^0(h^0)$, which is determined by (A2.21), may be represented in the form

$$\bar{A}_{\alpha}^0 = -\partial G(h^0)/\partial h_{\alpha}^0\,.\tag{A2.52}$$

Therefore we obtain from (A2.51)

$$(\varepsilon_{\alpha}\varepsilon_{\beta} - 1)\frac{\partial^3 G(x)}{\partial x_{\alpha} \partial x_{\beta} \partial x_{\gamma}}\bigg|_{x=0} = 0\,,$$

$$(\varepsilon_{\alpha}\varepsilon_{\beta} - 1)\frac{\partial^4 G(x)}{\partial x_{\alpha} \partial x_{\beta} \partial x_{\gamma} \partial x_{\delta}}\bigg|_{x=0} = 0\,,\tag{A2.53}$$

or

$$(\varepsilon_{\alpha}\varepsilon_{\beta} - 1)s_{\alpha\beta\gamma} = 0\,,\tag{A2.54a}$$

$$(\varepsilon_{\alpha}\varepsilon_{\beta} - 1)s_{\alpha\beta\gamma\delta} = 0\tag{A2.54b}$$

because

$$\left.\frac{\partial^3 G(x)}{\partial x_\alpha \partial x_\beta \partial x_\gamma}\right|_{x=0} = -u_{\alpha\lambda}^{-1} u_{\beta\mu}^{-1} u_{\gamma\nu}^{-1} s_{\lambda\mu\nu} \,,$$

$$\left.\frac{\partial^4 G(x)}{\partial x_\alpha \partial x_\beta \partial x_\gamma \partial x_\delta}\right|_{x=0} = u_{\alpha\lambda}^{-1} u_{\beta\mu}^{-1} u_{\gamma\nu}^{-1} u_{\delta\pi}^{-1} [-s_{\lambda\mu\nu\pi} + (3)s_{\lambda\mu\sigma} u_{\sigma\tau}^{-1} s_{\tau\nu\pi}] \qquad \text{(A2.55)}$$

($u_{\alpha\beta}, s_{\alpha\beta\gamma}, s_{\alpha\beta\gamma\delta}$ are the derivatives of $F(A)$ at the point A^0). Since $\varepsilon_\alpha \varepsilon_\beta s_{\alpha\beta\gamma} = s_{\alpha\beta\gamma}$, due to the time reversibility, (A2.54a) signifies that

$$(\varepsilon_\gamma - 1)s_{\alpha\beta\gamma} = 0 \,, \qquad \text{(A2.56)}$$

i.e. that $s_{\alpha\beta\gamma}$ is not equal to zero only if all parameters $B_\alpha, B_\beta, B_\gamma$ are time-even. Formula (A2.54b) means that the $s_{\alpha\beta\gamma\delta}$ differ from zero only if $B_\alpha, B_\beta, B_\gamma B_\delta$ have the same time parity.

The relations for impedances follow from the relations (A2.47). In fact, using [(5.6.6) v.1] we have

$$G_{1,23} = -G_1 Q_{1,23} G_2 G_3 \,, \qquad \text{(A2.57a)}$$

$$G_{1,234} = G_1(-Q_{1,234} + P_{(234)} Q_{1,25} G_5 G_{5,34}) G_2 G_3 G_4 \,. \qquad \text{(A2.57b)}$$

Substituting (A2.57a) into (A2.47c) and applying [(5.3.46) v.1], we arrive at

$$\varepsilon_{\alpha_3} Q_{2,13}^{\text{t.c.}}\Big|_{\omega_3=0} = Q_{1,23}\big|_{\omega_3=0} \cdot \qquad \text{(A2.58)}$$

In an analogous manner, from (A2.47b,57b) and from the equation

$$G_{12,34}^{\pm(2)} = G_1 G_2 Q_{12,34}^{\pm(2)} G_3 G_4 \,,$$

which is equivalent to [(5.7.47) v.1], we obtain

$$Q_{12,34}^{\pm(2)}\Big|_{\omega_4=0} = i\hbar \Gamma_2^\pm \varepsilon_{\alpha_4} [Q_{3,124} - P_{(124)} Q_{3,15} G_5 Q_{5,24}]^{\text{t.c.}}\Big|_{\omega_4=0} \cdot \qquad \text{(A2.59)}$$

Further, the substitution of (A2.57b) into (A2.47d) gives

$$\varepsilon_{\alpha_3}\varepsilon_{\alpha_4}[Q_{2,134}^{\text{t.c.}}\Big|_{\omega_3=\omega_4=0} - Q_{2,15}^{\text{t.c.}} G_{6,5} Q_{6,34}^{\text{t.c.}}\Big|_{\omega_3=\omega_4=0}$$

$$- P_{34} Q_{2,35}^{\text{t.c.}}\Big|_{\omega_5=0} G_{6,5} Q_{6,14}^{\text{t.c.}}\Big|_{\omega_4=0}]$$

$$= Q_{1,234}\big|_{\omega_3=\omega_4=0} - Q_{1,25} G_{5,6} Q_{6,34}\big|_{\omega_3=\omega_4=0}$$

$$- P_{34} Q_{1,35}\big|_{\omega_3=0} G_{5,6} Q_{6,24}\big|_{\omega_4=0} \cdot \qquad \text{(A2.60)}$$

Still, according to (A2.47c),

$$\varepsilon_{\alpha_3} Q_{2,35}^{\text{t.c.}}\Big|_{\omega_3=0} = Q_{5,23}\big|_{\omega_3=0} \,,$$

$$\varepsilon_{\alpha_4} Q_{6,14}^{\text{t.c.}}\Big|_{\omega_4=0} = Q_{1,64}\big|_{\omega_4=0} \cdot \qquad \text{(A2.61)}$$

Therefore

$$\varepsilon_{\alpha_3}\varepsilon_{\alpha_4}Q_{2,35}^{\text{t.c.}}\Big|_{\omega_3=0}\, G_{6,5}Q_{6,14}^{\text{t.c.}}\Big|_{\omega_4=0}$$
$$= Q_{1,64}\big|_{\omega_4=0}G_{6,5}Q_{5,23}\big|_{\omega_3=0} \tag{A2.62}$$

and the terms containing P_{34} in (A2.60) cancel out. Using formulas of the type [(5.2.21,22) v.1] we can write

$$G_{6,5} = G'_{\alpha_6,\alpha_5}(\omega_6)\delta(\omega_6 + \omega_5)\,, \tag{A2.63a}$$
$$Q_{6,34} = Q'_{\alpha_6,\alpha_3\alpha_4}(-\omega_6,-\omega_3)\delta(\omega_6 + \omega_3 + \omega_4) \tag{A2.63b}$$

with

$$G'_{\alpha_6,\alpha_5}(\omega_6) = \int \exp(-i\omega_6 t_{65})G_{\alpha_6,\alpha_5}(t_6,t_5)dt_{65}\,,$$
$$Q'_{\alpha_6,\alpha_3\alpha_4}(\omega_6,\omega_3) = \int \exp(-i\omega_6 t_{64} - i\omega_3 t_{34})$$
$$\times\, G_{\alpha_6,\alpha_3\alpha_4}(t_6;t_3,t_4)dt_{64}dt_{34} \tag{A2.64}$$

(the minuses in (A2.63b) are caused by the contravariant nature of the impedance). From (A2.63) we have

$$Q_{2,15}^{\text{t.c.}}G_{6,5}\varepsilon_{\alpha_3}\varepsilon_{\alpha_4}Q_{6,34}^{\text{t.c.}}\Big|_{\omega_3=\omega_4=0}$$
$$= \int Q_{2,15}^{\text{t.c.}}G'_{\alpha_6,\alpha_5}(\omega_6)\delta(\omega_5+\omega_6)\varepsilon_{\alpha_6}Q'_{\alpha_6,\alpha_3\alpha_4}(\omega_6,0)$$
$$\times\,\delta(\omega_6)d\omega_5 d\omega_6$$
$$= Q_{2,15}^{\text{t.c.}}\Big|_{\omega_5=0}\varepsilon_{\alpha_5}G'_{\alpha_5,\alpha_6}(0)Q'_{\alpha_6,\alpha_3\alpha_4}(0,0)\,. \tag{A2.65}$$

Analogously, by virtue of (A2.63) we get

$$Q_{1,25}G_{5,6}Q_{6,34}\big|_{\omega_3=\omega_4=0} = Q_{1,25}\big|_{\omega_5=0}\,G'_{\alpha_5,\alpha_6}(0)Q'_{\alpha_6,\alpha_3\alpha_4}(0,0)\,. \tag{A2.66}$$

It remains to use (A2.47c) in order to see that the expressions on the right-hand sides of (A2.65) and (A2.66) are equal to each other, so that the terms with $Q_{6,34}^{\text{t.c.}}$, $Q_{6,34}$ cancel out as well. Thus (A2.60) takes the form

$$\varepsilon_{\alpha_3}\varepsilon_{\alpha_4}Q_{2,134}^{\text{t.c.}}\Big|_{\omega_3=\omega_4=0} = Q_{1,234}\big|_{\omega_3=\omega_4=0}\,. \tag{A2.67}$$

It is just as simple as (A2.47d).

In conclusion of this section let us check the validity of the relations (A2.58,67) for the case of the inertialess phenomenological equations and of the simple inclusion of external forces. Writing the additional cubic term in [(6.1.3) v.1], we have

$$\dot{A}_\alpha = l_{\alpha,\beta}(x_\beta - h_\beta) + \tfrac{1}{2}l_{\alpha,\beta\gamma}(x_\beta - h_\beta)(x_\gamma - h_\gamma)$$
$$+\tfrac{1}{6}l_{\alpha,\beta\gamma\delta}(x_\beta - h_\beta)(x_\gamma - h_\gamma)(x_\delta - h_\delta) + \ldots\,. \tag{A2.68}$$

Isolating the terms linear in $(x - h)$, we obtain

$$x_\beta - h_\beta = c_{\beta\alpha} \left[\dot{A}_\alpha - \tfrac{1}{2} l_{\alpha,\beta\gamma}(x_\beta - h_\beta)(x_\gamma - h_\gamma) \right.$$
$$- \tfrac{1}{6} l_{\alpha,\beta\gamma\delta}(x_\beta - h_\beta)(x_\gamma - h_\gamma)(x_\delta - h_\delta)$$
$$\left. - \ldots \right]. \tag{A2.69}$$

where $\| c_{\beta\alpha} \| = \| l_{\mu,\nu} \|^{-1}$ (the matrix $l_{\mu,\nu}$ is assumed to be non-degenerate). Let us substitute the equation (A2.69) for $(x - h)$ into its right-hand side by iterations. This gives

$$h_\beta = x_\beta - c_{\beta\alpha}\dot{A}_\alpha + \tfrac{1}{2}c_{\beta\alpha}l_{\alpha,\sigma\tau}c_{\sigma\gamma}c_{\tau\delta}\dot{A}_\gamma\dot{A}_\delta + \tfrac{1}{6}c_{\beta\alpha}(l_{\alpha,\sigma\tau\pi}$$
$$- 3l_{\alpha,\sigma\mu}c_{\mu\nu}l_{\nu,\tau\pi})c_{\sigma\gamma}c_{\tau\delta}c_{\pi\varepsilon}\dot{A}_\gamma\dot{A}_\delta\dot{A}_\varepsilon + \ldots . \tag{A2.70}$$

Here

$$x_\beta = \frac{\partial F(A)}{\partial A_\beta}$$
$$= u_{\beta\alpha}(A_\alpha - A_\alpha^0) + \tfrac{1}{2}s_{\beta\alpha\gamma}(A_\alpha - A_\alpha^0)(A_\gamma - A_\gamma^0)$$
$$+ \tfrac{1}{6}s_{\beta\alpha\gamma\delta}(A_\alpha - A_\alpha^0)(A_\gamma - A_\gamma^0)(A_\delta - A_\delta^0) + \ldots \tag{A2.71}$$

if

$$F(A) = \tfrac{1}{2}u_{\alpha\beta}(A_\alpha - A_\alpha^0)(A_\beta - A_\beta^0)$$
$$+ \tfrac{1}{6}s_{\alpha\beta\gamma}(A_\alpha - A_\alpha^0)(A_\beta - A_\beta^0)(A_\gamma - A_\gamma^0)$$
$$+ \tfrac{1}{24}s_{\alpha\beta\gamma\delta}(A_\alpha - A_\alpha^0)(A_\beta - A_\beta^0)(A_\gamma - A_\gamma^0)(A_\delta - A_\delta^0)$$
$$+ \ldots . \tag{A2.72}$$

Substituting (A2.71) into (A2.70) and comparing the result with the equation

$$h_1 = Q_{1,2}(A_2 - A_2^0) + \tfrac{1}{2}Q_{1,23}(A_2 - A_2^0)(A_3 - A_3^0)$$
$$+ \tfrac{1}{6}Q_{1,234}(A_2 - A_2^0)(A_3 - A_3^0)(A_4 - A_4^0) + \ldots , \tag{A2.73}$$

which is equivalent to (A2.20), we obtain the impedances

$$Q_{1,2} = (u_{\alpha_1\alpha_2} - c_{\alpha_1\alpha_2}p_1)\delta(t_1 - t_2)$$
$$= -c_{\alpha_1\gamma}(p_1\delta_{\gamma\alpha_2} + d_{\gamma\alpha_2})\delta(t_1 - t_2), \tag{A2.74a}$$
$$Q_{1,23} = [s_{\alpha_1\alpha_2\alpha_3} + c_{\alpha_1\rho}l_{\rho,\sigma\tau}c_{\sigma\alpha_2}c_{\tau\alpha_3}p_2p_3]$$
$$\times \delta(t_1, t_2, t_3), \tag{A2.74b}$$
$$Q_{1,234} = [s_{\alpha_1\alpha_2\alpha_3\alpha_4} - c_{\alpha_1\rho}(l_{\rho,\sigma\tau\pi} - P_{(\sigma\tau\pi)}l_{\rho,\sigma\mu}c_{\mu\nu}l_{\nu,\tau\pi})$$
$$\times c_{\sigma\alpha_2}c_{\tau\alpha_3}c_{\pi\alpha_4}p_2p_3p_4]\delta(t_1, t_2, t_3, t_4). \tag{A2.74c}$$

By virtue of (A2.74b) the relation (A2.58) gives (A2.54a) and owing to (A2.74c) the relation (A2.67) gives (A2.54b).

Note that at $m \geq 2$ the general formulas

$$\int Q_{1,2\ldots m} dt_2 \ldots dt_m = \left. \frac{\partial^m F(A)}{\partial A_{\alpha_1} \partial A_{\alpha_2} \ldots \partial A_{\alpha_m}} \right|_{A=A^0} \tag{A2.75}$$

are valid. They can easily be derived from the equations

$$\int G_{1,2\ldots m} dt_2 \ldots dt_m = - \left. \frac{\partial^m G(x)}{\partial x_{\alpha_1} \partial x_{\alpha_2} \ldots \partial x_{\alpha_m}} \right|_{x=0}. \tag{A2.76}$$

In the case (A2.68,71) we have

$$\int Q_{1,2\ldots m} dt_m = \delta(t_1, \ldots, t_{m-1}) \int Q_{1,2\ldots m} dt_2 \ldots dt_m$$

as is seen from (A2.74). That is why (A2.58,67) and (A2.50) proved to be equivalent in this particular case.

A2.3 Cubic FDRs of the Third Kind for $Q_{1,23} \neq 0$

When the three-subscript functions do not vanish, additional terms appear in the formulas of Sect. 5.7 of Vol.1. Here we will write down the new terms denoting the old ones by dots .... Thus instead of [(5.7.5) v.1] we have

$$
\begin{aligned}
\langle \mathcal{E}_1, \mathcal{E}_2 \rangle &= \ldots \tfrac{1}{2} Q_1 Q_{2,56} \langle B_1, B_5 B_6 \rangle + \tfrac{1}{2} Q_{1,56} Q_2 \langle B_5 B_6, B_2 \rangle \\
&\quad + \tfrac{1}{4} Q_{1,56} Q_{2,78} \langle B_5 B_6, B_7 B_8 \rangle \\
&= \ldots + Q_1 Q_{2,56} \langle B_1, B_5 \rangle \langle B_6 \rangle + Q_{1,56} Q_2 \langle B_5, B_2 \rangle \langle B_6 \rangle \\
&\quad + Q_{1,56} Q_{2,78} \langle B_5, B_7 \rangle \langle B_6 \rangle \langle B_8 \rangle.
\end{aligned} \tag{A2.77}
$$

Substituting [(5.7.2) v.1] into (A2.77) and taking into account [(5.6.45b) v.1] gives (instead of [(5.7.6) v.1])

$$
\begin{aligned}
L_{12,34} &= \ldots + Q_1 Q_{2,56} G_{15} G_{6,34} + Q_{1,56} Q_2 G_{52} G_{6,34} \\
&\quad + P_{34} [Q_1 Q_{2,56} G_{15,3} G_{6,4} + Q_{1,56} Q_2 G_{52,3} G_{6,4} \\
&\quad + Q_{1,53} Q_{2,74} G_{57} G_3 G_4],
\end{aligned} \tag{A2.78}
$$

where ... has the same meaning as in (A2.77). On the other hand, using [(5.7.18), (5.6.68) v.1], we obtain

$$L_{12,34} = \ldots + Q_{12,5} G_{5,34} \tag{A2.79}$$

instead of [(5.7.28) v.1]. Equating (A2.78) and (A2.79) and using [(5.3.6,89), (5.4.48), (5.6.74b) v.1], after cancellations we arrive at

$$
\begin{aligned}
Q_{12,34} &= i\hbar [-\Gamma_2^- Q_{1,234} - \Gamma_1^+ Q_{2,134} + (\Gamma_1 + \Gamma_2) Q_{5.12}^{\text{t.c.}} G_5 Q_{5,34} \\
&\quad + P_{12} P_{34} (\Gamma_2 + \Gamma_{13}) Q_{1,35} G_5 Q_{4,25}^{\text{t.c.}}] \\
&\quad + Q_{12,34}^{(2)}
\end{aligned} \tag{A2.80}
$$

with the same function $Q_{12,34}^{(2)}$ satisfying the relations [(5.7.44) v.1]. The expression on the right-hand side of (A2.80) differs from the expression in [(5.7.36) v.1] by the terms containing $Q_{l,mn}, Q_{l,mn}^{\text{t.c.}}$. Using the equations

$$Q_{12,34} = Z_{12,34} p_3 p_4 \,,$$
$$Q_{1,234} = Z_{1,234} p_2 p_3 p_4 \,,$$
$$Q_{1,23} = Z_{1,23} p_2 p_3 \,,$$
$$p_1 G_{1,2} = Y_{1,2} \,,$$

we can rewrite (A2.80) in another form:

$$
\begin{aligned}
\beta Z_{12,34} = \ & \Theta_2^- Z_{1,234} + \Theta_1^+ Z_{2,134} + (p_1 \Theta_2 \\
& + p_2 \Theta_1) p_{12}^{-1} Z_{5,12}^{\text{t.c.}} Y_5 Z_{5,34} - P_{12} P_{34} (p_{13} \Theta_2 \\
& + p_2 \Theta_{13}) p_4^{-1} Z_{1,35} Y_5 Z_{4,25}^{\text{t.c.}}
\end{aligned}
\tag{A2.81}
$$

with $p_{12} = p_1 + p_2$, $p_{13} = p_1 + p_3$. The other cubic FDRs of the third kind can be obtained by the same method for the case of non-zero three-subscript functions.

A2.4 The Cubic Kirchhoff-Type Relations for $U_{1,23} \neq 0$

Substituting (3.2.2) into the equation

$$h_1 + \mathcal{E}_1 = Z_{1,2} J_2 + \tfrac{1}{2} Z_{1,23} J_2 J_3 + \tfrac{1}{6} Z_{1,234} J_2 J_3 J_4 \tag{A2.82}$$

(see [(5.6.26,49) v.1]) and using iterations we can obtain (for $S_{12} = I_{12}$)

$$
\begin{aligned}
g_1^r = \ & -2^{-1/2} M_{1,2} \mathcal{E}_2 + U_{1,2} g_2^a + \tfrac{1}{2} U_{1,23} (g_2^a + 2^{-1/2} \mathcal{E}_2)(g_3^a + 2^{-1/2} \mathcal{E}_3) \\
& + \tfrac{1}{6} U_{1,234} (g_2^a + 2^{-1/2} \mathcal{E}_2)(g_3^a + 2^{-1/2} \mathcal{E}_3)(g_4^a + 2^{-1/2} \mathcal{E}_4) \,,
\end{aligned}
\tag{A2.83}
$$

where

$$M_{1,2} = I_{1,2} - U_{1,2}$$
$$\| U_{1,2} \| = \| I_{1,2} \| - 2 \| Z_{1,2} + I_{12} \|^{-1} \,,$$
$$U_{1,23} = 2^{-3/2} M_1 Z_{1,23} M_2 M_3 \,,$$
$$U_{1,234} = \tfrac{1}{4} M_1 Z_{1,234} M_2 M_3 M_4 - P_{(234)} U_{1,25} M_5^{-1} U_{5,34} \,. \tag{A2.84}$$

We will use (A2.83) for calculating $\langle g_1^r, g_2^r \rangle$. In doing so the terms quadratic and cubic in $\mathcal{E}_l$ can be omitted.
Applying [(5.6.99) v.1] and the formula

$$\langle J_1 \rangle_{g^a} = 2^{-1/2} M_1 g_1^a - 2^{-3/2} U_{1,23} g_2^a g_3^a - \cdots \,, \tag{A2.85}$$

obtained from (3.2.2b), (A2.83), we arrive at

$$
\begin{aligned}
\langle g_1^r, g_2^r \rangle_{g^a} = \ & \tfrac{1}{2} M_1 M_2 \{ Z_{12} + Z_{12,5} [2^{-1/2} M_5 g_5^a \\
& - 2^{-3/2} U_{5,34} g_3^a g_4^a] + \tfrac{1}{4} Z_{12,34} M_3 M_4 g_3^a g_4^a \} \\
& - \tfrac{1}{2} M_2 U_{1,35} g_3^a [Z_{52} + 2^{-1/2} Z_{52,4} M_4 g_4^a] \\
& - \tfrac{1}{2} M_1 U_{2,35} g_3^a [Z_{15} + 2^{-1/2} Z_{15,4} M_4 g_4^a] \\
& - \tfrac{1}{4} M_2 U_{1,345} Z_{52} g_3^a g_4^a - \tfrac{1}{4} M_1 U_{2,345} Z_{15} g_3^a g_4^a \\
& + \tfrac{1}{2} U_{1,35} U_{2,46} Z_{56} g_3^a g_4^a \,.
\end{aligned}
\tag{A2.86}
$$

Hence

$$
\begin{aligned}
U_{12,34} &= \tfrac{1}{4} M_1 M_2 Z_{12,34} M_3 M_4 - 2^{-3/2} M_1 M_2 Z_{12,5} U_{5,34} \\
&\quad - 2^{-3/2} P_{34} [M_2 U_{1,35} Z_{52,4} + M_1 U_{2,35} Z_{15,4}] \\
&\quad - \tfrac{1}{2} M_2 U_{1,345} Z_{52} - \tfrac{1}{2} M_1 U_{2,345} Z_{15} \\
&\quad + \tfrac{1}{2} P_{34} U_{1,35} U_{2,46} Z_{56} \,.
\end{aligned}
\tag{A2.87}
$$

We now take into account FDRs of the third kind [(5.6.101a,b) v.1], (A2.81) and then express $Z_{1,2}, Z_{1,23}, Z_{1,234}$ in terms of $U_{1,2}, U_{1,23}, U_{1,234}$ by using (A2.84). After cancellations we get

$$
\begin{aligned}
\beta U_{12,34} &= \beta U^{(2)}_{12,34} - \Theta_2^- U_2 U_{1,234} - \Theta_1^+ U_1 U_{2,134} \\
&\quad - P_{34} \Theta_{13}^+ U_{1,35} U_{2,45} + (p_1 \Theta_2 + p_2 \Theta_1) p_{12}^{-1} U^{\text{t.c.}}_{5,12} (U_5 \\
&\quad + I_5)^{-1} U_{5,34} - P_{12} P_{34} (p_{13} \Theta_2 + p_2 \Theta_{13}) p_4^{-1} \\
&\quad \times U_{1,35} (U_5 + I_5)^{-1} U^{\text{t.c.}}_{4,25} \,.
\end{aligned}
\tag{A2.88}
$$

Here

$$
U^{(2)}_{12,34} = \tfrac{1}{4} M_1 M_2 Z^{(2)}_{12,34} M_3 M_4
$$

is the function satisfying the relations (3.3.32). Comparing (A2.88) with (3.3.30), we see that new terms containing $U_{k,lm}, U^{\text{t.c.}}_{k,lm}$ appear in (A2.88). The other cubic FDRs are not given here.

A3. Linear Transformation of Internal Parameters and Derivatives of Quasi-frenergy

Let the new internal parameters are

$$
B'_\rho = l_{\rho\alpha} B_\alpha \,.
\tag{A3.1}
$$

Here the matrix $\hat{L} = \| l_{\rho\alpha} \|$ may be degenerate and even nonsquare. The number of B'_σ may be less than the number of the original parameters B_α. From (A3.1) we obtain

$$
\langle B'_\rho, \dots, B'_\omega \rangle = l_{\rho\alpha} \dots l_{\omega\varepsilon} \langle B_\alpha, \dots, B_\varepsilon \rangle \,.
$$

Hence owing to (2.3.17)

$$
\varphi'_{\rho_1 \dots \rho_s} = l_{\rho_1 \alpha_1} \dots l_{\rho_s \alpha_s} \varphi_{\alpha_1 \dots \alpha_s} \,.
\tag{A3.2}
$$

Here $\varphi_{\alpha_1 \dots \alpha_s}$ is determined by (2.4.1) and analogously

$$
\varphi'_{\rho_1 \dots \rho_s} = \partial^2 \Phi'(x') / \partial x'_{\rho_1} \dots \partial x'_{\rho_s} \quad \text{at} \quad x' = 0 \,.
$$

At small κ transformation (2.4.8) is valid and therefore we can use (2.4.14) and

$$
\varphi_{\alpha\beta\gamma\delta} = \psi^{-1}_{\alpha\varepsilon} \psi^{-1}_{\beta\zeta} \psi^{-1}_{\gamma\eta} \psi^{-1}_{\delta\vartheta} \left\{ \psi_{\varepsilon\zeta\eta\vartheta} - 3[\psi_{\varepsilon\zeta\lambda} \psi^{-1}_{\lambda\mu} \psi_{\mu\nu\vartheta}]_{\text{sym}} \right\} \,.
$$

Applying these equations and the reversed ones with primed and non-primed matrices, we can find from (A3.2) how the new derivatives are expressed in terms of the old ones:

$$\|\psi'_{\rho\sigma}\| = \|l_{\rho\alpha}l_{\sigma\beta}\psi^{-1}_{\alpha\beta}\|^{-1}, \tag{A3.3}$$

$$\psi'_{\rho\sigma\tau\upsilon} = \psi'_{\rho\varphi}\psi'_{\sigma\chi}\psi'_{\tau\psi}\psi'_{\upsilon\omega}l_{\varphi\alpha}l_{\chi\beta}l_{\psi\gamma}l_{\omega\delta}\psi^{-1}_{\alpha\epsilon}\psi^{-1}_{\beta\zeta}\psi^{-1}_{\gamma\eta}\psi^{-1}_{\delta\vartheta}\psi_{\epsilon\zeta\eta\vartheta} \tag{A3.4}$$

if all $\psi_{\alpha\beta\gamma} = 0$. When $\hat{L}^{-1}$ exists, we have

$$\psi'_{\rho_1...\rho_s}l_{\rho_1\alpha_1}\cdots l_{\rho_s\alpha_s} = \psi_{\alpha_1...\alpha_s},$$

so that $\Psi'(B') = \Psi(\hat{L}^{-1}B')$. Formulas (A3.3,4) are useful when $\hat{L}^{-1}$ does not exist.

Now we apply (A3.3,4) to the case when $B(\mathbf{r}) \equiv B_1(\mathbf{r})$ play the role of B_α, i.e. $\mathbf{r}$ plays the role of α. Let

$$B'_1 \equiv A = \frac{1}{cS}\int_L B(\mathbf{r})\varphi(\mathbf{r})d\mathbf{r} \tag{A3.5}$$

with

$$c = \frac{1}{S}\int_L \varphi^2(\mathbf{r})d\mathbf{r}.$$

Here L is the whole region having the area S. Equation (A3.5) has been obtained with the help of the formula

$$B(\mathbf{r}) = A\varphi(\mathbf{r}) \tag{A3.6}$$

(see Sect. 2.5.4), i.e. (A3.5) would be valid if (A3.6) were true. In reality (A3.5) is true and (A3.6) is untrue when fluctuations of $B(\mathbf{r})$ are taken into consideration.

For (2.5.44) we have

$$\psi(\mathbf{r}, \mathbf{r}_1) = \left\{g(R_c - R) + h[(\partial/\partial\mathbf{r})^2 + k_c^2]\right\}\delta(r - r_1),$$
$$\psi(\mathbf{r}_1, \ldots, \mathbf{r}_n) = 24d(\mathbf{r}_1, \ldots, \mathbf{r}_n).$$

Then

$$\int \psi^{-1}(\mathbf{r}, \mathbf{r}_1)f(\mathbf{r}_1)d\mathbf{r}_1 = \left\{g(R_c - R) + h\left[\left(\frac{\partial}{\partial\mathbf{r}}\right)^2 + k_c^2\right]\right\}^{-1}f(r)$$

with arbitrary $f(\mathbf{r})$ and

$$\int \psi^{-1}(\mathbf{r}, \mathbf{r}_1)\varphi(\mathbf{r}_1)d\mathbf{r}_1 = g^{-1}(R_c - R)^{-1}\varphi(\mathbf{r}). \tag{A3.7}$$

We apply (A.3.3,4,7) and find the new frenergy

$$\Psi'(A) = \tfrac{1}{2}\psi'_{11}A^2 + \tfrac{1}{24}\psi'_{1111}A^4.$$

Calculations confirm the previous result (2.5.53).

References

Chapter 1

1.1 W.Bernard, H.B.Callen: Rev. Mod.Phys. **31**, 1017 (1959)
1.2 R.L.Stratonovich: Zh. Eksp. Teor. Fiz. **39**, 1647 (1960)
1.3 G.N.Bochkov, Yu.E.Kusovlev: Zh. Eksp. Teor. Fiz. **72**, 283 (1977)
1.4 C.W.Gardiner: *Handbook of Stochastic Methods for Physics, Chemistry and the Natural Sciences* (Springer, Berlin, Heidelberg 1983)
1.5 G.N.Bochkov, Yu.E.Kusovlev: Fluctuation-dissipation relations and stochastic models in nonequilibrium thermodynamics. Preprints Nos 138,139 (NIRFI, Gorkii 1980) [In Russian]

Chapter 2

2.1 I.Prigogine: *Étude Thermodynamique des Processus Irreversible* (Desoer, Liege 1947)
2.2 I.Prigogine: *Introduction to the Thermodynamics of Irreversible Processes* (Thomas, Springfield 1955)
2.3 P.Glansdorff, I.Prigogine: *Thermodynamic Theory of Structure, Stability and Fluctuations* (Wiley, New York 1971)
2.4 R.L.Stratonovich: Izv. VUZ Radiofizika **25**, 779 (1982)
2.5 S.Kullback: *Information Theory and Statistics* (Wiley, New York 1959)
2.6 O.Taussky: J. Soc. Indust. Appl. Math. **9**, 640 (1961)
2.7 M.Marcus, H.Minc: *A Survey of Matrix Theory and Matrix Inequalities* (Allyn and Bacon, Boston 1964)
2.8 M.Abramowitz, I.A.Stegun: *Handbook of Mathematical Functions with Formulas, Graphs and Mathematical Tables* (NBS, New York 1964)
2.9 N.N.Bogolubov, Yu.A.Mitropol'skii: *Asymptotic Methods in the Theory of Nonlinear Oscillations* (Hindustan Publ. Corp., Delhi 1961)
2.10 R.L.Stratonovich: Izv. VUZ Radiofizika **23**, 942 (1980)
2.11 G.Nicolis, I.Prigogine: *Self-Organization in Non-equilibrium Systems* (Wiley, New York 1977)
2.12 R.Landauer, J. Stat. Phys. **9**, 351 (1973)
2.13 G.N.Bochkov, Yu.E.Kusovlev: Izv. VUZ Radiofizika 21, 1467 (1978)
2.14 *Thermodynamics of Biological Processes* I.Lamprecht, A.I.Zotin eds (Walter de Gruyter, Berlin 1978)
2.15 G.N.Bochkov, Yu.E.Kusovlev: Zh. Eksp. Teor. Fiz. **76**, 1071 (1979)
2.16 I.Csiszar: Magyar Tud. Acad. Mat. Kutato Int. Közl 8, 85 (1963)
2.17 I.Csiszar: Stud. Sci. Math. Hung. **2**, 299 (1967)
2.18 F.Schlögl: Phys. Reports (Rev. Sect. of Phys. Rev.) **62**, 267 (1980)
2.19 R.Thom: *Stabilité Structurelle et Morphogénèse* (Benjamin, New York 1972)

2.20 H.Haken: *Synergetics. An Introduction* (Springer, Berlin Heidelberg 1978)
2.21 W.Ebeling: *Strukturbildung bei irreversiblen Prozessen. Eine Einführung in alle Theorie der dissipativer Structuren* (Teubner, Leipzig 1976)
2.22 W.Ebeling, H.Engel-Herbert: Rostocker Phys. Manuskripte **2**, 23 (1977)
2.23 W.Ebeling: Phys. Lett. **A68**, 430 (1978)

Chapter 3

3.1 L.D.Landau, E.M.Lifshits: *Course of Theoretical Physics Vol.5 Statistical Physics* (Pergamon, London 1958)
3.2 D.N.Klyshko: Dokl. Akad. Nauk SSSR **244**, 563 (1979)
3.3 D.N.Klyshko: *Photons and Nonlinear Optics* (Gordon & Breach, New York 1988)
3.4 R.L.Stratonovich: Dokl. Akad. Nauk SSSR **245**, 354 (1979)
3.5 R.L.Stratonovich: Dokl. Akad. Nauk SSSR **257**, 83 (1981)

Chapter 4

4.1 R.Zwanzig: Phys. Rev. **124**, 983 (1961)
4.2 C.W.Gardiner: *Handbook of Stochastic Methods for Physics, Chemistry and the Natural Sciences* (Springer, Berlin, Heidelberg 1983)
4.3 M.Abramowitz, I.A.Stegun: *Handbook of Mathematical Functions with Formulas, Graphs and Mathematical Tables* (NBS, New York 1964)
4.4 H.Mori: Progr. Theor. Phys. **33**, 423 (1965)
4.5 R.Balian, Y.Alhassid and H.Reinhardt: Phys. Lett. Ser. C Physics Reports. 131, 1 (1986)

Subject Index

Springer Series in Synergetics

Editor: Hermann Haken

Synergetics, an interdisciplinary field of research, is concerned with the cooperation of individual parts of a system that produces macroscopic spatial, temporal or functional structures. It deals with deterministic as well as stochastic processes.

Springer-Verlag
and the Environment

We at Springer-Verlag firmly believe that an international science publisher has a special obligation to the environment, and our corporate policies consistently reflect this conviction.

We also expect our business partners – paper mills, printers, packaging manufacturers, etc. – to commit themselves to using environmentally friendly materials and production processes.

The paper in this book is made from low- or no-chlorine pulp and is acid free, in conformance with international standards for paper permanency.